THAMES BARRIER DESIGN

Proceedings of the conference held in London
on 5 October, 1977

THE INSTITUTION OF CIVIL ENGINEERS, LONDON, 1978

PRODUCTION EDITOR: Bonny Harding

ISBN: 0 7277 0057 X

Published by the Institution of Civil Engineers, and produced and distributed by
Thomas Telford Ltd, PO Box 101, 26-34 Old Street, London EC1P 1JH

Printed by Inprint of Luton (Designers and Printers) Ltd

Contents

Contents

Opening address

P.A.COX, FICE*

This Symposium was conceived to bring together all the threads of the extensive study, research and detailed design which led to the start of the construction work on the Thames Barrier in 1974. The time-scale of the project is quite impressive. The Waverley Committee was set up after the floods in 1953 and the first major contract was let in 1973. The project is due to be finished in 1982. This timescale is long, even by modern standards, for getting major projects under way.

When this Symposium was conceived apprehension was expressed by a number of people who felt that, perhaps before the project had been constructed and all the ideas tested, it was inviting a fair degree of criticism from those who might have approached the problem in a different way. However, with the timescale of this project, it seemed inappropriate to defer a presentation of this wealth of useful information until such time when many of the people concerned would have either retired or left and moved to other projects. The long period of discussion enabled a scheme to be produced which optimized all the requirements which had to be considered - the engineering structural design, ground conditions, site availability, navigation problems and, an important point in relation to those, the construction feasibility of building a structure of this type in a waterway still extensively used.

There were also the environmental considerations. One other option was the raising of the banks throughout London. In the earlier stages this might almost have been acceptable environmentally. Today it almost certainly would not be. The papers deal with the detailed studies by the consulting engineers and many research organizations. There are other research organizations which played an important part in the development of the design. During all of this development stage, when the guidance was coming from the GLC and initially the Department of the Environment and later the Ministry of Agriculture, Fisheries and Food, those who were involved in the development work found a great deal of understanding from these organizations who were obviously concerned to have the right scheme developed, but equally concerned to ensure that the money expended was being used properly.

During the later stages of the development work other aspects of the project were being developed. These were the contractual arrangements for carrying out the work. The decision was taken to split the project into a series of major contracts. These included preliminary site clearance and the construction of the river wall on the south bank site, the major civil works contract and the contracts for the supply of the gates, the operating machinery and the services.

There were further subsidiary contracts for the supply of particular components. These were placed ahead of the major contracts so that there should not be any delay in the latter, and included the supply of the bearings, the trunnions on which the gates rotate, the support structures built into the piers for the trunnions and the timber roofs. Using this method of placing the contracts it was possible to maintain control over the programme and also allow the phasing of the design work to take advantage of the research projects which were still under way when the order was given for the contracts to be prepared.

---

*Partner, Rendel Palmer and Tritton

1. Historical background to the Thames Barrier

R. W. HORNER, TD, BSc, FICE, FIWE, FIPHE, MIMechE, MInstWPC*

FLOOD HISTORY

1. The Thames Barrier is now being built to solve a problem which has beset London for the whole of its recorded history. From the earliest times, low lying areas alongside the Thames have been subject to flooding as a result of exceptional high water levels in the southern North Sea.

2. The Anglo-Saxon Chronicle records a surge tide in 1099: "This year also, on the festival of St. Martin**the sea flood sprung up to such a height and did so much harm as no man remembered that it ever did before. And this was the first day of the new moon".

3. Stow in his History of England writes: "In the Year 1236 the River Thames overflowing its banks caused the marshes about Woolwich to be all a sea wherein boats and other vessels were carried with the stream, so that besides cattle a great number of inhabitants there were drowned, and in the great Palace of Westminster men did row with wherries in the midst of the Hall".

4. Pepys in his diary for December 7 1663 wrote: "There was last night the greatest tide that ever was remembered in England to have been in this river: all Whitehall having been drowned".

5. The levels reached by these earlier tides must be a matter of conjecture, but by 1791 a tide is estimated to have reached a level at London Bridge nearly 4.27 m (14 ft) above Ordnance Datum Newlyn. In 1881, almost a century later, the exceptional surge tide reached 4.91 m (16.11 ft), a level almost equalled by another exceptional high water the next year. The 1881 level was not exceeded until 1928, when a level of 5.2 m (17.09 ft) was reached at London Bridge, to be exceeded again in 1953 when a level of 5.41 m (17.75 ft) was reached.

6. These levels are shown graphically against time in Fig. 1, and an adverse change of about 0.73 m (2.4 ft) per century is evident. Dr. J.R. Rossiter in 1968 analysed high water

and low water heights for all tides for the period 1934-66 recorded by the Southend and Tower Pier tide gauges (ref. 1). He prepared estimates of the tidal parameters, the change of high waters at Southend being 363 ± 76 mm (1.19 ± 0.25 ft) per century, and at Tower Pier 774 ± 116 mm (2.54 ± 0.38 ft) per century. At Tower Pier the mean tide level was rising by 433 ± 82 mm (1.42 ± 0.27 ft) per century, and **mean tidal amplitude by 344 ± 70 mm** (1.13 ± 0.23 ft) per century.

7. An explanation of these results is that due to the increase of the volumes of the world's oceans, plus a possible sinking of south east England, mean level of the Thames is rising at over 305 mm (1 ft) per century, and in addition the tidal range at London Bridge is increasing by at least 305 mm (1 ft) per century.

8. The situation has therefore come about in which a large urban area, including both industrial and residential development, is at risk from inundation by high water levels in the southern North Sea. A look at the history of London shows that this has been the inevitable result of the development of an urban complex on this particular site. The original settlement was probably pre-Roman, and came about because of the concentration of trade routes at a point nearest to the sea where the River Thames could be readily crossed. This provided the shortest route between the Continent and the Midlands and North of England. The construction of the first London bridge, and as a result the restriction of movement of sea-going ships up-stream of that point, led to the development of a port. By the beginning of the 17th Century, the available high ground on the north side of London Bridge was fully built on, and already further building had to take place on the marshy areas to the east of the City. As the City expanded, both industrial and residential developments took place. With the enormous growth of the urban area in the 19th Century, many square miles of the low lying marshy areas on the North and South banks of the river were built on.

9. During the same period as London developed, the steady change referred to above had been taking place in the estuary. The more temperate conditions following the end of the last Ice Age had led to the recession of the glaciation which had covered the country as far south as the

---

*Deputy Director, Department of Public Health Engineering, Project Manager, Thames Barrier, Greater London Council

**Early in November

Thames Barrier Design. Institution of Civil Engineers, London, 1978, 3-8

3

Thames Valley. The melting of the ice cap led to a rise in sea levels, and the removal of the large mass of ice from the northern part of the British Isles is thought to have caused a change in equilibrium and to have led to a rotation of the country about an axis aligned approximately Bristol - Newcastle, the north-west rising and the south-east sinking (ref. 2).

10. The increase of the tidal range which has been referred to above, cannot be attributed to any one factor but is probably due to the cumulative effect of a number of factors. These include the effect of greater water depths in the southern North Sea and the outer estuary resulting from the rise in sea level, dredging operations to improve navigational channels increasing the speed of movement of the tidal wave, and improved flood defences in the lower estuary restricting the spread of the incoming tide over the saltings.

11. From consideration of these factors, the adverse change over the years can be understood, but it is also necessary to explain why, from time to time, high waters (or surge tides) occur which exceed the predicted tide level by considerable amounts (Fig. 2)

12. These variations in the normal pattern of tidal movement are always associated with the presence of low air pressure and gale force winds which affect the surface of the sea. In the past, the largest surges have been produced by depressions which have moved in from the Atlantic, passing the North of Scotland in an easterly direction, and then turning on a south-easterly course across Southern Scandinavia to Northern Germany. The particular track and speed of movement of the depression has a marked influence on the heights of the surge in the Thames Estuary. The surge is generated in the deep waters of the Atlantic by the low air pressure of the depression, is amplified by being carried into the shallower waters of the Continental Shelf, and is driven down the converging funnel of the southern North Sea by the strong northerly winds blowing on the western flank of the depression. The movement of such depressions in from the Atlantic is a fairly common occurrence as can be seen from the weather maps in the daily papers, but fortunately it is rare for a depression to move on a course and at a speed to produce the largest surge effects.

13. In the case of the 1953 high water, the tide was elevated some 1.98 m (6.5 ft) above its predicted level. In the Thames, high water levels as much as 2.38 m (7.8 ft) above predicted have been recorded. On the Elbe, at Hamburg in 1962 a surge tide was recorded which was 4.0 m (13.2 ft) above its predicted level. Such a surge, is more dangerous if it occurs on a high spring tide than on a much lower neap tide. It is interesting to note that the exceptional tide referred to in the Anglo-Saxon Chronicle in 1099 occurred on "the first day of the new moon" which would be a period of spring tides, but not of course the top of the spring tides of that particular moon.

14. The incidence of exceptional high waters (surge tides) is best treated as a matter of pure chance, since interaction between meteorological conditions in the North Sea, and astronomical tides appears unlikely. This approach was adopted by Dr. H.E. Hurst in December 1957 in an unpublished paper (ref. 7) and by Commander Suthons in his paper on high sea levels on the East and South Coast of England (ref. 3).

FLOOD PROTECTION

15. Reclamation of the salt marsh on the shores of the estuary has taken place since the earliest time, and development for industrial and residential use increased the loss which resulted from flooding and therefore provided an incentive for improved flood defences. These flood defences were however the responsibility of the riparian owners, and the unsatisfactory situation arose, whereby the work of a conscientious owner who maintained his defences in good condition, would be of little value if his next door neighbour was dilatory and allowed his to fall into disrepair.

16. The high waters and the consequent floods of 1874 and 1875 led to the passing of the first Thames Flood Act of 1879 (ref. 4), which gave authority to the Metropolitan Board of Works to prescribe statutory flood defence levels and to require riparian owners to raise their defences to these levels. In the event of failure of an owner to carry out his responsibilities, the MBW had authority to carry out the work for him and to recover the cost.

17. The level prescribed by the MBW, which in the words of their Chief Engineer Sir Joseph Bazalgette, was based on the level reached by the highest recorded tide plus a margin of 6 inches to 1 foot, provided a satisfactory defence until 1928 when overtopping occurred. The 1928 surge tide which coincided with a period of high upland flow, 283 m³/s (10,000 ft³/s), showed that a single level for the whole central London area flood defences was not adequate, and higher levels were necessary for the western part of the estuary. These revised levels were prescribed in 1930, ranging from 5.54 m (18.17 ft) at Hammersmith to 5.18 m (17.0 ft) at Woolwich.

METHODS OF FLOOD DEFENCE

18. The traditional method of flood defence, which has evolved over the centuries, has been by the raising of flood banks along the estuary. Even Teddington Weir, built in 1811, does not exclude high surge tides or even high spring tides from the river upstream, which are largely contained by the river banks.

19. In the Central London area, where the statutory levels apply, arrangements have proved generally satisfactory in the past, but as banks were raised to higher and higher levels, limitations began to appear. The Waverley Committee, set up by the Government to report on the floods of 1953, suggested as one of their recommendations: "that, in regard to the Thames, investigations should be undertaken urgently into the

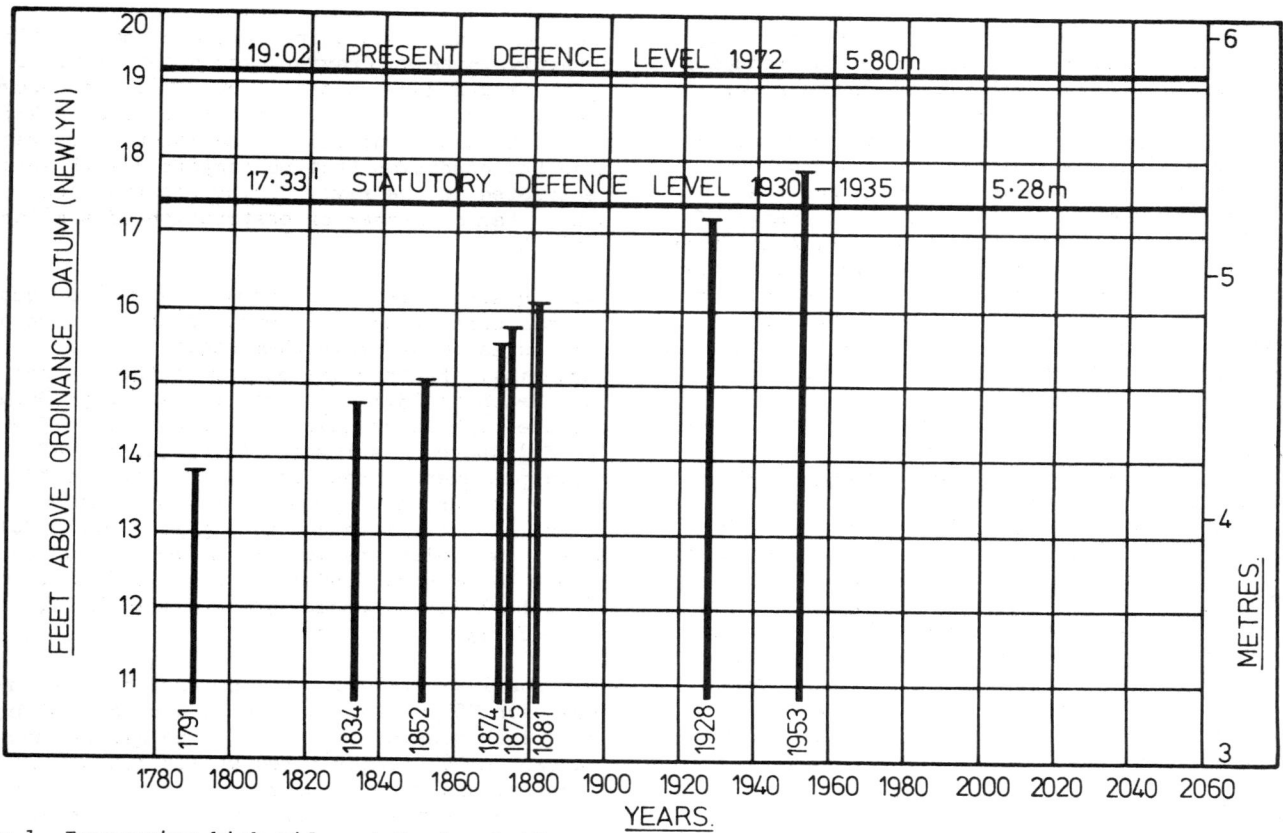

Fig.1. Increasing high tides at London Bridge

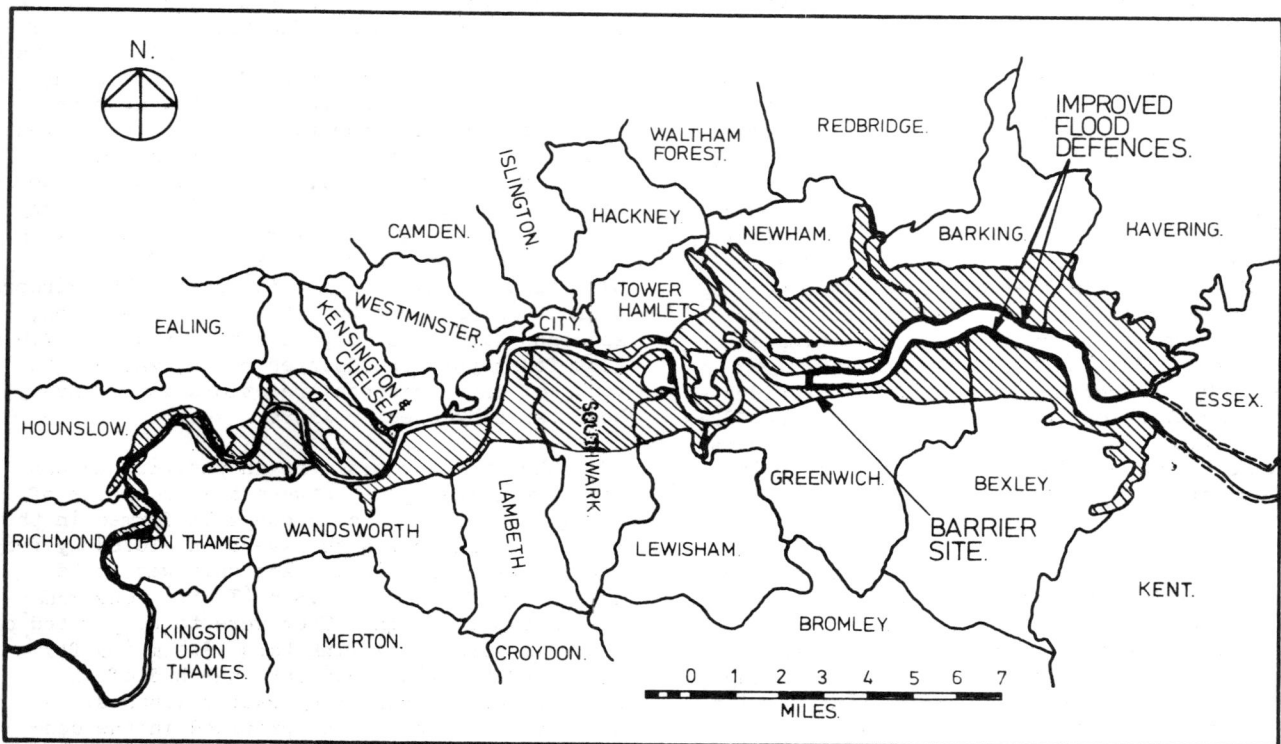

Fig.2. Area affected by a surge flood tide

possibility of providing a suitable structure, capable of being closed, as a means of reducing the maximum water levels higher up the river at the time of a surge" (ref. 5). The use of a barrier in the form of a shaped piece of hardboard was first demonstrated in a hydraulic model of the Thames which had been built and operated by the PLA. The officer in charge of the investigation at the time was Mr. Alan Price who is now at the Hydraulics Research Laboratory. The use of storm gates to protect dock entrances is fairly common practice, but this was the first case where this form of protection had been suggested for the Thames Estuary.

20. In contrast, barrages to aid navigation on the Thames have been discussed as far back as the 1780's. The idea was revived strongly at the beginning of this century, and a book by a group of authors led by Thomas Barber, "The Port of London and a Barrage", was published in 1904. The idea behind these proposals was to build a structure to impound the upper estuary as an aid to navigation and amenity. Following the 1928 high water, the idea was revised with the added bonus of flood protection, and was pursued strongly in the 1930's, but was turned down on the grounds of risk of blocking of the river as a result of an air attack.

BARRIER PROPOSALS

21. Following the Waverley Committee recommendation, a Thames Technical Panel was set up to investigate the feasibility of a barrier (ref. 6). The panel included Chief Engineers of all the bodies with interests along the river, and technical representatives of other bodies concerned.

22. A consultative group on navigation was also set up (ref. 7), composed of members of the Technical Panel and navigational experts from Trinity House, Port of London Authority and the Admiralty. A site in the centre of Long Reach was selected, and two clear openings of 152 m (500 ft) were specified with a vertical clearance over high water spring tides of 67 m (220 ft) and two smaller openings of 76 m (250 ft) with 15.2 m (50 ft) clearance. The structure was to be capable of containing a surge tide 1.8 m (6 ft) higher than the 1953 level. On the recommendation of the President of the Institution of Civil Engineers, two firms of Consulting Engineers were appointed, Messrs Rendel, Palmer & Tritton, and Sir Bruce White, Wolfe Barry & Partners.

23. Site investigations, and model experiments were carried out and preliminary designs prepared. Three different forms of barrier were suggested. These were designs making use of drop gates, gates suspended from retractable girders, and gates suspended from swing bridges. These proposals were put to a Steering Committee set up in 1961, but were turned down by the Navigational interests, who considered that the oil installations in the centre of Long Reach, with the need to turn 30-40,000 ton tankers in

the area, ruled out this site, and that an alternative site at Crayfordness about 1½ miles upstream was more suitable (ref. 7). As however the site was on a bend, a clear opening of 427 m (1400 ft) was required. In addition there was also the condition that no temporary work should be erected in the river during construction other than a dredger or obstruction of similar mobility.

24. These conditions imposed massive problems for the Consulting Engineers, and the design of structures to overcome them required much ingenuity. Further the bottom of the channel in Long Reach is 13.7 m (45 ft) below Newlyn Datum, and the high water levels to be contained 7 m (23 ft) above this level. A dam capable of rapid placement across the channel in about 1 hour in currents up to 3 m/s was required, 427 m (1400 ft) long and nearly 21.3 m (70 ft) high. Two solutions were produced, a low level scheme by Messrs. Rendel, Palmer & Tritton, and a high level scheme by Sir Bruce White, Wolfe Barry & Partners.

25. The low level scheme was based on two girders each 259 m (850 ft) long and housed in dry docks each side of the river. The girders were mounted on 164 rubber tyred wheels 1.5 m (5.0 ft) diameter, and were to be run out across the river on a concrete track constructed under water. Sluices were to be provided in the girders to be closed when the structure was in position.

26. The high level scheme required two girders each 343 m (1126 ft) long, 27.5 m (90 ft) high and 36.6 m (120 ft) wide, which would have been run out across the river from dry docks in the north and south banks. The bottom of each of the girders was to be formed by a flotation tank 2.4 m (8 ft) deep, the top of the tank being 1.8 m (6.0 ft) above O.D.N. so that at most stages of the tide the tank would be submerged. The space above the tank up to the maximum design flood level of 7.0 m (23 ft) O.D.N. would have been closed by vertical sluices. The dry docks where the girders were to be housed when not in use, were to be 347 m (1140 ft) by 67 m (220 ft) by 13 m (43 ft) deep. The girders would be hauled out by cables led to winches, and each girder would pass out over a pier 7.9 m (260 ft) beyond the dock entrance, where both vertical and horizontal reactions would have been taken by steel rollers.

27. When the girders were in position across the river, the space between the underside of the girders and the concrete sill formed in the bed of the river was to have been closed by triangular barrier units. Each unit was to be 12.2 m (40 ft) wide, 21.3 m (70 ft) long and 12.2 m (40 ft) high. They were to be pivoted off the main girders by link bars 19.8 m (65 ft) long, and to swing down onto the bed of the river under control of hydraulic winches. A steel abutment was to be provided in the concrete sill to take the thrust of the water against the barrier unit, and also the thrust on the girders transmitted by the link bars.

Fig.3. Artist's impression of Thames Barrier

REPORT BY PROFESSOR BONDI

28.   The technical problems of barriers at the Crayfordness site caused the Government to look at the whole scheme afresh, the high cost of the project raising questions of benefit cost. The problem had so many possibilities and aspects that it was thought that a review by someone with a powerful intellect and·no preconceptions, would help in putting the whole into focus.

29.   Professor Bondi,* who was then Professor of Mathematics at King's College London was asked to report (ref. 7). In his report (ref. 8), he dealt with the benefit cost aspect by taking the view that the kind of disaster in London that could be caused by an exceptional surge tide, was a risk that could not be tolerated, irrespective of the odds. He considered the advantages of a barrage, but suggested that if this was not practical, then sites between Woolwich and Long Reach should be investigated to allow the construction of a less expensive barrier than those proposed for Crayfordness.

THE GREATER LONDON COUNCIL INVESTIGATION

30.   Following the receipt of Professor Bondi's report, the Government requested the Greater London Council, the statutory flood prevention authority for London, to undertake a detailed investigation into both barrier and barrage schemes for a limited number of sites upstream of Crayfordness. The Government agreed to pay

half the cost. The GLC agreed to undertake the investigation, provided that the scope was widened to cover all practical schemes.

31.   To direct the investigation, a Policy Committee was formed, Chaired by Lord Kennet, then Joint Parliamentary Secretary of the Ministry of Housing and Local Government, with representatives from Government Departments, Hydraulics Research Station, Port of London Authority, Chamber of Shipping, Kent and Essex River Authorities and the Greater London Council (ref. 9). The investigation was controlled in detail by a Steering Committee, Chaired by the Director of Public Health Engineering, GLC, Mr. S.H. Dainty, with similar representation to the Policy Committee. Working Parties were formed to advise on Navigation, Shipping, Pollution and Siltation, Ground Water, Oceanographical and Meteorological, Civil Engineering and Amenity aspects (ref. 9).

32.   The two firms of Consulting Engineers who had submitted reports at the earlier stage agreed to form a joint team to carry out desk studies of barrier and barrage structures at their various sites. An investigation into the cost of raising the existing flood defences to the tidal limit, as an alternative to the barrier and barrage proposals, was carried out by the Rivers Branch of the Department of Public Health Engineering, GLC.

---

* Now Sir Hermann Bondi, Chief Scientific Adviser to the Ministry of Defence.

33. A comprehensive research programme was initiated to cover all aspects ranging from hydraulic and siltation studies by the Hydraulics Research Station, mathematical model studies by the Institute of Coastal Oceanography and Tides (now part of the Institute of Oceanographical Sciences), hydraulic characteristics of various gate designs by the British Hydromechanics Research Association, and detailed studies by many other bodies.

34. To manage and co-ordinate the various studies, a Project Team was appointed by the GLC, with a Project Manager, Deputy Project Manager and a Research Engineer. Technical and Administrative support was provided by the Rivers Branch, DPHE, GLC, under the direction of the Chief Engineer, Mr. C.J.S. Anderson.

35. The investigation initially considered the standard of defence to be adopted, and a variety of types of defence. Besides the traditional method of the raising of flood banks along the estuary, barrier and barrage structures were considered. The term 'barrier' is used in this paper to describe a movable structure, normally kept open and not causing any material change in the tidal flow of the river, but which can be closed to hold up an incoming surge tide. The term 'barrage' is used to mean a permanent structure across the river with sluices and possibly with locks.

36. Other types of defence were also considered in various degrees of detail. The team was bombarded with suggestions by the general public covering a wide variety of schemes, and periodic "brain storming sessions" were indulged in by the Project Team to try and discover an elegant, practical and economic solution which had so far not been considered. These included constriction of the outer estuary, and overspill of the incoming surge tide onto marshland areas on the shores of the lower estuary.

37. Consideration of the various possible means of providing the necessary defence, led to the choice of a barrier as the best solution to the problem considering factors of standard of protection provided, speed of construction, cost, environment, impact and effects on navigation and river regime.

38. The second stage of the investigation concentrated on the site and form of the barrier. The final decision was in favour of the site on the western half of Woolwich Reach, the structure being provided with four main openings each 61 m (200 ft) in width. This proposal formed the basis of the scheme for which Parliamentary Powers were sought by the GLC in the Bill presented to Parliament in 1972. The Thames Barrier Act received the Royal Assent in August 1972 (Fig. 3).

ACKNOWLEDGEMENTS

39. The Author thanks the Director of Public Health Engineering, and the Greater London Council for permission to publish this paper. Special thanks are tendered to Colonel S. K. Gilbert of the Department of the Environment who was a constant and valuable source of wisdom, advice and encouragement during the Investigation Stage, and who continues to take a keen interest in the project up to the present day. My particular thanks too to Mr. Brian Hall, now Divisional Manager of the Essex Rivers Division of the Anglian Water Authority, who made a major contribution to the project as Research Engineer and subsequently Assistant Project Manager, on the GLC team responsible for the Investigation. At the same time, sincere thanks are tendered to all those men and women, politicians, engineers and administrators, who have contributed so much to the Project and made possible the progress which has been achieved to date.

REFERENCES

1. ROSSITER J.R. and LENNON G.W. (1968). An intensive analysis of shallow water tides. Geophys J. R. Astr. Soc., 16, 275-293.

2. DUNHAM K.C. The Evidence for Subsidence of South East England. Phil. Trans. R. Soc. Lond., A272, 79-274, 1972.

3. SUTHONS C. T. (1963). "Frequence of Occurrence of Abnormally High Sea Levels on the East and South Coasts of England". Proc. I.C.E., 25, 433-450.

4. Metropolis Management (Thames River Prevention of Floods) Amendment Act 1879.

5. Departmental Committee on Coastal Flooding (1954). Report HMSO London, May 1954. Cmd. 9165.

6. Ministry of Housing and Local Government (1960). "Technical Possibilities of a Thames Barrier". HMSO London, March 1960, Cmnd.956.

7. GILBERT S.K. The Thames Barrage. Inst. of Civil Engrs. Vernon Harcourt Lecture 1969.

8. BONDI Professor H. London Flood Barrier, Report to Ministry of Housing and Local Government 1967.

9. Thames Flood Prevention, First Report of Studies. Greater London Council. Oct., 1969.

2. Sites and schemes considered

R. W. HORNER, TD, BSc, FICE, FIWE, FIPHE, MIMechE, MInstWPC*

A. H. BECKETT, MBE, BSc, FICE†

R. C. DRAPER, FIED‡

INTRODUCTION

1.   When the Government asked the GLC to carry out an investigation into the London flood problem in January 1968, they requested the examination of the following schemes (ref. 1):

(a) The construction of a movable flood barrier in either Halfway Reach (Dagenham) or Woolwich Reach.

(b) The construction of a fixed barrage at some point above the entrance to India and Millwall Docks with provision for the passage of shipping.

2.   The GLC agreed to carry out the investigation on the basis that they would examine the sites and schemes requested, but also wished to be free to examine any other schemes which might be considered possible solutions. The work was started early in 1968 with a considerable degree of urgency. This had a marked influence on the manner in which the investigation was carried out. There were three main areas requiring appraisal:

(i) A decision on the standard of protection which should be provided. This required an assessment of the overall problem, the area, population, and investment in the shape of residential and industrial property etc. at risk, enabling a cost of damage from surge tides of various magnitudes to be assessed. On this basis a standard of protection was evolved, which was a comprise between the optimum benefit cost ratio and the standard which was desirable to give a sense of complete security to people who lived and worked in the flood area.

(ii) A review of possible solutions and an assessment of their advantages and disadvantages.

(iii) The selection of a site and the general outline design of the preferred solution.

3.   Work in these three areas was carried out in parallel, although initially maximum

*Deputy Director, Department of Public Health Engineering, Project Manager, Thames Barrier, Greater London Council
†Partner, Sir Bruce White, Wolfe Barry and Partners
‡Senior Design Engineer, Rendel, Palmer and Tritton

emphasis was of course on (i), moving as the investigation proceeded to (ii), and then finally to concentrate on (iii). This was essential in view of the urgency of the whole investigation, and the need to arrive at the final solution as quickly as possible.

4.   The system of committees, and working parties set up to control and advise the Investigation Team have been described in the paper on the Historical Background. Initially the Oceanographical and Meteorological, Navigational and Civil Engineering Working Parties were the most active. In the second stage the Pollution and Siltation, Ground Water, and Amenity Working Parties became involved.

GENERAL PATTERN OF THE INVESTIGATION

5.   It was decided at the outset, that a physical model of the Thames Estuary was necessary to investigate the pattern of movement up river of surge tides reaching various levels at Southend, and with different profiles of rate of rise of water against time. Such a model would take time to construct, and therefore the Hydraulics Research Station was requested to undertake the construction of a model as rapidly as possible. Following consultations with HRS it was decided to build the model to a horizontal scale of 1/600 and a vertical scale of 1/60. It was necessary to cover the estuary from Southend to Teddington a distance of 62 miles, and this resulted in a model 116 m (380 feet) long. To contain the model in this length even, entailed the alteration of hand of the Chiswick Bend, which effectively located Teddington Weir somewhere in the Wandsworth area, thus making the lay-out more compact. It was expected that the model would be needed for some time, and due to shortage of space at HRS, it was decided to construct it in a store shed rented from the Central Electricity Generating Board on the site of the Didcot Power Station. Construction started in August 1968 and testing and proving was completed by the end of January 1969 (ref. 2).

6.   It was decided to make use of a mathematical model of the Thames Estuary developed from initial work carried out by Messrs J.R.H. Otter and A.S. Day (ref. 9) in respect of the barrier, by the Liverpool Tidal Institute (now Institute of Oceanographic Sciences) which was complementary to the HRS model. This model had the advantage that the boundary was some 71 km (44 miles)

seaward of Southend. Barrier closure effects moving downstream would be reliable well beyond Southend without any interfence from the boundary (ref. 3).

7.   At the same time the Working Parties and the Investigation Team were busy setting up the framework for the studies. The Oceanographical and Meteorological Working Party advised on the manner of the occurrence of surges, the probability of water levels of various heights, and the adverse changes likely to take place with the passage of time.

The Navigational Working Party advised on the size and number of the main shipping openings which should be provided for barrage and barrier structures at the various points in the estuary, ranging from a single 427 m (1400 ft) opening at Crayfordness to three 46 m (150 ft) openings at Cannon Street.

8.   The Joint Consulting Engineers produced outline schemes for barriers and in certain cases, barrages, for the various sites. The Siltation and Pollution Working Party advised on studies necessary for the schemes to evaluate the siltation and pollution effects. The Ground Water Working Party deliberated on the possible effects on ground water levels and movements resulting from barrier and barrage schemes, and made recommendations as to a pattern of bore holes to investigate levels of the ground water. The cost of raising existing flood defences from Teddington to Southend by 0.3 m to 1.8 m (1 to 6 feet) in steps of 0.3 m (1 ft) was evaluated by a team set up by the Rivers Branch of the Department of Public Health Engineering.

9.   In all some 41 schemes for barrages and barriers, at six sites from Cannon Street Railway bridge to Crayfordness, were investigated by the Joint Consulting Engineers, and estimates of costs and time of construction were prepared. In addition, two schemes were investigated involving a 183 m (600 ft) wide canal through the neck of the Isle of Dogs, with a barrier structure controlling the surge tide. It may be of interest to mention that Professor Schofield of Manchester College of Science and Technology suggested that a flexible 'Parachute' device be investigated as a possible barrier, and this idea was tested by HRS. However it was not considered sufficiently reliable or permanent enough for the duty to be performed.

10.   Details of the main sites are given in Table 1 (see also ref. 4); details of the work of the various organisations involved in the Investigation are given in the Appendix.

EVOLUTION OF THE SELECTED SCHEME

11.   The study of the area of low lying land at risk from tidal flooding was carried out by the Barrier Team and was largely a matter of review of literature, examination of records, contour plans etc., to assemble the required information. A good deal of material was available from the earlier studies referred to in Paper No.1, and from the Ministry of Housing and Local Government.

12.   The decision on the standard of protection to be provided was a matter of judgement supported by benefit cost studies. Initially the Thames Barrier Panel figure of 1.8 m (6 ft) above the 1953 high water was adopted (ref. 5), as this appeared a reasonable first shot considering the area at risk and the return period of such a high water. This gave a level of 6.8 m (22.25 ft) at Crayfordness and 7.0 m (23.00 ft) at Woolwich. Engineering studies were thus able to proceed on this basis. Subsequently this standard was revised to that of the adverse trend of high waters up to the year 2030 A.D. This gave a level of 6.4 m (21.00 ft) at Crayfordness, and therefore the earlier assumption allowed a free board of 0.4 m (1.3 ft) which was a reasonable basis for the design of hard top defences.

13.   The probabilities of high waters as calculated by Suthons (ref. 6), were revised in conjunction with Commander Suthons to allow for the adverse change which had been identified by Rossiter (ref. 7). Levels throughout the estuary for a given probability with average upland flow, were derived from the figures for Southend plus the differential of an average spring tide for the particular point. This method was adopted as it was considered that levels at upriver points such as Tilbury and Woolwich were based on historical records, and therefore the probabilities derived there from these would be reduced by overspill in the outer estuary which had occurred in the past, but which would be eliminated in the future by improved flood defences downriver. Once a general standard had been adopted for the lower estuary, high waters would be contained, and the model studies carried out even at that stage showed that differences between Southend and points in the upper estuary would be comparable with the normal high spring tide differences.

14.   Some consideration was given to schemes on the lines of the Dutch Delta Plan for the Eastern Scheldt. Possible sites were between Margate and Clacton and also between Shoeburyness and Sheerness, and again between Canvey Island and Cliffe Marshes. These schemes were all rejected mainly on the grounds of the large navigational opening required for the large vessels using the outer estuary, but also on the grounds that time for construction would be excessively long. Navigation openings of 762 m (2,500 ft) were required for these sites, and this would need a retractable barrier. The difficultues of construction of such a structure even at the Canvey Island site would be considerable.

15.   Besides the barrier and barrage schemes and the traditional bank raising method, consideration was given to two other methods of providing protection. One of these was that of providing a constriction of the outer estuary, which would be sufficient to reduce the tidal range within the estuary itself. Old London Bridge, which was demolished in 1834, acted in this manner. On the removal of this bridge and the construction of Rennie's Bridge, tidal range up river was increased by about 1.4 m (4.5 ft), high waters being higher by about 0.3 m (1 ft) and low waters

Table 1

SITE	LENGTH OF FLOOD DEFENCES DOWNSTREAM (KILOMETRES) ⊗	MINIMUM SIZE OF MAIN OPENING (METRES)	FOUNDATION	TYPE OF GATE	WIDTH OF WATERWAY H.W.O.S.T. (METRES)	DEPTH H.W.O.S.T. TO BED (METRES) *	CROSS SECTIONAL AREA OF WATERWAY (SQ.METRES) †
CANNON STREET	44	46	LC	1 2 3	213	11.0	1950
BLACKWALL REACH	35	61	W&R	1 2 3 4	350	13.4	3900
WOOLWICH REACH	30	61	C&TS	1 2 3 4	550	13.7	4830
HALFWAY REACH	21	305	C	3 4	670	15.5	8180
CRAYFORDNESS	14	427	C	3 4	790	17.4	10130
GRAVESEND	0	427	C	3 4	853	20.4	12360

GATE TYPES
1 DROP GATE
2 RISING SECTOR
3 DRUM GATE
4 RETRACTABLE

FOUNDATIONS
LC LONDON CLAY
W&R WOOLWICH AND READING BEDS
C CHALK
TS THANET SAND

⊗ THIS FIGURE IS THE ESTIMATED LENGTH OF FLOOD DEFENCES BOTH SIDES OF THE RIVER BETWEEN THE PARTICULAR SITE AND THE GRAVESEND SITE (TO THE NEAREST KILOMETRE)

* DEPTH TO DEEPEST PART OF BED

† APPROXIMATE ONLY

H.W.O.S.T. TYPICAL MEAN SPRING TIDE (2.86m HIGHWATER AT SOUTHEND)

Table 2. Comparative cost of flood protection with barrier: £million

SCHEME	TOTAL COST	STRUCTURE	DEVEL- OPMENT	BANKS UP STREAM[1]	BANKS DOWN STREAM[2]	LAND	COMPEN- SATION	SHIPPING DELAY
				BREAKDOWN OF COST				
BANK RECONSTRUCTION	65		–	–	55	–	10	–
CANNON STREET (or up-stream) Barrier & banks downstream.	46.5	4.5	–	2.0	33.0	1.0	6.0	–
BLACKWALL REACH Barrier & banks down-stream.	47	15.0[X]	–	2.0	24.5	1.0	4.5	–
WOOLWICH REACH Barrier & banks down-stream.	45	22.5[3]	–	2.0	14.5	1.0	2.5	2.5
HALFWAY REACH Barrier & banks down-stream.	50	29.0[X]	2.0	8.0	7.0	1.0	1.5	1.5
CRAYFORDNESS Barrier & cut off banks upstream banks raised 1-2ft	48	33.0	2.0	8.0	2.0	0.5	0.50	2.0
GRAVESEND Barrier & cut off banks upstream banks raised 1-2ft	53[4]	36.0[X]	2.0	8.0	2.0	0.5	0.50	4.0

NOTE
This table has been ab-stracted from the first report of studies. Costs are comparative only & are at 1969 prices.

NOTES
1 An element of cost has been allowed to provide improved protection from fresh water flow.
2 Surge warnings. Late closure is assumed to be avoided for levels above 15.00 OD at Southend.
3 Woolwich Barrier. The Consultants' estimates increased by £1m to allow for substrata revealed by subsequent site investigation.
4 Gravesend Barrier. Increased protection would save some £3.0 m and reduce this to £50 m.

Estimates marked X are subject to further review.
Shipping delay. If upstream bank raising is omitted then the effect rises as follows.: Halfway reach £ 2.0 m, Cray-fordness £ 2.0 m, Gravesend £ 10.0 m.

12

lower by 1.1 m (3.5 ft). Model tests on both the physical and mathematical hydraulic models (ref. 3), showed that this constriction must be of the order of an 85% reduction of area in order to be effective. Very high velocities would be experienced in the remaining 15% area, and navigation would be difficult if not impossible!

16. **The alternative which was given some** extended study, was the possibility of overspill in the outer estuary to reduce the levels upstream. In the case of the 1953 high water, the Hydraulics Research Station, on the evidence of the tests carried out on the first physical model of the Thames Estuary, considered that breaching and overspill had had the effect of reducing levels at London Bridge by 0.12 m (5 in). There were two very serious problems with this method of dealing with surge tides. The first of these was the practical problem of getting sufficient area which could be held available for flooding when the emergency arose, and the second one was the effect on levels upstream, should the designated area become filled, (ref. 3), before the peak of the surge tide had passed the overspill point. When this occurred, there was a resonance effect which resulted in enhanced levels upstream. In view of these two very serious disadvantages, it was felt that this **scheme was not practical.**

17. The situation was therefore reached fairly early in 1969, when it was apparent that the solution to the problem of the protection to the low lying areas of London against dangerously high waters in the estuary, would lie with either the traditional method of raising the flood defences throughout the whole estuary, or with the construction of a barrier or barrage in the general area Crayfordness to Cannon Street Railway Bridge.

DETAILS OF SCHEMES CONSIDERED

18. Table 1 gives a summary of the barrier and barrage schemes considered. In addition the raising of flood defences up to and beyond Teddington Weir by up to 1.8 m (6 ft) was evaluated.

19. An analysis of cost (Table 2) showed that the bank raising proposals would be the most expensive, and also would take the longest time to construct. In addition there were strong objections on environmental grounds, and there would obviously be very considerable opposition to this proposal. Of the barrier or barrage schemes, the siltation tests on half tide operation of a barrier in either Woolwich or Blackwall Reaches showed a considerable siltation effect, and the effect from a barrage would be much greater.

20. Almost all the schemes considered were of the movable type; a fixed barrier with a locking system would not have proved acceptable to the Port of London Authority and other interests and was therefore never seriously considered.

21. In 1956 the Joint Consultants were appointed to look into the practicability of constructing a movable flood barrier across the

River Thames at Long Reach. The chosen position in the river had a straight length of approximately 3 miles which would facilitate easy navigation through a barrier.

22. Three basic types of barrier were considered in general terms, (a) a vertical lift barrier, (b) a swing bridge barrier, and (c) a horizontally retracting barrier. The Report containing these proposed schemes was assessed by the Navigational Working Party, and discarded in favour of a barrier providing a clear open of 427 m (1400 ft). The requirement of a clear opening of this size posed the Joint Consultants with an unprecedented problem. Of the schemes so far considered only the retractable design could be developed to suit the requirement. In 1956 the Joint Consultants presented a new Report proposing two possible retractable barrier schemes.

23. <u>High Level Retractable Barrier</u>

One was a high level scheme, where the supporting truss remained above the water level and where closure was affected by hinged gates slung below. The supporting trusses were to cantilever some 214 m (700 ft) from support systems at each side of the river.

24. <u>Low Level Retractable Barrier</u>

The second proposal was for two similar types of trusses riding on wheels running on a track, pushed out from launching ways at each bank. The trusses were to be partially submerged with the top chords above surge level. Internal sliding gates would be closed when the trusses met at mid-river. On assessment, reservations were made regarding their practicability and reliability.

In 1968 the Joint Consultants were re-commissioned to undertake a further design study. The choice of sites, however was narrowed down to three, the original site, Crayfordness and Woolwich Reach or Cannon Street. The Crayfordness site had the attraction of a good chalk foundation close to the river bed, and for this reason a further attempt was made to develop a barrier at this site.

25. <u>The Drum Gate Barrier</u>

The drum gate barrier was developed for the Crayfordness site. Thirteen gates were proposed, each 30.5 m (100 ft) across and shaped like a wedge of cheese, giving a total clear opening of 457 m (1500 ft). The point of the wedge was hinged at bed level, and foundations were designed to house the remainder of the wedge in a complementary shaped recess below bed level, when not in use. The structures were to be hollow and buoyant and would be operated by dewatering or flooding the foundation cavity. One of the main difficulties that would be encountered would be in constructing the foundation cavities. They were to be deep, approximately 41 m (135 ft) below river bed level. The second difficulty, which eventually caused the scheme to be discarded, was that of providing an efficient seal around the gate structure at bed level and maintaining it.

The Cannon Street site did not prove attractive, due to the considerable lengths of bank downstream that would require raising and improving. In addition the effect of barrier closure at this position on high tides downstream under certain circumstances, would be to increase them by as much as 0.9 m (3 ft) by the reflected wave. Of all the intermediate sites the best possible were in Woolwich Reach.

26. The following general range of Barrier gate schemes were investigated for these sites:

(a) Vertical Lifting Types

27. In these schemes overhead gates are raised and lowered between towers, the gates being lowered to the bed of the river to effect closure. The gates span between the piers, and no part of the water pressure load is taken by the sill. Due to the very large deadweight of the gates it is necessary for them to be counterweighted, but an imbalance occurs when they enter the water due to the buoyancy effect. This was overcome on the earlier designs by holding the main support structure above water level and lowering hinged flap gates from the main structure. However, many options are open to overcome this effect, not least, that of using a variable counterweight system.

This type of barrier was considered to be the most suitable and reliable proposition.

(b) Swing Bridge Type

28. This type takes the form of horizontally balanced swing spans carrying drop gates which are suspended underneath the spans and lowered onto a sill in the bed of the river.

29. The structural design problems of this type of barrier are similar to those of the lifting spans, with the exception that suitable horizontal bearings combined with vertical pivot bearings have to support the horizontal water load. The main disadvantages are from a navigational point of view. In order to provide adequate protection to the spans from possible damage by collision when open, the piers would have to be to dimensions greater than the breadth of the spans.

This type of barrier had many precedents structurally and was regarded as a feasible proposition.

(c) Horizontal Retractable Types

30. In these schemes the barrier is formed of units which are moved across the river on tracks laid on a sill in the bed of the river. Each unit is self-stable and no piers are required. The methods of operation are by winches or by rack and pinion gear at the forward end of the cambers in which the structures are normally housed. Stop logs or similar are provided for maintenance purposes.

31. The major disadvantage of this type is the

necessity of a track, on which the barrier units move, in the bed of the river. The track would always be liable to damage and obstruction by debris.

32. The alternative method to this is to cantilever from each shore. If they were to be kept above water level the considerable cantilever spans would require counter-balances as a means of holding the tail ends down. On the other hand if the spans were partially submerged an effective weight saving could be achieved, but only at the expense of allowing the tidal flow to apply a horizontal force whilst the structure is pushed out. The subsequent horizontal deflection could be considerable. The launching of cantilever spans of the size necessary for a barrier with a tidal stream pressing against it, is outside present engineering experience.

These types of barrier, although considered, do not provide a confident method of flood protection.

(d) Submerged Types

33. In these schemes the gates of the barrier normally lie in the bed of the river when not in use, and are raised vertically and brought into position when required by means of buoyancy tanks or hydraulic jacks. The number of piers can be limited with this type and it has the advantage that the load due to water pressure is transmitted direct to the foundations.

34. It is considered that with all these types, maintenance would be difficult. The gates, lying on the river bed, would always be vulnerable to damage and obstruction which would not be readily detected. However, it will be seen later that it is from this category of barrier types that the ultimate design was chosen.

(e) Bascule Type

35. This scheme comprises numerous bascule bridge leaves, supported on piers and abutments, which swing downwards and lock together to form a continuous structure connecting piers and abutments across the river. Gates are rolled out from each bank, suspended from the underside of the bascule leaves and passing through transverse openings in the piers, thus closing the river.

The type was rejected for many practical reasons, not least of all the amount of room required on each shore for the gates.

(f) Floating Type

36. Various types of floating barrier have been suggested. These types consist of floating units which are manoeuvred into position across the river and sunk onto a floor in the river bed to form the barrier, either in the form of an arch or spanning between piers. It is considered that it would be extremely difficult to control and move large floating units into position and sink them accurately onto a sill, in the fast currents which occur in the Thames.

(g) Vertical Hinged Gate Types

37. The hinged gate type may be sub-divided into the mitre and straight sill structures. The gates are usually semi-buoyant to reduce reaction on the hinges. It is considered that these types suffer from the same disadvantages as the floating type, in that control of the gates whilst being rotated into the barrier position in a flowing current would be difficult, and large slamming forces would result just before closure.

FURTHER STUDIES

38. The Joint Consultants continued their study of the problem to find a type of gate which would satisfy all the design parameters. This further investigation considered the Drum Gate and the Drop Gate, both of which have been previously described, and a new type of gate called the "Rising Sector Gate" invented by Mr. R.C. Draper of Rendel, Palmer & Tritton.

The "Rising Sector Gate", due to limitation of design stresses in relation to the requirements, indicated a limiting width of 61 m (200 ft).

SELECTED SCHEME

39. Following the publication of the First Report of Studies in October 1969, both the GLC and the Government backed the proposal to protect London against tidal flooding by the construction of a barrier in Woolwich Reach. About this time the Port of London Authority announced the closure of the Surrey Docks and this information had a significant effect on the size of opening selected for the structure.

40. Initial proposals had been for a structure with a central opening for navigation of 137 m (450 ft) and two side openings of 61 m (200 ft). The closing of the Surrey Docks and indicated that the trend of the closure of upriver docks was continuing, and therefore it would be logical to assume that the West India Docks would close in the foreseeable future. It therefore appeared likely that the number of large vessels using the estuary above Woolwich Reach would decrease as the years passed. With this background, there seemed to be little justification for the provision of the large 137 m opening, which entailed the use of a drop gate weighing several thousand tons. Sub-division of this central opening into two 61 m (200 ft) openings would enable the four main openings to use identical gates of whatever design might be selected.

41. The situation was therefore reached by the autumn of 1970, when general informed opinion was tending to favour a barrier in Woolwich Reach as the right solution for the London Tidal Flood Problem, with four main 61 m openings for the existing 183 m navigational channel.

42. Consideration of cost and also time of construction indicated that the most economic solution, and also the scheme requiring the shortest time for construction would be a barrier site in Woolwich Reach.

43. For comparison, details of the latest estimates (March 1977) of the cost of the Flood Protection works within this London excluded area are given in Table 3.

ACKNOWLEDGEMENTS

44. The Authors thanks the Director of Public Health Engineering, The Greater London Council and Sir Bruce White, Wolfe Barry & Partners and Rendel, Palmer & Tritton, for permission to publish the paper. Any opinions stated are those of the Authors and do not necessarily represent Council Policy or that of the Joint Consultants.

TABLE 3

(Costs shown are at March 1977 levels)

	£
Investigations and design studies	480,128
Interim bank raising*	3,915,700
Thames Barrier	207,042,326
Downstream Bank raising	45,985,000
Contribution to Downstream Water Authorities flood defences under Barrier Act of 1972	27,229,000
	284,652,154

*Flood defences have been temporarily raised some 18" throughout London to provide an improved standard of defence in the short term pending the completion of the permanent defences.

The Ministry of Agriculture Fisheries and Food contribute 75% of the cost of the works from central government funds, the remaining 25% being met from the Greater London Council finances.

In the initial stages a contribution of 50% towards investigation costs was made by the Department of the Environment (then the Ministry of Housing and Local Government).

REFERENCES

1. Greater London Council Minutes of Meeting dated 26th March 1968.

2. Thames Flood Prevention, First Report of Studies. Greater London Council - October 1969 Appendix 2.

3. Thames Flood Prevention, First Report of Studies. Greater London Council - October 1969 Appendix 3.

4. Thames Flood Prevention, First Report of Studies. Greater London Council - October 1969 Appendix 5a.

5. Ministry of Housing and Local Government (1960). "Technical Possibilities of a Thames Barrier". HMSO London, March 1960, Cnd. 956.

6. SUTHONS C.T. (1963). "Frequency of Occurrence of Abnormally High Sea Levels on the East and South Coasts of England". Proc. I.C.E., 25, 433-450.

7.   ROSSITER J.R. and LENNON G.W. (1968).
An intensive analysis of shallow water tides.
Geophys J.R. Astr. Soc., 16, 275-293.

8.   Thames Flood Prevention. Second Report of
Studies. Greater London Council. 1971.

9.   OTTER, J.R.H., (late Partner of Rendel,
Palmer & Tritton) and DAY, A.S. Tidal Flow
Computations. The Engineer, 29 January 1960,
177.

APPENDIX

BRIEF SUMMARIES OF THE WORK OF VARIOUS
ORGANISATIONS INVOLVED IN THE INVESTIGATION
UP TO 1 JULY 1972

Government Departments

44.   The investigation was carried out at the
request of the Ministry of Housing and Local
Government, who supported the work with a 50%
grant aid.

Following the passing of the Barrier Act in 1972,
responsibility was passed over to the Ministry
of Agriculture, Fisheries and Food. The
Ministry of Transport and the Board of Trade
were represented on the Policy and Steering
Committees, and the Board of Trade on the
Navigation Working Party.

Greater London Council

45.   Initially, the investigation into methods
of protecting London against tidal flooding was
the responsibility of a special team, set up
within the Department of Public Health Engineering. This team became part of Rivers Branch
in 1970 as the result of a reorganisation of
the Department.

In addition to providing administrative support
to the investigation team, Rivers Branch staff
also contributed the following studies to the
work:

46.   Rivers Branch, Department of Public Health
Engineering prepared estimates of the cost of
raising the existing flood defences along both
banks of the river from Teddington Weir east-
wards to Crayfordness in steps of 1 foot up to
6 feet. These were subdivided into sections so
that the cost of the Barrier and associated bank
raising downstream could be assessed for a wide
variety of sites. Benefit cost analyses were
prepared showing the benefit cost ratios of
schemes protecting the London area against floods
of varying return period. Outline studies were
also carried out on various schemes, such as an
outer estuary barrage, river construction at
Canvey Island, combined new river channel and
barrier at Broadness, and a combined new river
channel and barrier at the Isle of Dogs.

Methods of implementing overspill schemes for the
Kent Marshes were also studied. Borehole contracts for ground water studies were let and
river traffic surveys on shipping and lighterage

were carried out. Studies of locking times and
capacities were also made.

47.   The Architect to the Council advised the
Joint Consulting Engineers on the architectural
treatment appropriate for the finally selected
structure. The Director of Mechanical and
Electrical Engineering advised on the mechanical and electrical engineering aspects of the
structure.

48.   The Department of Planning and Transporta-
tion carried out an evaluation of the relative
cost and benefits of the alternative sites and
types of barrier in the Woolwich and Blackwall
Reaches of the River Thames. A study was also
carried out into the possibility of constructing
a bridge across the river in conjunction with the
Barrier.

49.   Other Departments of the Council which
contributed their specialist knowledge and
expertise to the Investigation were those of
the Solicitor, the Parliamentary Officer, the
Valuer, and the Parks Department.

The Joint Consulting Engineers

50.   The two firms of Consulting Engineers
experienced in marine works, Messrs Rendel,
Palmer & Tritton, and Sir Bruce White, Wolfe
Barry & Partners had been commissioned by the
Department of Housing and Local Government to
produce schemes for the flood defence barriers
at selected sites in the Thames Estuary.

It was decided to continue to make use of the
experience and expertise of these firms in the
studies carried out by the GLC Organisation and
a joint team was set up. In the first stage of
the investigation preliminary schemes were drawn
up for some 6 sites totalling 41 different
structures (ref. 4). In the second stage of the
investigation, a more detailed study was made of
a drum gate barrier at a site at Crayfordness,
and a Rising Sector Gate structure at a site in
Woolwich Reach 18½ miles and 8½ miles downstream
of London Bridge respectively (ref. 8).

Research Organisations

(a)   Hydraulics Research Station

51.   The station constructed and operated 1/600
and 1/300 physical models of the Thames Estuary
from Southend Pier to Teddington Weir.
Investigations were carried out into the maximum
levels recorded in the Estuary by surge tides
reaching the level recorded at Southend for the
1953 tide and levels 2.4 and 6 feet above this
level. The effect of various surge profiles
(water levels against time), and the effects
of Barrier operation for either full or partial
closure were studied. Tests were carried out
to ascertain the extent of the area flooded
following overspill of the banks at 1930 statutory levels, and the effect of increasing these
levels by 1 foot and 2 feet. Siltation effects
resulting from flood and tide control operation
were studied. A Report was made of the performance of the "parachute" barrier proposed by
Professor Schofield.

(b) The Institute of Oceanographic Sciences

52. The Institute developed mathematical models for the Thames Estuary and southern North Sea, and used these models to investigate levels throughout the estuary, resulting from an input of the observed 1953 levels at Southend. Investigation of the effects of overspill onto Cliffe Marshes were also carried out, and the effect of temporary narrowing of the river at the seaward end of Canvey Island. The result of Barrier closure at various locations in the estuary, particularly Long Reach and Woolwich, and at different intervals during the tidal cycle were ascertained. As a check, a similar model was used to investigate the effect of closure of the existing Richmond half-tide barrier and the model results compared with those actually measured in the river.

(c) British Hydromechanics Research Association

53. The Association constructed a flume at their laboratory at Cranfield to test out a 1/20 scale gate for instability during closure. Tests were run to simulate all possible conditions during the raising of the gates and forces acting of the gates were measured. Further scour tests were carried out on 1/40 and 1/100 scale two-dimensional models. A study was also carried out into the behaviour of a model drum gate.

(d) Institute of Geological Science

54. The Institute carried out an investigation to determine the effects of a barrage or barrier in raising water levels in the strata forming the Thames Flood Plain. The distribution and thickness of the water bearing deposits within the Flood Plain Terrace were determined by examining some 1000 existing drilling records. 39 bore holes were drilled principally along three lines of section across the Flood Plain. Water levels were monitored with float operated recorders. The work is continuing. A study has been made of the effect of ground water pumping, by the London Transport Executive.

(c) Water Pollution Research Laboratory*

55. Using a mathematical model, the Water Pollution Research Laboratory studied the effect on salinity, temperature and chemical conditions of the estuary, of the operation of a barrier at Woolwich as a tide control structure, closing at half tide levels on the ebb tide, and opening again at a similar level on the flood tides. These studies were subsequently amplified to cover simple flood barrier operations.

Other Organisations

56. The Port of London Authority, the Kent and Essex River Authorities and the Chamber of Shipping of the United Kingdom were represented on the Policy and Steering Committees. The Chamber of Shipping carried out a study into the cost of delay to shipping, of closure of the river at both the centre of Woolwich Reach and at Crayfordness for varying periods from 6 to 12 hours.

Messrs. Cremer and Warner as consultants to the PLA were represented on the Pollution and Siltation Working Party and carried out an investigation into the effect of a barrage scheme at London Bridge on oxygen levels upstream.

---

*   Now the Water Research Centre.

3. The site chosen

R. W. HORNER, TD, BSc, FICE, FIWE, FIPHE, MIMechE, MInstWPC*

INTRODUCTION

1.   As described in the previous Papers, by the autumn of 1970, the Investigation into the problem of tidal flooding of the low lying areas of London had reached a stage where the best solution appeared to be provided by a barrier sited in Woolwich Reach.

2.   The ideal barrier site would have to satisfy the following requirements:

(a) Sound foundations for the massive structure necessary to resist the thrust resulting from the differential head across the gates.

(b) Long straight approaches from either direction for navigation.

(c) Land available on both sides of the river for construction and for the permanent shore installations.

(d) Minimum navigation openings, thus providing an economic and reliable structure.

(e) Minimum overall construction time for the project, including the barrier and the raising of banks downstream.

(f) Minimum overall cost of the combined project.

(g) Site in area where the environmental impact of the barrier would be acceptable.

CHOICE OF SITE

3.   The merits of the various sites against these requirements are summarised in Table 1. Comments on the various sites are as follows:

Cannon Street This site provided a practical and economic solution. The site was limited in space for construction purposes, but this problem could be overcome at a cost. The bulk of the work involved would have been on raising flood defences downstream which would give rise to difficulties.

Blackwall Reach This site was not very attractive with problems over foundations, approaches and land acquisition. The river is narrow here, some 350 m (1150 ft) in width, and this would have made difficult the dredging of an adequate diversion channel required during construction. This site appeared to have little to recommend it.

Woolwich Reach West This site was satisfactory for foundations, approaches, and acquisition of land. In addition the size of the main openings in the structure could be kept down to a reasonable figure. Ample space existed on the north side of the river for dredging a diversion channel while the southern half of the structure was being constructed, and the work in raising banks downstream was not excessive.

Woolwich Reach East This site had good foundation conditions, but approaches were not as good as the Woolwich Reach West site and land acquisition would have been fairly difficult. On balance the saving of about 2 miles of flood defences in comparison with the West site was more than offset by the disadvantages.

Halfway Reach The disadvantage of this site was the requirement of the large main opening for shipping of 305 m (1000 ft) and the poor foundations. In addition the presence of the Ford Motor Works on the north bank would have involved complications.

Crayfordness This site possessed good foundations and land was reasonably available for construction. The disadvantages were the large navigation opening which was necessary, and the poor approaches since the site was on a bend of the river.

Gravesend Foundations here were good, and approaches satisfactory. The main navigation opening required was large, and the site would have been very congested. The PLA considered that they might have to dredge this reach to a greater depth to allow for larger ships going to the Tilbury Docks, and this would have given rise to difficulties in the design of the structure.

From the above it will be seen that the Woolwich Reach West emerges as the most satisfactory site from an overall consideration of the main factors.

* Deputy Director, Department of Public Health Engineering, Project Manager, Thames Barrier, Greater London Council

THE CHOSEN SITE

4.   The main advantage of the Woolwich Reach sites, was that these were the points furthest downstream where a 61 m (200 ft) opening could be used. Downstream of the Royal Docks, the PLA insisted on much larger openings.

5.   The use of a 61 m (200 ft) opening had two considerable advantages. First, either a drop gate, or a Rising Sector Gate of moderate dimensions could be used. Second, the 61 m (200 ft) opening was sufficiently small that total failure of one of the main gates would not allow sufficient water through the opening to create excessively dangerous flood conditions upstream. If therefore the 183 m (600 ft) navigational channel were catered for by providing four main openings, as long as three could be closed under any circumstances, a satisfactory degree of reliability would have been attained. Operationally the situation therefore changed dramatically from efforts to achieve 100% reliability, to a target of not less than 75% reliability. Of course, from the design aspect, every effort was still made to reach 100% reliability, leaving the possibility of tolerable conditions with the failure of one gate as a safety factor.

6.   The main opening factor was paramount, but it is interesting to note how conveniently the Woolwich Reach West site was advantageous in other respects. Foundation conditions were reasonably good, approaches were good, land was available on both north and south banks, and the width of the river was adequate to allow for the dredging of a diversion channel for use of shipping during the construction period.

7.   The use of four main openings of 61 m (200 ft) was generous, but had the advantage that two openings would be available for inward bound traffic and two openings for outward bound. The provision of only three main openings was considered, but it was felt that the use of the central opening for inward bound traffic on the flood tide, and outward bound traffic on the ebb tide would entail a risk of confusion and collision. In respect of the overall cost of the structure, the cost of the additional pier involved, in substituting two smaller openings for the fourth main opening, would certainly have outweighed the economy possible on the gates and machinery.

8.   It was decided that all the four main gates would have sills at the same level and therefore the gates would be identical. The model tests detailed in Paper No.5 indicated that this arrangement might suffer from some siltation in the approaches to the northernmost main gate, and also to a lesser extent in the case of the most southern main gate. Experience so far during the construction stage suggests that this should not be too serious a problem.

Table 1

SITE	Foundations	Approaches	Land	Navigational Opening (metres)	Overall Construction Time (years)	Total Cost (Y)	Environ- mental Acceptibi- lity
Cannon Street	S	S	D	46	7	88	S
Blackwall Reach	P	P	P	61	6	89	P
Woolwich Reach - West	G	G	G	61	5	85	G
Woolwich Reach - East	G	S	D	61	5	85	G
Halfway Reach	P	S	D	305	7	94	G
Crayfordness	G	P	S	427	7	90	G
Gravesend	G	S	D	427	7	100	S

Remarks: S - Satisfactory   Y - Percentage of the most
           P - Poor   expensive overall scheme
           G - Good
           D - Difficult

9. The level of the sills was fixed to conform with the "ruling depth" for this reach of the river. This gave a level of -9.25 m (-30.4 ft) below Ordnance Datum Newlyn for the top of the sill structure.

10. To avoid problems with the regime of the river, the requirement was laid down that the area of waterway at half tide at the Barrier Site should not be reduced by more than 25% by the structure. It was a fortunate aspect of the chosen site, that the existing area of waterway is rather greater than the net "regime" area for that particular point in the estuary, and therefore a figure as high as 25% could be accepted.

11. The minimum width of pier to provide adequate strength and stability, and also sufficient space to house operating machinery, was fixed as 11 m (36 ft). It was a requirement by the navigation authorities that the overall length of the piers should be kept to a minimum, and a dimension of 61 m (200 ft) was aimed at. In the final design, this had to be increased to 65 m (213 ft) but this was to allow for the cut waters. In developing the layout the minimum width of the pier was 11 m over a length of 55 m (180.5 ft).

12. Besides the main 61 m (200 ft) openings, the PLA considered that for navigation, it would be desirable to provide a subsidiary 31.5 m (103 ft) opening on the north side, adjacent to the main spans, and a similar opening on the south. The level of the top of the sill for these openings was required to be at -4.6 m (-15.1 ft) O.D.N. level. It was decided that the remaining openings for the structure could conveniently be made 31.5 m (103 ft) in width, and consideration of the regime requirement gave a top of sill level of zero O.D.N. for these openings.

CONCLUDING DESIGN REQUIREMENTS

13. The stage was thus reached where the main design requirements for the structure had largely been determined. There remained the important consideration of the head across the structure for which it should be designed. It was felt from operational aspects, that the authority responsible for decision to close the structure, should not be inhibited from setting the closure operation in motion due to constraints of the head which the structure could accept. It was at that time considered desirable to close at about low water if the decision could be taken at that stage, in order to minimise effects downstream.

The decision was therefore taken that the structure should be capable of resisting the full differential head of a low water closure with a maximum surge tide. In practice of course it will be possible to allow water through the structure and there is every chance that this onerous condition will not be met.

14. A further aspect that influenced the design was the need for the main gates to remain closed over the low water period should a surge

tide lasting for some 18 or 24 hours occur. In these circumstances there might be reluctance to open the main gates and then re-close. For this reason, it was considered desirable that the gates should be able to hold back water from upstream over the low water period. The design requirements as put to the Consulting Engineers for the design of the final structure are detailed below:

Design Requirements

15. The main factors which dictated the design of the Barrier were as follows:-

a. Flood Defence Criteria

Three conditions of loading were considered as follows:-

(i) Water levels on the downriver side 6.9 m (22.5 ft) O.D.N. with water level on the upriver side -1.5 m (-5.0 ft) O.D.N. under normal conditions with the gate crest in the closed position at minimum level of 6.9 m (22.5 ft) O.D.N.

(ii) Water level on the downriver side 7.2 m (23.5 ft) O.D.N. with water level on the upriver side -2.7 m (-9.0 ft) O.D.N. in abnormal conditions with 0.3 m (1.0 ft) overspill flowing over the gate.

(iii) Water level on the downriver side -1.8 m (-6.0 ft) O.D.N. with water level on the upriver side 4.3 m (14.0 ft) O.D.N.

b. Navigational Requirements

(i) 61 m (200 ft) clear width main navigational openings.

(ii) In addition two subsidiary openings of 31.5 m (103 ft) clear width to be provided.

(iii) Piers to be as narrow as possible within considerations of structural strength.

(iv) Sills of the main openings to be at a level which does not restrict navigation in the Reach, or lead to excessive siltation.

c. Miscellaneous Conditions

(i) The gate closure operation to require less than one hour from start to full closure.

(ii) The structure to have the maximum possible reliability in operation.

(iii) Structure to be as unobtrusive as possible in the environment of Woolwich Reach.

ACKNOWLEDGEMENT

15. The Author thanks the Director of Public Health Engineering, and the Greater London Council for permission to publish this paper. Any opinions stated are those of the Author and do not necessarily represent Council policy.

Discussion on Papers 1–3

MR BECKETT

1. The river is not a rigid body yet its whole motion is to be changed by the application of a braking effect at one position only in its length. The river mass can be measured in millions of tons. When we close the barrier on the upstream-flowing river, we are, in effect, trying to stop its movement throughout its length, but not all at once. Immediately upstream of the barrier, water level will be depressed in the form of a trough. This will extend to the tidal limit, return as a wave and, in due course, create a to-and-flow oscillation which decays but leaves levels rising gradually due to the inflow of fresh water from up river. Downstream of the barrier the obstruction to the river movement will cause a piling up against the structure and this piling up effect will travel downstream, converting some of the kinetic energy of the moving river into potential energy stored in its raised level. Thus there will be a general upheaving of the whole river proceeding seawards, the height of which depends on the river depth, its velocity and the speed at which the front of the upheavement travels seawards.

2. On the upstream side of the barrier there is a trough moving away from it and lowering the level whereas on the seaward side there is an elevation or upheaving moving seawards. The height of the upheaving can be calculated from the volumetric formula in which the speed of the crest multiplied by the height of the upheaving equals the depth of river multiplied by the change of velocity, the change being brought about by the barrier itself. Thus if we let some of the water flow through the barrier we reduce the height of the upheaving.

3. The shape of the front of the upheaving depends on the speed at which barrier closure takes place. Slow closure will produce a flattish gradient. The process of raising water level downstream will cease when the front of this reflected wave reaches the rapidly widening estuary whereupon the super-elevated river will become unstable, producing a reversion to kinetic energy, and seaward flow followed by a subsidence or trough running back to the barrier. This trough will reduce the level of the river as it runs back to the barrier and, if carefully timed to coincide with high water, reduce the risk of flooding.

4. With careful timing barrier closure and control of flow through the structure it will be possible to avoid the aggravation of flood risk downstream and even improve conditions by timing the returning trough to coincide with high water where flood defence levels may be critical. Experience in barrier operation is, of course, essential but all the devices necessary to make it possible are to be incorporated in the structure.

COLONEL S.K. GILBERT (Adviser to Department of the Environment)

5. All along there have been those for the Thames Barrier and saving London and those for the Port of London Authority and saving London's shipping. Some wanted a structure in the river and others did not. In 1959 agreement on a 500 ft opening at Long Reach was reached for which the consulting engineers produced proposals. However, they were unacceptable to the Port of London Authority who proposed an alternative site at Crayfordness where the required opening was 1400 ft.

6. In 1965 there was further disagreement. Gains in efficiency, through containerisation, moved the project down river from London to Tilbury and in effect freed part of the upper reaches for a barrier with 200 ft gaps. These difficulties explain the delays which occurred between the initiation and the start of construction of the barrier.

MR A.R. YOUNG (Partner, Robert Cuthbertson and Partners)

7. In 1968, I had the privilege of being a member of the design team set up to carry out feasibility studies for the Greater London Council who were responsible for London's flood problem. The terms of reference were fairly wide. At that stage there were several sites to be considered. The choice was narrowed down to Woolwich and various types of structure were considered. One suggestion from a member of the general public was a parachute idea which comprised a folded, flexible membrane on the bed of the river. Tubes passed between the fold which, when inflated, caused the membrane to form a parachute. This device was held in place by another membrane placed flat on the bed of the river. The parachute was attached to this membrane and there were also ties from the top of the parachute down to the membrane. Underneath the foundation membrane was a drainage system which was pumped out thereby causing suction, pressing the whole thing down on to the bed of the river. It was an ingenious device but one could not imagine what would take the enormous horizontal forces involved.

8. Another idea came from someone who thought that the Thames could be cured from flooding by straightening it out, improving the gradient and increasing the cross-sectional area. He must have been thinking about upland flooding because this certainly would not have prevented the propagation of a surge tide into the heart of London.

9. The idea of the rising sector gate as a solution was put forward by Mr Draper in 1970 and another report of studies was submitted to the Greater London Council. This was accepted and in order to convert this idea into a reliable structure a tremendous amount of work had to be done.

DR D.E. WRIGHT (Sir William Halcrow and Partners)

10. The decision to build the barrier ranks as one of the single largest investment decisions of its kind that have been made in recent years. There is no doubt that the decision to invest in works which will reduce the risk of flooding is correct, but the consequential point concerns the degree of protection which it is appropriate to provide. This directly affects the scale of the project, its costs and its anticipated benefits which lie essentially in the reduction of future anticipated losses.

11. Perhaps the single parameter which best characterizes the scale of this project is the elevation of the surge for which protection is to be provided and hence the height of the barrier and the flood defence walls required on both sides of the estuary downstream of it.

12. The need for an economic input to the overall justification of the project is recognized by the Authors. Sir Hermann Bondi's view (Paper 1, paragraph 2) was that a disaster of the kind caused by an exceptional surge was a risk that could not be tolerated, irrespective of odds. In paragraph 2 of Paper 2, the Authors note that the standard evolved was a compromise between the scheme with an optimum benefit-cost ratio and the one desirable to give a sense of complete security. They also state that the decision on the standard to be provided was a matter of judgement supported by benefit-cost studies. Despite Sir Hermann's remark, the standard of protection that is to be provided by the barrier is still finite. Were yet higher standards of protection considered in the process of arriving at the present scheme?

13. Although most flood alleviation schemes are on a similar scale and less complex than the barrier project, could the Authors give more information about the methods used in the benefit-cost studies, the point to which the analysis was carried, and the nature of the compromise between the optimum economic scheme and the one required for complete security?

MR G. COLE (Ministry of Agriculture, Fisheries and Food)

14. I do not believe one should say that interaction does not occur. Suthons,[1] using Jenkinson's method,[2] worked with observed sea levels because they were readily available and convenient to use.

15. Although the inverted barometer is important with small surges, on a big surge what counts is the sweeping of the water down the North Sea at approximately the same velocity as the tide. Thus, in the Waverley Report[3] it is stated that a surge is mainly due to strong winds blowing forward the surface waters of the sea, although small accumulations of water may also be produced by changes in atmospheric pressure. Suthons stated: "It was well known that, roughly speaking, if the pressure fell by 1 inch of mercury there would be a rise of sea level of about 1 foot...". The surge of 1953 reached nearly 9 ft at Southend.

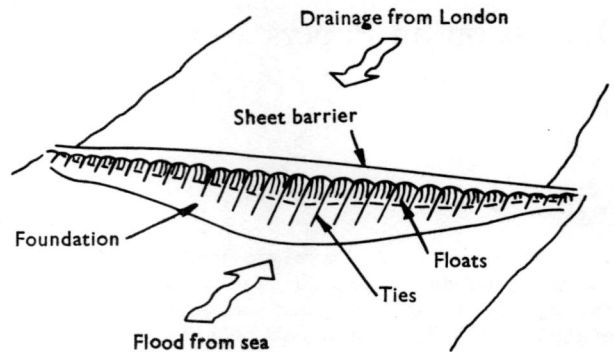

Fig.1

MR J.M. HOLLOWAY (Department of Public Health Engineering, Greater London Council)

16. The GLC's Thames Barrier project team looked at the possibilities of tidal power generation associated with the barrier in Woolwich Reach. The mean tide amplitude of the Thames there is about 5.4 m and the average tidal discharge can be assumed to be of the order of 2800 cumecs on both the ebb and the flood tide.

17. The greatest power output from a run of the river scheme would be obtained from generating on both the ebb and flood tides, using bulb type turbines similar to those installed at La Rance in France. The installed capacity would be limited to about 30 MW and on the basis of ten hours' generation every twenty-four hours the annual output would be only about 88 million units.

18. The installation of power generation turbines would convert a movable barrier into a barrage and shipping would then have to be locked through. A lock of the size required to handle the volume of shipping would increase the capital and operating costs of the works considerably.

19. Essentially the generation of power would entail the replacement of the four main barrier gates by perhaps 32 turbine generator units and in estimating the cost of power generation the cost of the four main gates can be set as a credit against the turbine-generator costs. Even so, with the additional cost of a shipping lock, the cost of tidal power intermittently generated at the site becomes totally uneconomic and it was considered that tidal power generation could not be justified for inclusion in the project.

PROFESSOR A.N. SCHOFIELD (University Engineering Department, Cambridge)

20. A flood from the sea causes a horizontal thrust of tens of thousands of tonnes against a flood barrier. In other barrier designs this thrust is resisted by base friction and lateral soil pressure against massive concrete caissons which rest on the river bed. My proposed parachute[4] (Fig.1) would only have been anchored to drainage tubes lying on the river bed (Fig.2), which would have formed a strong but lightweight grillage. However, above this grillage of strong tubes a sealed pavement was to extend across the river bed, forming a continuous membrane below which there would have been continual pumping of water from the drains. The total weight of water

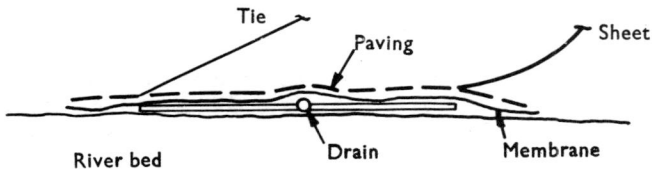

Fig.2

above the membrane would therefore have generated a large vertical effective stress between the grillage and the soil, so that large frictional forces would have been available to resist horizontal movement of the grillage. By employing simple principles of hydrostatics and soil mechanics this flexible but strong underwater structure, made from only a few thousand tonnes of modern materials, could have resisted tens of thousands of tonnes of horizontal thrust, by exploiting the weight of hundreds of thousands of tonnes of overlying flood water. The flood would have been made to hold itself back by 'sitting on its own coat tails'.

21. Until reading Paper 2 I was unaware that there had been any adverse technical report on such a parachute barrier. The models which I tested proved entirely reliable except under conditions of heavy overflow, in which case other fabric dams are also known to flutter. However, that adverse condition was not intended to occur, as the closure operation was envisaged to take place rapidly and without overflow. The location envisaged was in Long Reach, and river banks downstream would have had to be raised to contain the reflected wave, to allow closure just before river banks upstream began to be at risk. The parachute was not designed to meet sophisticated requirements such as tide control on an ebb tide (Fig.3). It was not considered impermanent because plastics material survives much better than metal out of sight on the sea bed. The parachute was simply considered to be an economical and reliable design with a good probability of effecting closure safely in a rare storm surge.

22. It is interesting to consider in this connection the costs listed in Table 3 of Paper 2. In 1970 I argued[5] the overriding need for interim bank raising by following the discounting methods that were set out in reference 6. My simple approach was to guess that the risk in 1970 ran at about £100 million a year, corresponding to a one in twenty risk of a £2000 million loss in any one season of risk. An alternative way of guessing the risk was to suppose that the 2000 year return period flood might cause £2000 million loss, that the 200 year flood might select and inundate a quarter of the area of risk and that a 20 year flood might inundate a sixteenth of the area. Adopting such an approach one could guess that the premiums for insurance against such risks could not be less than those shown in Table 1.

23. On that basis the risk in 1970 ran at about £10 million a year. The 1972 interim bank raising cost less than £4 million and was economically justified because it eliminated the 1:20 year risk. However, permanent works to eliminate longer term risks could never have been justified on such cost-benefit calculations. Clearly in London[7] as in Holland[8] in 1953, people at risk

from tidal flooding placed a much higher value on permanent major works than appeared justified by engineering economic calculations using discounting methods. It appears that once people agree that there is any really significant but unfamiliar risk, they will pay a stiff premium for a sound scheme which they feel confident will virtually eliminate all such risks - people will only enter into calculation of engineering economics in circumstances in which they are familiar with and are prepared to live with a risk. The premium in this case was very stiff; the cost of the present permanent barrier and of all associated works downstream of it may well be a hundred times as much as the cost of the interim works - much more than the cost of a parachute barrier.

24. The relative level of the tidal defences is expected to continue to fall over the next century (although part of the fall might stop if the aquifer below the London clay were to be recharged, which would end the present continuing underdrainage). It may not be possible to raise the river banks again; centrifugal model tests show that there is a limit to bank raising because of the thickness and softness of the alluvium on which some of the banks rest. If further bank raising is not possible then at a future date there may be no alternative to the construction of a second barrier a long way downstream, say in Gravesend Reach. A parachute barrier could well be appropriate in the next century, and in view of the difficulty there was in making the parachute barrier concept credible this time it may be appropriate to start some research and trial construction now so that such structures are better understood next time.

MR J.H. FLEMING (Partner, Sir M. MacDonald and Partners)

25. Comparison of various solutions for the 30 m wide Hull Tidal Surge Barrier indicated that the most economical gate was a vertical lifting gate, and that the adoption of a turn-over type enabled the tower height to be kept within the limits prescribed and yet provided sufficient clearance for navigation. It would therefore appear that

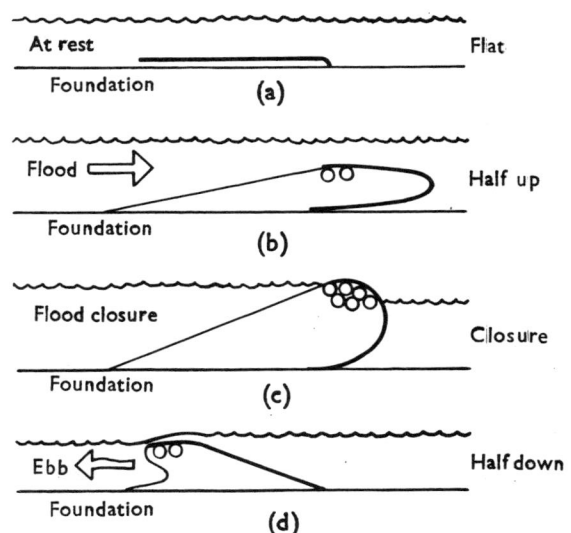

Fig.3

a similar solution could have been appropriate for the 31.5 m navigation openings in the Thames Barrier, and simple lifting gates would have served for the other 31.5 m openings.

26. Since a vertical lifting gate was also considered for the original 137 m opening, and was the favoured solution in that case (paragraph 27 of Paper 2), it would seem that a similar system would be technically feasible for the 61 m openings finally adopted.

27. A big factor in adopting the rising sector gate was the need for the structure to be as unobtrusive as possible in the environment of Woolwich Reach. While considering the impact of a tidal barrier on the natural beauty of Woolwich Reach, let us imagine a proposal to demolish Tower Bridge because it is not unobtrusive in the environment of the Tower of London! Tall engineering structures are not objectionable per se.

28. Can the Authors give a cost comparison between the solution adopted and a structure with tall towers and lifting gates? It is understandable that there were local pressures against building a tall structure, and all concerned deserve great credit for the manner in which they were able to accommodate these views, but the operational advantages of lifting gates in a tidal barrier are self-evident, and it is likely that there would be a cost advantage in adopting them for some, or all, of the openings. If a price had to be paid for unobtrusiveness, then this price should be recorded.

MR J.R. O'DONNELL (Chief Engineer, Marine Services, Port of London Authority)

29. The primary responsibility of the PLA in connection with this project was to exercise its statutory duty relating to navigation on the Thames.

30. The width of the entrance lock to India and Millwall docks is 80 ft and that to the Royal Docks 100 ft and the PLA were responsible for ensuring that large sea-going vessels transitting these locks were not endangered in navigating underway through the barrier wherever it was to be placed and in all weathers and conditions.

31. The tidal Thames is a waterway with a public right of navigation and for the two principal sites discussed, namely Long Reach and Woolwich Reach, the deep sea vessels passing these locations would be up to 680 ft long and 86 ft beam and 550 ft long and 73 ft beam respectively. Furthermore, the larger vessels invariably move near high water and, because there is a tendency for increased vessel movement inwards at weekends to commence discharge on Monday and there is a corresponding heavy outward movement at the end of a working week when ships have been discharged, the barrier design had to accommodate a maximum movement situation. Such a typical traffic movement in a tidal cycle would be 7 vessels of over 3000 t, 13 vessels between 1000 and 3000 and 13 under 1000 t, as well as over 200 other craft such as colliers, coastal tankers, ballast supply vessels, tugs with tows and other local and self-propelled craft.

32. To avoid danger from a collision not only between vessels but also between a vessel and the barrier, the widest possible navigation openings need to be provided and a mariner would need

a navigation width of at least three times the beam of his vessel in order to transit without mishap in all states of tide and weather conditions. If one major ship opening were provided then a total distance between piers of six ship widths would be required with two side openings of lesser width for the many other movements.

33. The PLA were not anti-barrier if only because dock areas would be subject to flooding in the event of a large surge without such protection, but they had to ensure that the right of free relatively unhindered navigation was provided to keep alive that activity on which London was primarily founded. In fact, the imposed decision of the early 1970's for four 200 ft openings with two narrower and shallower side openings of 100 ft for a Woolwich Reach Barrier fell short of the requirements of the PLA. However, since then, the movements of the larger vessels have reduced in numbers and the associated risks are also reduced.

34. Very large vessels are frequent visitors to London and it is not part of the PLA's role to reduce the capacity of Britain's most important water highway or to permit its reduction.

35. The air drafts required by the PLA were not primarily for ocean-going vessels but mainly for the movement of the PLA's heavy lift floating crane 'Mammoth' which worked at Tilbury as well as the Upper Docks. This crane had an air draft of 200 ft, to which must be added clearances and height of tide above OD in order to arrive at an acceptable level for any structure across the tideway.

MR HORNER, MR BECKETT and MR DRAPER

36. Dr Wright raised the point of benefit-cost and standard of defence being provided. Total cost of flood defence for the London area is made up of the barrier cost and the cost of associated flood defence banks downstream. The barrier cost is comparatively insensitive to standard of protection decided upon, the main cost being that of providing a structure and operating it. The strength of the gates and the size of the piers are determined by the differential head across the structure. Early closure on a 100 year return period high water would only give the same differential across the gates as a slightly later closure, by about one quarter hour, against a 1000 year return period tide.

37. On average, bank costs on industrial frontages are roughly proportional to the height by which the level must be raised. The difference of bank costs in the area of GLC responsibility between the 1000 and 100 year standard would be about 30%, say £14 million.

38. At the time the benefit-cost study was carried out the value of the barrier and the associated raising of the banks downstream in the London area was £61 million (1970 prices). The annual cost of flood damage was some £5 million, which gives a benefit-cost ratio of about unity for a discount rate of 10%. A lower standard than the 1000 year return period would have favoured a solution to the problem by raising banks throughout the estuary, as an alternative to the construction of the barrier. This solution was of course adopted in the interim bank work of 1971-72, which provided a 60 year

return period standard. Due to inflation, costs
of the flood protection works had risen considerably, but the cost of flood damage would rise in
a similar ratio, so that the conclusions reached
as a result of the earlier studies should still
be correct.
39. The discounted cash flow technique tends to
favour low cost protection against short return
period events. A high benefit-cost ratio means
more frequent flooding. This of course is not
what the public want. As a result of the flooding on the tributaries of the Thames in the sum-
mer of 1977, the cry has been heard again and
again that action must be taken to prevent
a recurrence of this kind of event. The additional cost on the whole project in the GLC area
of less than 10% to raise the standard of protection from the 100 year return period to the 1000
year return period is therefore readily accepted.
40. The 1000 year return period standard of protection, with allowance for adverse change up to
the year 2030, was considered to be the highest
level to which existing flood defences downstream
of the barrier site could be raised in view of
the cost of overcoming the engineering difficulties. The barrier itself of course could resist
a higher surge quite satisfactorily with a
slightly later closure, the design differential
head across the structure not being exceeded.
41. We agree with Mr Cole that there is reason
to believe that interaction between surge and
tide occurs, and that the wind stress component
of the surge is probably more important than
pressure effect in those cases.
42. Professor Schofield's interesting and ingenious proposal for a parachute barrier was given
serious consideration. His view that if the
adverse trend affecting the Thames Estuary continued, then in say another 100 years or so a
more ambitious scheme in the outer estuary would
be necessary to provide for the flood defence of
the area, was one with which few engineers would
disagree. Under these circumstances, the parachute barrier might well prove to be an effective form of flood defence. With reference to
the favourable benefit-cost aspect of the interim bank raising, in addition to containing
the one in 60 year risk, it also had the benefit
of much reducing the scale of flooding resulting
from even higher water levels.
43. In answer to Mr Fleming (paragraph 28) the
drop gate scheme was for a 450 ft central span
and therefore was not strictly comparable with
the rising sector gate design which was 17%
cheaper. The rising sector gate was much favoured by the planning authorities, who considered the 'visual intrusion' on the vista along
Woolwich Reach less objectionable. However, a
serious difficulty with the drop gate at this
particular site was the risk of vibration or
'bouncing' of the gate as it passed through the
rapidly moving flow of water. Because of the
shape of the Thames Estuary, closure of the gate
generated a considerable differential head across
the structure within a few minutes. This produced a high efflux velocity under the gate
which, impinging on the lateral bracing of the
gate, might cause it to bounce. In order to
eliminate this phenomenon the gate would have to
be lowered at slack water, low tide, which in
turn places an unacceptable operating restriction

on barrier closure times. The surge warning
would have to be given very early in order to
comply with this and would therefore destroy the
flexibility that has been built into the warning
system.
44. The situation where a drop gate is sited
at the entrance to a creek, such as the site at
Hull, is entirely different since there is no
possibility of the build up of a large differential across the gate during closure.
45. Mr O'Donnell mentioned the navigational
requirements which imposed such onerous conditions on the earlier designs of the barrier.
Final design for the barrier was inevitably a
compromise between navigational requirements and
engineering factors.

MR TAPPIN

46. Mr Fleming mentioned the subject of drop
gates for larger spans and referred to the Hull
Barrier. The Hull Barrier gate is designed to
span approximately 30 m and while it must resist
a similar maximum differential water head to
that related to the Thames Barrier, the depth
from surge level to sill level is considerably
smaller. This results in the hydrostatic load
per unit length on the Thames Barrier being over
three times that of the Hull structure.
47. For the span and loading requirements of
the Hull Barrier, a vertical skin-plate stiffened
by horizontal plate girder forms a very satisfactory and economical structure. However, this
structural form does not extrapolate to larger
spans where heavier loads must be resisted. The
plate girders must be replaced by deep horizontal trusses and the gate becomes a considerably
larger structure with consequent problems relating to the operating machinery and support structures.
48. This form of gate was considered but rejected, partly for the reasons above and partly
because of the problem of hydrodynamically induced vibrations.

REFERENCES

1. SUTHONS C.T. Discussion. Conference on North
   Sea Floods. Institution of Civil Engineers,
   London, 1954, 38-39.

2. JENKINSON A.F. Frequency distribution of the
   annual maximum (or minimum) values of meteorological elements. Q. Jl R. Met. Soc., Apr.,
   1955, 81, 158-171.

3. DEPARTMENTAL COMMITTEE ON COASTAL FLOODING.
   Report. HMSO, London, 1954, Cmd 9165.

4. SCHOFIELD A.N. A parachute to tame the tide.
   New Scientist, 1970, 8 Oct., 66-68.

5. SCHOFIELD A.N. Cost of preventing Thames tidal
   flood. Nature, 1970, 227, 1203-1209.

6. INSTITUTION OF CIVIL ENGINEERS. An introduction to engineering economics. Institution of
   Civil Engineers, London, 1969.

7. BOWDEN Lord V. Letter. The Times, 1970,
   26 Nov.

8. UNITED NATIONS. Flood problems in deltaic
   areas. Proc. 4th Reg. Tech. Conf. Wave
   Resource Development, Colombo, Ceylon, 1960.
   Economic Commission for Asia and the Far East,
   Bangkok, 1962.

4. Site investigation and geotechnical considerations

P. G. FOOKES, PhD, FIMM, FGS*

P. L. MARTIN, MSc, MICE†

INTRODUCTION

1. The Paper describes the final geotechnical
investigation carried out prior to tenders being
invited for the construction of the Barrier and
discusses the results of the investigation,
particularly in relation to the design of the
foundations to the piers and abutments. The
geological and engineering characteristics of
the alluvial deposits and bedrock at the Site
are described, and the criteria adopted in the
selection of founding levels are discussed.

2. The northern end of the Barrier will be
founded in Thanet Sand and the southern end in
Chalk. The investigations indicated that the
Thanet Sand is reworked near its upper surface
and showed that there is a general absence of
solution and solifluction features in the Chalk.
Chalk at the southern end of the Site is frost-
shattered to a depth of a few metres below the
surface but where the Chalk is overlain by
Thanet Sand frost-shattering was found to be
absent. Several small faults were located at
the Site.

GEOLOGY OF THE SITE

3. The regional position of the Site in its
solid geological setting is shown in plan and
cross-section on Fig. 1. The basic geology of
the London area is fairly simple with the Thames
running over a gentle synclinal basin composed
of Tertiary sediments overlying Cretaceous Chalk,
the angle of dip of the beds usually being less
than 30⁰ to the horizontal. North and south of
this basin the Chalk outcrops at the edge of the
syncline. Fig. 1 shows the axes of the larger
fault and fold structures in the area. The
principal trends present are:-

(i) north/south monoclines

(ii) east-west folds (anticlines and synclines)

(iii)north-east/south-west folds and faults.

4. Wooldridge (ref. 1) has attributed these
fault and fold trends to continuing movement on
existing structural lines deep in the underlying
hard rock basement (the Devonian and older), and
to movements during the Tertiary mainly related

to compressive forces emanating from the growing
Alpine mountains in southern Europe.

5. The area around the Site has been disturbed
by tectonic movements. The nearest known large
fault is the "Greenwich Fault", which is a normal
fault belonging to group (iii) with a maximum
downthrow of some 30 m to the north-west. A
small synclinal fold parallel to the fault occurs
a little to the south of the Site and has exposed the London Clay in its axis, with the
underlying older Tertiaries and the Chalk
successively outcropping northwards between the
fold and the Site.

6. From the maps and records of the Institute
of Geological Sciences, local literature, visits
to local exposures, and data from previous site
investigations in the area carried out during
studies to select the Site (Paper 3), the probable distribution and sequence of the soils and
bedrock at the Site was determined. These investigations indicated that just north of the
river, at the Site, alluvial deposits overlie
Thanet Sand which in turn overlie the Chalk,
whilst just south of the river alluvial deposits
rest directly on the Chalk.

7. The chalk was expected to be part of the
Upper Chalk, which is fissured and jointed and
contains flints, and to range in age from
approximately the thick macrofossil zones of
Terebratulina lata to Micraster coranguinum.
The boundary between the Chalk and the Thanet
Sand is usually marked by the Bull Head Bed,
which is up to half a metre thick and generally
consists of irregular, greenish-coated flints
embedded in silty and sandy clay.

8. The Thanet Sand, the lowest local member of
the Tertiary, rests unconformably on the eroded
Upper Chalk surface and was expected to consist
of a dense fine sand, light grey to buff in
colour, or dark greenish where the unoxidised
mineral glauconite was present, with possibly
some silty layers and occasional lenses of clay.
It was considered that, as a result of erosion,
the boundary between the Thanet Sand and the
overlying alluvial deposits would be irregular
and possibly ill-defined.

9. The alluvial deposits were expected to consist of Flood Plain Gravel, which is the most
recent of the terrace deposits that line the
Thames Valley, overlain by Recent Alluvium and
river bed sediment.

*Consultant, Rendel, Palmer and Tritton
†Senior Soils Engineer, Rendel, Palmer and Tritton

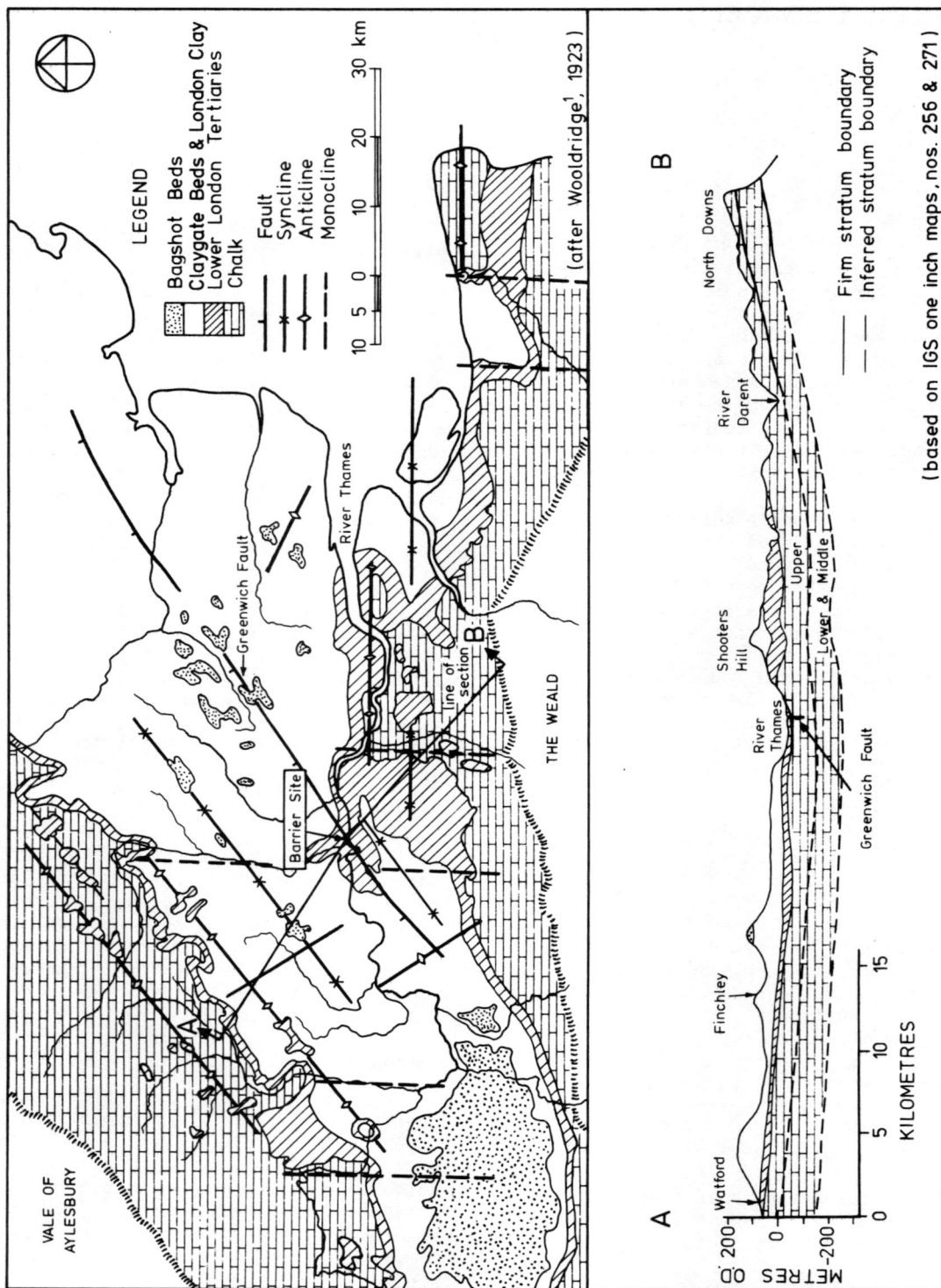

Fig.1. Plan and section showing barrier site in relation to geological structures in the London Basin

10. The important geological events of the Quaternary Ice Ages (Table 1) had a profound effect on the engineering properties of all near-surface materials. In the London Basin during the Middle Pleistocene, glaciers of the Anglian Glaciation displaced the course of the Thames, then further north than today, from about Watford and Finchley southward (Fig. 2). However, the glaciers of the subsequent Wolstonian (next to last) and Devensian (last) glaciations, which occurred during the Upper Pleistocene, did not quite reach as far south as the London Basin, and the Thames cut for itself a series of channels in response to the low glacial sea levels. It is interesting to note that at the Site the bed of the Thames now lies approximately along the axis of the London Basin syncline (Fig. 1).

11. During the warmer interglacial periods (the Hoxnian and the Ipswichian), when sea levels were higher, extensive aggradation of alluvium took place with the development of various terraces and including finally, the lowest, the Flood Plain terrace. Subsequently, during the current post-glacial period, terrace deposits continued to be laid down to form the Recent Alluvium.

12. During the cold glacial periods, deep frost-shattering of the chalk took place together with the development of 'Coombe Rock' - extensive solifluction sheets of broken and moved chalk debris. On wetting and working, this material, in particular, easily forms 'putty chalk'.

13. In addition to the frost-shattering, solifluction and other cold climate phenomena affecting the chalk, extensive development of pipes and other solution features also occurred. Their formation was not necessarily related only to the cold periods (ref. 3) but may also have occurred during warmer periods, and might possibly be continuing at the present time (ref. 4). It was considered, therefore, that the chalk at the Barrier Site could have a frost-shattered surface, possibly even covered with solifluction debris, and perhaps containing various solution features, narrow pipes or deep dish-shaped depressions. In addition to the features particular to the Chalk, the possibility of small undiscovered faults and of buried channels existed. One of the principal requirements of the site investigation was, therefore, to establish the presence or absence of these features.

SITE INVESTIGATION

14. After various studies, detailed elsewhere (ref. 5) and discussed in preceding papers of this Symposium, a preliminary site investigation was carried out in Woolwich Reach during 1970. This work was performed by Foundation Engineering Ltd., who sank seven boreholes and carried out in-situ testing at the locations G1 to G7 shown in Fig. 3. Following the selection of the site for the Barrier the detailed investigation was carried out.

15. The detailed investigation as initially envisaged was to have included the sinking of small diameter shafts down to and into the chalk to permit in-situ inspection, plate loading tests and the taking of bulk and undisturbed samples. However, it was concluded that due to potential problems associated with groundwater, the effect of disturbance to the chalk by the shaft construction, and the lack of assurance regarding the representability of the results as applied to the whole site, the expenditure and effort would not be justified. The proposal was therefore abandoned and it was decided that evaluation of the ground conditions at the Site would be based mainly on a boring and drilling programme with detailed core logging and correlation of standard penetration tests with results available from in-situ tests at other sites, together with the information obtained from borehole permeability tests, laboratory tests on samples, a micropalaeontological assessment of the chalk stratigraphy and structure, and studies on the existing literature and relevant histories.

16. The site investigation contract was let to Foundation Engineering Ltd., who carried out the work between December 1971 and May 1972. The investigation involved the sinking of seventy-three boreholes, fifty-four of which were over water. Sixty-eight of the boreholes (Nos. 1 to 24, 24A, 24B, 25 to 55, 60, 60A and 61 to 69) were directly related to the piers and abutments of the Barrier, the associated river walls and the onshore works, while the remaining five (Nos. D1 to D5) were in connection with the dredging for the proposed diversion channel. The locations of these boreholes are shown in Fig. 3, except those of D1, D2, D4 and D5 which lie outside the area. Because of a modification made, after the completion of the site investigation, to the design of the support system for the gates (for structural rather than for geotechnical reasons, as described in Paper 14), the boreholes at the northern end of the Site are located slightly to the west of the Barrier axis.

17. As a separate part of the work carried out under the contract, seventeen land boreholes were sunk on the north and south banks of the river for the installation of piezometers and recording equipment in connection with the regional groundwater investigations being carried out by the Institute of Geological Sciences. The locations of those boreholes sunk within the Site boundary (W12A, W13, W14A, W14B and W15) are also shown on Fig. 3.

18. The site investigation boreholes were sunk by shell and auger methods in the soils overlying the chalk and in the weaker chalk, and, where required, were continued by rotary core drilling. Standard penetration tests (S.P.T.'s) and 100 mm diameter open-drive sampling were carried out, as appropriate, in the shell and auger boreholes and representative disturbed samples were taken. An H double tube, swivel-type core barrel with a face discharge bit taking a 73 mm diameter core,

Table 1. Geological succession in the London Basin

PERIOD	EPOCH	STAGE	DEPOSIT (INCLUDING THAMES TERRACES)	AGE YEARS
QUATERNARY	HOLOCENE (= RECENT)	FLANDRIAN	Estuarine Alluvium	10,000
	PLEISTOCENE (Upper)	DEVENSIAN (cold/ glacial) low sea level	Main Solifluction Deposits 'Lower Flood Plain' Terrace	
		IPSWICHIAN (warm in- terglacial) high sea level	'Upper Flood Plain' Terrace	
		WOLSTONIAN (cold/ glacial)	Taplow Terrace Main Coombe Rock	
	(Middle)	HOXNIAN (warm/in- terglacial)	'Boyn Hill' Terrace	
		ANGLIAN (cold/ glacial	Chalky Boulder Clay Glacial Sand & Gravel	
	(Lower)	Uncertain	Pebble Gravels Plateau Gravels	2M
TERTIARY	EOCENE		Bagshot and Clay- gate Beds London Clay Lower London Tertiaries (in cludes Thanet Sand)	38M 55M
CRETACEOUS			Upper Chalk	

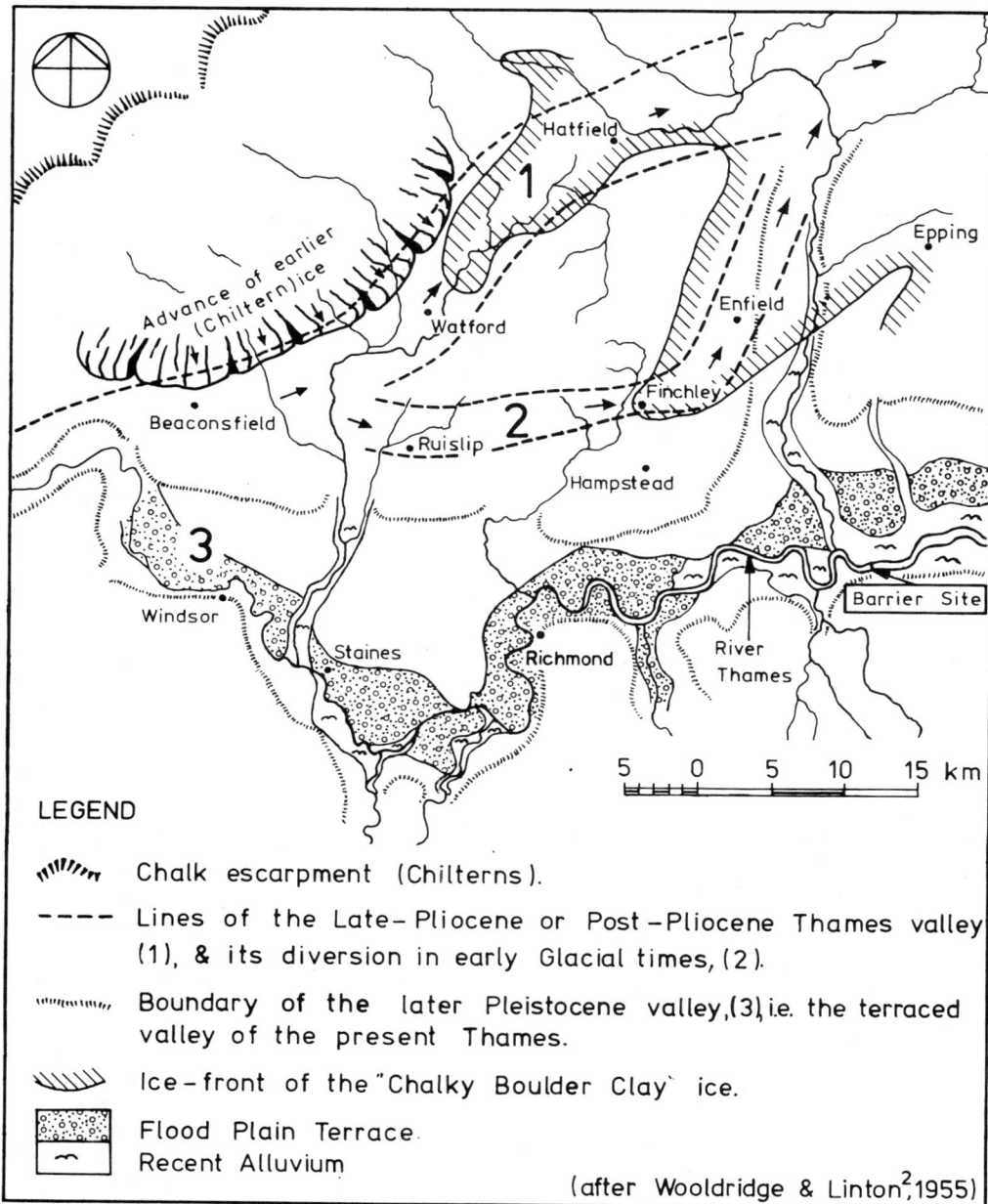

Fig.2. Plan showing present and former courses of Thames and distribution of superficial deposits

was used in ten of the eleven boreholes from which core was obtained (Nos. 1, 2, 3, 7, 9, 19, 25, 28, 31 and 45). In borehole 52, however, which was one of three boreholes drilled to a depth of about 100 m for micropalaeontological zoning purposes, an SF core barrel taking a 113 mm diameter core was used in order to obtain better core recovery. Fig. 4 shows simplified logs and the results of standard penetration tests for boreholes sunk at the upstream and downstream ends of the locations proposed for Pier 1 and Piers 4 to 9, indicating the variation in the ground conditions across the Site.

19. Tests were carried out to determine the permeability characteristics of the chalk, the Thanet Sand and the gravels by temporarily stopping the boreholes and carrying out rising and falling head tests using the borehole casing to seal off water in the soil above the test sections. It was appreciated that, because of the difficulty of sealing the boreholes effectively and the uncertainty about the natural piezometric level at the test depth, the values of permeability given by these tests would only be approximate. However, by carrying out a sufficient number of such tests it was considered that it would be possible to establish, with sufficient accuracy, the order of permeability of these deposits.

20. Tests of greater accuracy were carried out for chalk in borehole 52 using standpipe piezometers. Piezometers were installed in five other boreholes (Nos. 24B, 30, 48, 53 and 54) for use in the evaluation of permeability, to determine the influence of the tide on groundwater levels in the chalk and gravels, and to provide information regarding river wall design.

21. Advantage was taken of the presence of a disused sandpit next to Maryon Park, Charlton (about 1 km south of the Site) where Thanet Sand is exposed (ref. 6). Mechanical plant was used to excavate into and below the exposed face enabling it to be logged, in-situ density determinations to be made, and samples to be taken for laboratory testing. It had been hoped to be able to excavate deeply enough to expose the Thanet Sand/Chalk interface and the associated Bull Head Bed but the depth proved to be too great for this to be practicable.

22. In order to assess the restraint likely to be caused to the operation of the gates by the deposition of material between the gates and the sills, the investigation included sampling and testing of the surface material of the river bed. Sediment traps were installed for up to twelve weeks and subsequently recovered for undisturbed strength tests and grading tests on the deposited material. Vane tests were also carried out in the river bed at various locations.

23. In the laboratory, classification tests, particle size analyses, chemical and bacteriological tests and tests to determine strength and deformation properties were made on selected open-drive and core samples. Using prepared specimens of core, Mr. D.J. Carter of Imperial College carried out a micropalaeontological study of the chalk (ref. 7), in order to determine its stratigraphy in sufficient detail to detect the location of any possible faults and to help establish the presence and nature of any soliflucted or frost-shattered chalk.

24. Wimpey Laboratories Ltd., carried out a waterborne, acoustic survey to provide details of the sub-river bed features, in particular the chalk surface and structure. Two seismic profiling systems using sound energy at different frequencies, the 'pinger' (transducer source) and 'boomer' (mechanical percussive source), were used in trial surveys. The pinger proved more satifactory and was used for the main survey. Although survey records were obtained over the whole of the river width they could only be reliably interpreted in the southern part of the deep channel, possibly because of the energy absorption occurring elsewhere due to the thickness and organic content of the river bed sediment. An attempt was made to overcome the problem by using a more powerful boomer source but this was only marginally more successful.

25. A further small experimental geophysical trial was carried out, aimed at obtaining correlation between the detailed information from a 4.5 m diameter shaft and boreholes previously sunk in the Upper Chalk at the CEGB site at Littlebrook and from the boreholes at the site of the Barrier. The trial involved the logging of boreholes by means of down-the-hole equipment employing acoustic and radiation emissions. Seismograph Service (England) Ltd., undertook this work, which unfortunately was unsuccessful for reasons tentatively attributed to absorption losses in the chalk.

26. Studies were made of information available from other investigations and projects relevant to foundation construction on strata similar to those existing at the Site, particularly those concerning the nature and engineering characteristics of near-surface chalk. In addition, informative discussions were held with a number of civil engineering and specialist contractors regarding construction methods. Amongst the published records of design and construction of foundations on chalk those relating to the Medway Bridge (refs. 8, 9), the Woolwich Ferry terminals (refs. 11, 12), and the works at Tilbury (refs. 13, 14, 15, 16), were particularly relevant. Further information on the Medway project, in which problems were encountered with solution features, was made available by Mr. F.H. Hansen who was chief engineer to the main contractor for that work. Little published information was available on deep excavation and construction in Thanet Sand, presumably because of its relatively limited occurrence and the depth of overburden normally covering it.

DISCUSSION OF RESULTS OF THE SITE INVESTIGATION

27. During and after the geotechnical investigations, consideration was given to several possible methods of founding and constructing the piers and abutments. These included construction within deep, sheet-piled cofferdams, the

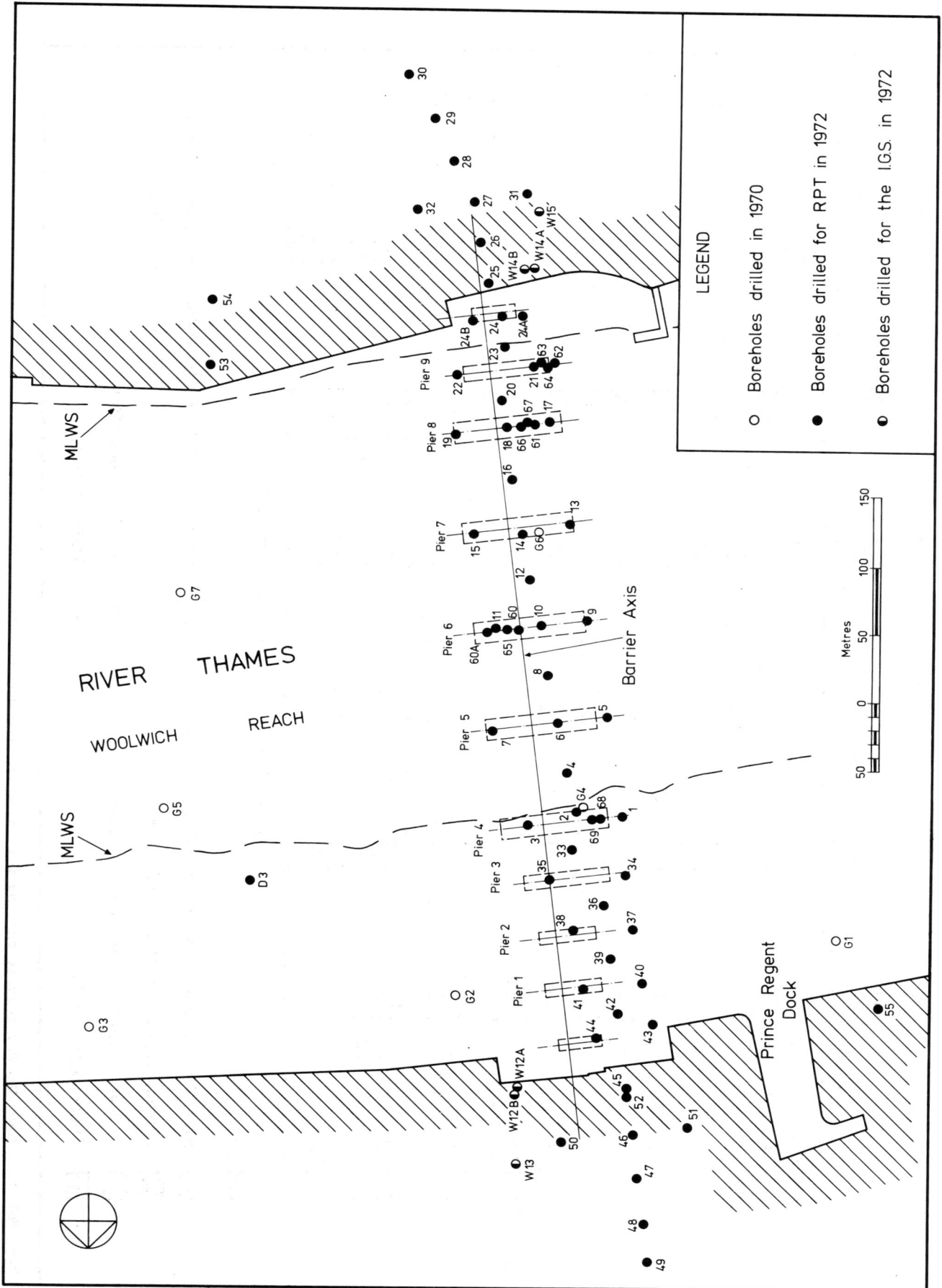

Fig.3. Plan of barrier site showing borehole positions

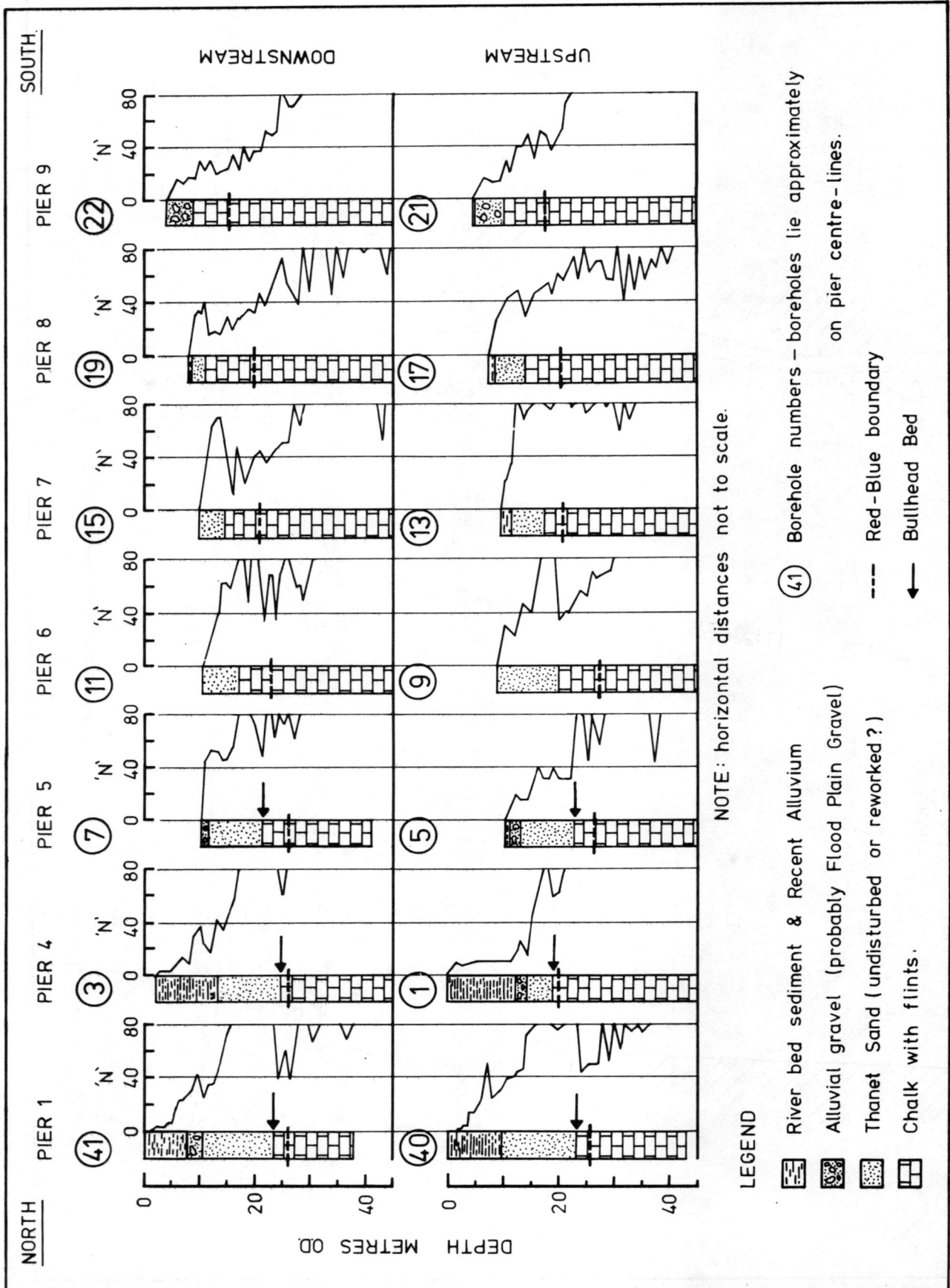

Fig.4. Borehole logs and SPT results showing variation in ground conditions across site

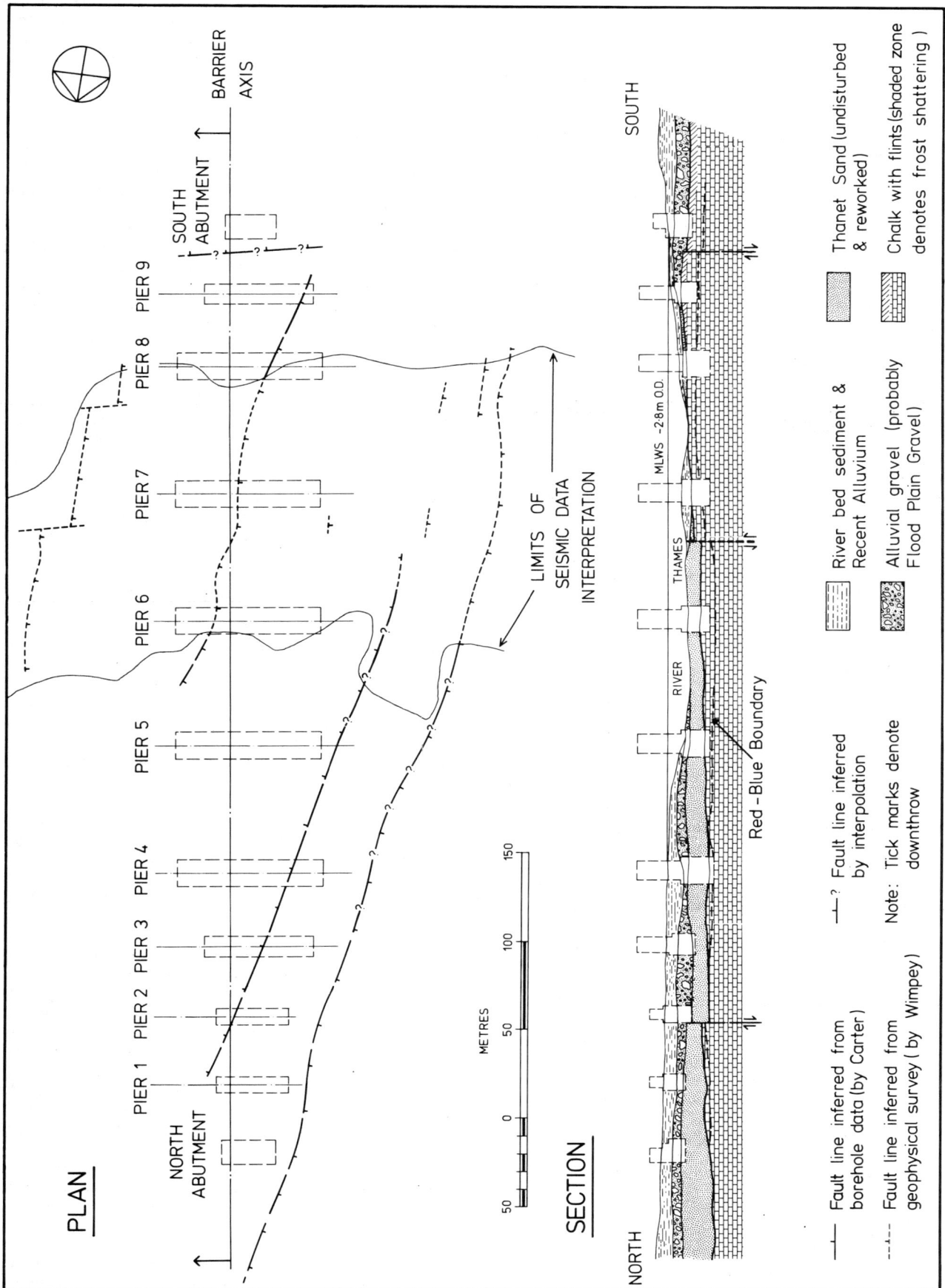

Fig.5. Geological cross section on barrier axis and plan showing faults

use of bored piles or diaphragm walls combined with shallower sheet-piled cofferdams, and construction by open-dredged caissons. The considerations governing the decision by the Consulting Engineers to adopt the first of these alternatives are given in Paper 14 of this Symposium and the discussion in the present paper of the results of the site investigation in relation to foundation design, refers only to this method of construction.

28. In Fig. 5 a cross-section on the Barrier axis shows the positions proposed for the piers and abutments in relation to the distribution of the alluvial deposits, the Thanet Sand and the Chalk. The following discussion of the results of the investigation of these deposits concentrates mainly on the information obtained about the Thanet Sand and the Chalk as it is this which is most relevant to the foundations of the piers and abutments.

Geological and Engineering Characteristics of the Alluvial Deposits

29. River bed sediment was found to comprise a mixed deposit of river-transported sediment ranging in particle size from clay to medium sand (Fig. 6). In the bacteriological tests carried out on this material the concentration of sulphate-reducing bacteria was found to be high and it was considered that the concrete sills should contain sulphate-resisting cement (as well as maximum possible corrosion protection of the steel gates housed in the sill recesses). The percentage of organic matter in all the samples tested averaged about 10%.

30. Recent Alluvium exists over much of the Site but is thickest beneath both banks of the river and is generally absent below the deepest part of the river bed. The material is variable, consisting of layers and lenses of silt, silty clay and peat, and this variability is reflected in the soil properties measured in the laboratory. The range of particle size distribution curves is shown in Fig. 6. In the peats and organic clays the liquid limit values were found to be between 300% and 400% with plasticity indices around 150%. The undrained shear strength was found generally to be between 5 and 60 kN/m^2 with sensitivity ranging from 2 to 10. Oedometer tests on samples 75 mm diameter by 12 mm gave values of coefficient of compressibility varying between 2×10^{-3} m^2/kN for peat and 0.2×10^{-3} m^2/kN for silty sandy clay.

31. The sulphur trioxide contents of the silts and silty clays were generally above 0.2% confirming the need for the use of sulphate-resisting cement in the sills. Because of their massive nature it was not, however, considered necessary to use sulphate-resisting cement in the piers and abutments.

32. Alluvial gravels were encountered in the boreholes sunk on the north and south banks of the river and the range of particle size distribution curves determined for this material is shown in Fig. 6. Although maps of the Institute of Geological Sciences show the deposit to be

Flood Plain Gravel, erosion by the river has caused local disturbance and it is believed that much of the material, particularly on the north bank, has been re-deposited. This was evident from the considerable amount of relatively fine-grained granular material which was encountered in the boreholes.

33. Permeability tests carried out in the boreholes gave a considerable range of values. Two tests gave results of the order of 300×10^{-6} m/sec which is typical for sandy gravel, whilst other tests gave results between 0.3 and 20×10^{-6} m/sec which, although of the order expected for the finer material contained within the gravel, could possibly be a reflection of the limited accuracy of the testing technique. From the results of the tests it was anticipated that open excavation would involve considerable de-watering problems unless an effective cut-off or sealing treatment were provided.

34. Because of the variability in thickness and nature of the alluvial gravels and the variation in consistency of the material from medium dense to dense shown by the standard penetration tests, the deposit was not considered to be suitable for the foundation of the piers and abutments. It was not anticipated, however, that there would be particular difficulty in excavating or driving piles through the material.

Geological and Engineering Characteristics of the Thanet Sand

35. The boreholes showed that the Thanet Sand deposit is somewhat wedge-shaped in cross-section, the material being absent at the southern end of the Site and increasing to a thickness greater than 10 m at the northern end. The boundary of the outcrop appears to be in the vicinity of Pier 8 (Fig. 5). Towards the edge of the outcrop, however, the absence of the Bull Head Bed and the change in character of the material from sand to clay and silt indicates that the material has probably been extensively reworked or redeposited by nature.

36. North of the location of Pier 6, the Bull Head Bed was encountered in every borehole, and the overlying Thanet Sand was shown by the standard penetration tests to be generally a very dense material. A small reduction in standard penetration test value in the upper 3 m of the stratum, immediately below the Flood Plain Gravel, was observed indicating that the surface material had suffered some disturbance either by weathering or reworking. Below the river bed the standard penetration test results showed that at least the top few metres of the deposit and occasionally its full thickness is significantly less dense than the material on the north bank under the Flood Plain terrace or the Recent Alluvium (Fig. 4), suggesting a more active development of weathering or reworking of the material.

37. The deposit, as anticipated, was found to consist, generally, of grey-green silty fine sand with the proportion of fine material tending to increase with depth. Grading curves for

this material are given in Fig. 6 which shows that the Thanet Sand sampled nearby at Maryon Park is slightly more uniform and coarser in grain size than much of the material at the Site, indicating that the material at the two sites is possibly from different stratigraphical horizons. Inspection of samples obtained using a split-spoon sampler provided some indication that the Thanet Sand is uncemented at the Site. The results of the rising and falling head tests indicated that the coefficient of permeability for the Thanet Sand lies between the limits of 0.5×10^{-6} and 5×10^{-6} m/sec, which is of the general order that can be anticipated for silty fine sand.

38. Because it was not considered feasible to recover undisturbed samples for density determination, tests were carried out to measure the maximum and minimum dry density of bulk disturbed samples of the material, and the results were then compared with values of in-situ dry density determined for the Thanet Sand at Maryon Park. The dry density range determined by the U.S.B.R. procedure (ref. 17), was 1260 to 1540 kg/m^3 and the in-situ dry density of the Maryon Park material was 1760 kg/m^3. The higher value for in-situ dry density compared with the range determined from the laboratory tests might be due to the difference in grain size of the materials from the two sites or because of the presence of slight cementing in the material at Maryon Park, or because of the heavy overconsolidation to which the in situ material has probably been subjected.

39. Using a 60 mm square shear box apparatus, drained tests were carried out on disturbed material for a range of normal pressures on specimens prepared to both a loose and dense state of compaction. These tests gave angles of shearing resistance of 34° and 43° for the loose and dense material respectively and values of critical density (defined as the density at which there is no change in volume during shear) of 1400 kg/m^3 and 1500 kg/m^3 at normal pressures of 70 kN/m^2 and 410 kN/m^2 respectively.

40. The values of shear strength obtained from the laboratory tests are comparable with those obtained from the relationship between shear strength and the standard penetration test value, N, given by Peck, Hanson and Thornburn (ref. 18). The lowest values of N determined in the investigation were of the order of 20 and, generally, N was greater than 60 below the "reworked" material, indicating an angle of shearing resistance of at least 33° and generally greater than 42°.

Foundations in the Thanet Sands

41. Of those abutment and pier locations where the thickness of the Thanet Sand is sufficient for it to be considered as a foundation stratum, it was only at the locations north of borehole 33 (North Abutment and Piers 1, 2 and 3) that the deposit was judged to be suitable. Over the northern part of the Site the investigation showed the material to be very dense, with N generally greater than 60, and indicated the absence of lenses of clay; from consideration of

the bearing pressures of the structures (up to 450 kN/m^2), the deposit was found to be adequate from consideration of both bearing capacity and settlement. Although piers founded on the material further south would have an adequate factor of safety against shear failure, calculations showed that settlements in this less dense material would be likely to exceed the differential settlement considered to be allowable.

42. The deformation modulus E used in the calculations was estimated from published empirical relationships between N and E for fine sands (refs. 19,20). Settlements were estimated from the deformation modulus and also from an alternative correlation which relates settlement directly with penetration resistance in fine sands (ref. 21). A wide range of values for E was obtained but the suitability of the foundation stratum was confirmed using a pessimistic value for assessment of the elastic settlements. Analysis showed that, for the North Abutment and Piers 1, 2 and 3, settlement would be within acceptable limits if they were founded on Thanet Sand sufficiently dense to give values of N greater than 40. In the final selection of the founding level for a structure, however, consideration would have to be given to the degree to which the material below founding level might be disturbed by the excavation process.

43. If excavation is carried out underwater, even if an excess head is maintained inside the cofferdam, disturbance is likely to occur to a greater depth than if the excavation is dewatered by a system of pumped wells. Therefore, if underwater excavation is adopted, strict measures will need to be taken to minimise disturbance of the formation and so maintain settlements within tolerable limits.

44. Because of the very dense nature of the material on the north bank it was anticipated that the driving of piles for the temporary works cofferdams would be difficult and would require the use of heavy equipment.

Geological Characteristics of the Chalk

45. The information obtained from the boreholes showed the upper surface of the chalk to be fairly flat and dipping very gently to the north-north-east. It was considered, from study of a contour plan of the chalk surface prepared from the information obtained from the boreholes and the geophysical survey, that large scale solution features (i.e. features tens of metres across) were probably absent at the Site. Although neither large nor small solution features were encountered by the boreholes, the existence of narrow pipes could not be discounted. The presence of relatively narrow pipes such as that shown in Fig. 7 was not indicated by the geophysical studies but, even if present, it was not expected that they would necessarily be found by geophysical methods. It was considered, however, that such features, if found during the excavation for the foundations to the Barrier structures, should be relatively small and could be investigated and treated as necessary at that time.

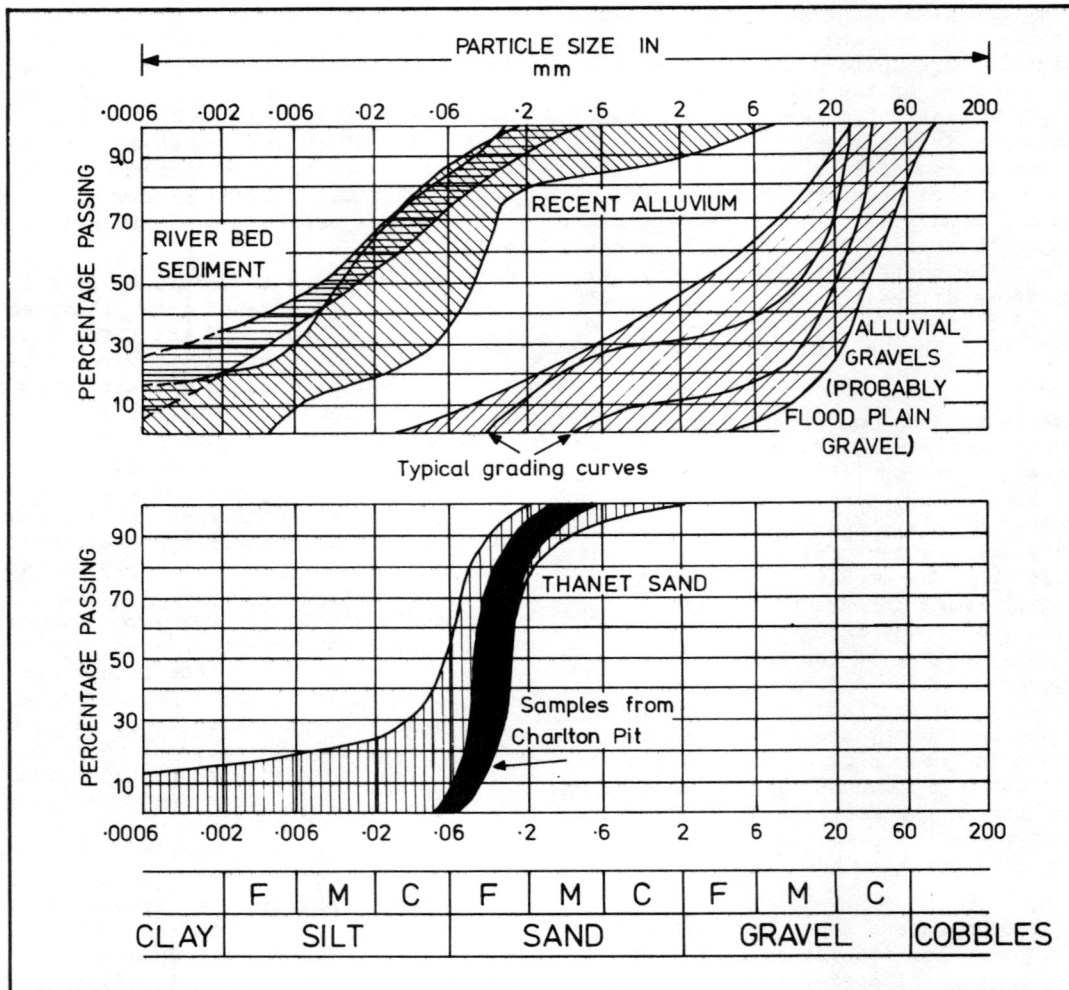

Fig.6. Particle size distribution of alluvial deposits and Thanet Sand

Table 2. Zones in the Upper Chalk encountered in the site investigation boreholes

STAGE	MACROFOSSIL ZONES	MICROFOSSIL ZONES (after Carter*)	approx thickness
SENONIAN (part)	Vintracinus socialis[+]		
	Micraster coranguinum	Red	9 m
		Blue) Orange)	57 m
	Micraster cortestudinarium	Violet	17 m
TURONIAN (part)	Holaster planus	Yellow	

* See Carter (ref. 7), 1977 for details of microfossil zones
+ Not identified in samples but possibly present at the Site

46. In the determination of the relatively thin microfossil zones of the chalk by Carter the occurrence of an easily recognisable abrupt change in the microfauna of the thick Micraster coranguinum zone was particularly useful (Table 2). This boundary was located in boreholes north of borehole 26 and, for convenience of reference, the zones above and below the boundary were called "Red" and "Blue" respectively. The so-called Red-Blue boundary which is shown for the boreholes depicted in Fig. 4 and also on the cross-section in Fig. 5, indicated that the chalk at the Site is only gently folded, being less sharply folded than the chalk nearby at Charlton and Greenwich.

47. Five faults were located by the studies, one trending N80°E and the others, aligned en echelon at a small angle to the Barrier axis, trending N10°E. The existence of one of the latter set had been indicated previously by the geophysical survey as a step in the chalk surface. With the exception of the fault located at the extreme northern end of the Site, the faults inferred from the micropalaeontological work and the geophysical survey are shown in plan and, where they cross the Barrier axis, in cross-section on Fig. 5. They are described by Carter as steep-angled normal faults with lateral displacements of up to 50 m which he interpreted as tears with a fairly small vertical throw of up to 5 m. Some slickensiding was observed in the cores obtained from several boreholes but this could not be correlated with the faults indicated by Carter's studies and it was concluded, therefore, that other less prominent faults might exist.

48. As part of the geotechnical evaluation a close inspection was made of the Upper Chalk exposed in the cliffs at Pegwell Bay, Kent which contain a series of small faults which were thought to be similar in character to those located at the Site by the micropalaeontological studies and the geophysical survey. The faults visible in the cliffs are steeply inclined and generally consist of a series of parallel surfaces in zones, typically between half a metre and one metre in width (Fig. 8). Inspection showed that the discontinuities adjacent to the fault zones and the faults themselves appeared to be tightly closed.

49. A study of the microfauna in the chalk at Pegwell Bay showed that it is probably from a different horizon from that at the Site, but the engineering properties of the Chalk from the two locations were judged to be similar. The inspection of the cliffs at Pegwell Bay was considered to be particularly valuable in providing a better understanding of the faulting and macrostructure of the chalk at the Site.

50. The site investigation boreholes showed the chalk to be creamy white in colour and to contain abundant flints, the presence of which made open-drive and core sampling difficult. The 73 mm diameter core taken using the H core barrel was generally badly broken up by flints but an improvement in core recovery was achieved for the borehole drilled using an SF barrel. Vertical and horizontal joints were apparent in the borehole cores.

51. The open-drive samples and all cores were logged and photographed. Fig. 9 shows photographs from the site investigation report of typical open-drive samples that have been extruded and split for examination. Fig. 10 shows a photograph of core recovered from borehole No. 52 from 57 to 60 m below ground level, together with the appropriate section of borehole record. The record shows not only the information given by the site investigation contractor but also the Consulting Engineer's field sketch log of the core recovered which was used in assessing the quality of the chalk.

52. Determination of the microfossil zones indicated that solifluction chalk was not present at the Site as no derived fauna were found. The presence and depth of frost-shattered chalk was established from consideration of the fracture spacing and iron-staining on joint surfaces in the open-drive samples and from observation of the disturbance to the sample. An important feature in the assessment of chalk quality was that the frost-shattered chalk, containing a matrix of remoulded chalk, was significantly less disturbed by the sampling than the hard chalk which tended to fracture into thin plates perpendicular to the axis of the sampling tube (Fig. 9).

Engineering Characteristics of the Chalk

53. The quality of the chalk was classified into grades on the basis of the S.P.T. values. The classification used was that proposed by Wakeling, based on his work on the Middle Chalk at Mundford (Norfolk) and on the Upper Chalk at Southampton and Northfleet (ref. 22). The Northfleet site is only 20 km from the Site and an investigation of the microfauna by Carter showed that the chalk at both sites belongs to a similar geological horizon. The S.P.T. range and the description appropriate to each grade is given in the Appendix. The results of standard penetration tests in Upper Chalk can vary widely depending on whether the cone penetrates intact chalk or enters a fissure or joint, or whether the tool is obstructed by a flint. Some judgement is required, therefore, in interpreting the results of the tests.

54. An important finding was that the chalk at the Site is generally sound and unweathered (Grades I to III) where it is protected by a cover of Thanet Sand, but where it is overlain by alluvial deposits at the southern end of the Site there is indication of frost-shattering and softening in the upper layers (Fig. 5). Occasional pockets of Grade V quality chalk were encountered. Although there was no evidence of frost-shattering of the chalk between Piers 7 and 8, it is known that frost-shattering of the chalk at the ground surface did occur in the London region and it is thought likely that the affected material has since been removed by erosion.

Fig.7. Narrow pipe in frost shattered chalk in a quarry 19km east of site

Fig.8. Fault in Upper Chalk at Pegwell Bay, Kent

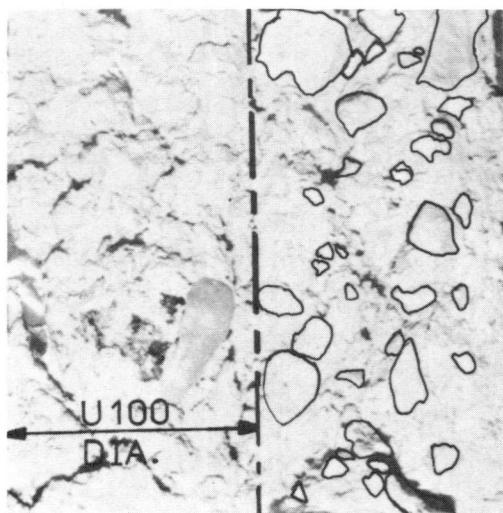

(a)

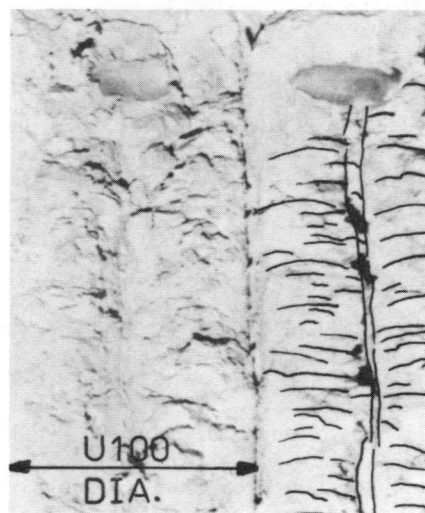

(b)

Fig.9. Comparison of extruded and split 100 mm dia. open-drive samples of frost-shattered and grade I chalk: (a) frost-shattered chalk, grade IV, from borehole 20, 2.7 m to 3.1 m below chalk surface; (b) grade I chalk from borehole 10, 20.0 m to 20.4 m below chalk surface

55. The results of standard penetration tests would be expected to reflect any reduction in quality of the chalk resulting from such processes as weathering or frost-shattering. N values plotted against depth below chalk surface and relative to the Red-Blue boundary are shown in Fig. 11. A comparison is made between the results of tests carried out in boreholes 15, 16, 20, 24, 24A, 24B, 25, 64 and 67 at the southern end of the Site, where the chalk is overlain by alluvial deposits (righthand graphs), and those carried out in boreholes further north, 2 to 8, 10, 33 to 36, 38 and 65, where the chalk is overlain by the Thanet Sand and therefore apparently not vulnerable to frost-shattering (left-hand graphs). For the chalk overlain by Thanet Sand, N was found not to be less than 35 and a wide scatter of results is evident with no significant trend of N with depth relative to either the chalk surface or the Red-Blue boundary. However, there is a trend of N value increasing with depth below the chalk surface for the chalk which is not protected by the Thanet Sand. There is no marked difference, however, in the N values above and below the Red-Blue boundary showing that the engineering properties of the two microfossil zones, as indicated by the S.P.T. results, are similar. It may be concluded, therefore, that the engineering characteristics of the near surface chalk which is not oberlain by Thanet Sand have been significantly affected by frost-shattering.

56. In every borehole, chalk of Grade II quality was encountered within a depth of 10 m below the top of the chalk. Another important finding was that the results of the standard penetration tests carried out in boreholes near faults were not significiantly different from the results of tests carried out elsewhere on the Site; this being generally consistent with the observations made at Pegwell Bay. However, there was some indication from standard penetration tests in two of the boreholes which lay close to a fault, that there could be a greater proportion of Grade III chalk in this region.

57. Permeability tests carried out in the boreholes showed, as expected, a considerable scatter of results with the majority of values of coefficient of permeability lying between 1×10^{-6} and 50×10^{-6} m/sec; the range of results possibly being, in part, a reflection of the method of testing. These tests were, however, carried out principally to detect the presence of open joints or fissures which would have considerably higher permeability than chalk having tightly closed discontinuities. On inspection of the results, it was concluded that such fissures had not been encountered at the test locations, which was encouraging, but it was considered that the presence of water-bearing fissures could not be entirely discounted. There was some small indication from the tests that permeability decreases with depth. The results of tests carried out in boreholes which were considered to be near to or in fault zones were not significantly different from the results obtained from other tests.

58. Tests of greater accuracy were carried out in borehole 52 in which standpipe piezometers were installed in sand pockets at three different levels in zones where the borehole core indicated the possibility of open fissures. The results showed, however, that the coefficient of permeability was not greater than 0.05×10^{-6} m/sec at the test sections.

59. Laboratory tests were carried out on the open-drive samples and 75 mm diameter core specimens to determine density, moisture content and strength and deformation characteristics. The dry densities and natural moisture contents of, respectively, 361 and 381 specimens obtained from the open-drive samples were determined. The values ranged between 1230 and 1680 kg/m^3 (average 1574 kg/m^3) and 15 and 36% (average 28%). Because of sample disturbance it is possible that the dry density in some cases was under-estimated and the moisture content over-estimated.

60. Drained triaxial compression tests on three specimens of Grade III chalk and one specimen of Grade II chalk, all approximately 100 mm diameter by 200 mm, taken with the open-drive sampler, indicated an angle of shearing resistance, ϕ' of about 40° and negligible effective cohesion (c'). The existence in pits, close to the Site, of near vertical faces of chalk of similar quality to that tested indicates, however, that the in-situ chalk has a significant cohesion value and it may be inferred, therefore, that the specimens tested may have been affected by sample disturbance, the measured values being more representative of the softened strength of the chalk.

61. Three axial compression tests were carried out on core specimens of approximately 75 mm diameter fitted with strain gauges to measure axial and lateral strain. The specimens gave average values of modulus of elasticity and Poisson's ratio of 9600 MN/m^2 and 0.26 respectively. The results of these tests gave some indication of the properties of intact chalk. The properties of the in-situ chalk are, of course, influenced by the joints and other discontinuities and, in an attempt to quantify the effect of these, an unconfined compression test was carried out on three specimens stacked axially. The modulus of elasticity of the composite specimen, which had a length of about 400 mm, was 1800 MN/m^2.

Foundations in the Chalk

62. Because the bearing capacity of the chalk is high in relation to the bearing pressure of the Barrier structures the depth of the foundations to Piers 4 to 9 and the South Abutment will be governed by the consideration of allowable settlement, unless lower founding levels are required for structural reasons. Settlement was expected to be due mainly to the immediate, "elastic" compression of the chalk, assuming that soft or loose material resulting from the excavation is removed from formation level before the placing of tremie concrete. Because the settlement characteristics of the in-situ

ENGINEER'S SKETCH LOG	CONTRACTOR'S RECORD								
	Depth m.B.G.L.	Core Recov- ered%	Solid Core			Frac. Index	R.Q.D.	Strata		Description of Strata
			Tot.	Max	Min			Grade	Legend.	
	−57	0 100	m	cm	cm		%			57·00−58·50 :− Bioturbated chalk, breaks horizontal but occasion- ally 70°. Open vertical fissure 57·00−57·25, 57·86−58·50 with second parallel fissure at 58·30 −58·50. Core broken 57·60−57·77, 58·00−58·12. Small flints at 58·06. Large flints at 58·06 and 57·60−57·70.
	−58		0·40	18	7	7·5	22			
								1		
	−59		0·15	15	−	6·7	10			58·50−60·00 :− Open vertical fissure at 58·50−59·15, 59·30− 59·53, 59·64−60·00 with second parallel fissure 58·50−58·63. Flints at 59·29, 59·55−59·60. Core broken 58·84−59·00, 59·15− 59·19, 59·53−59·64. Core worn by drilling 59·64−59·80.
	−60									

■ Core recovered   ▨ Sample recovered

NOTES: Solid Core means broken or unbroken core with complete cross section intact. Fracture Index is the no. of breaks per metre in the solid core. Max. & Min Core lengths refer to the solid core. Rock Quality Designation (R.Q.D.) is the sum of the lengths of pieces of solid core more than 10cm long as per cent of total run.

♣ Flint

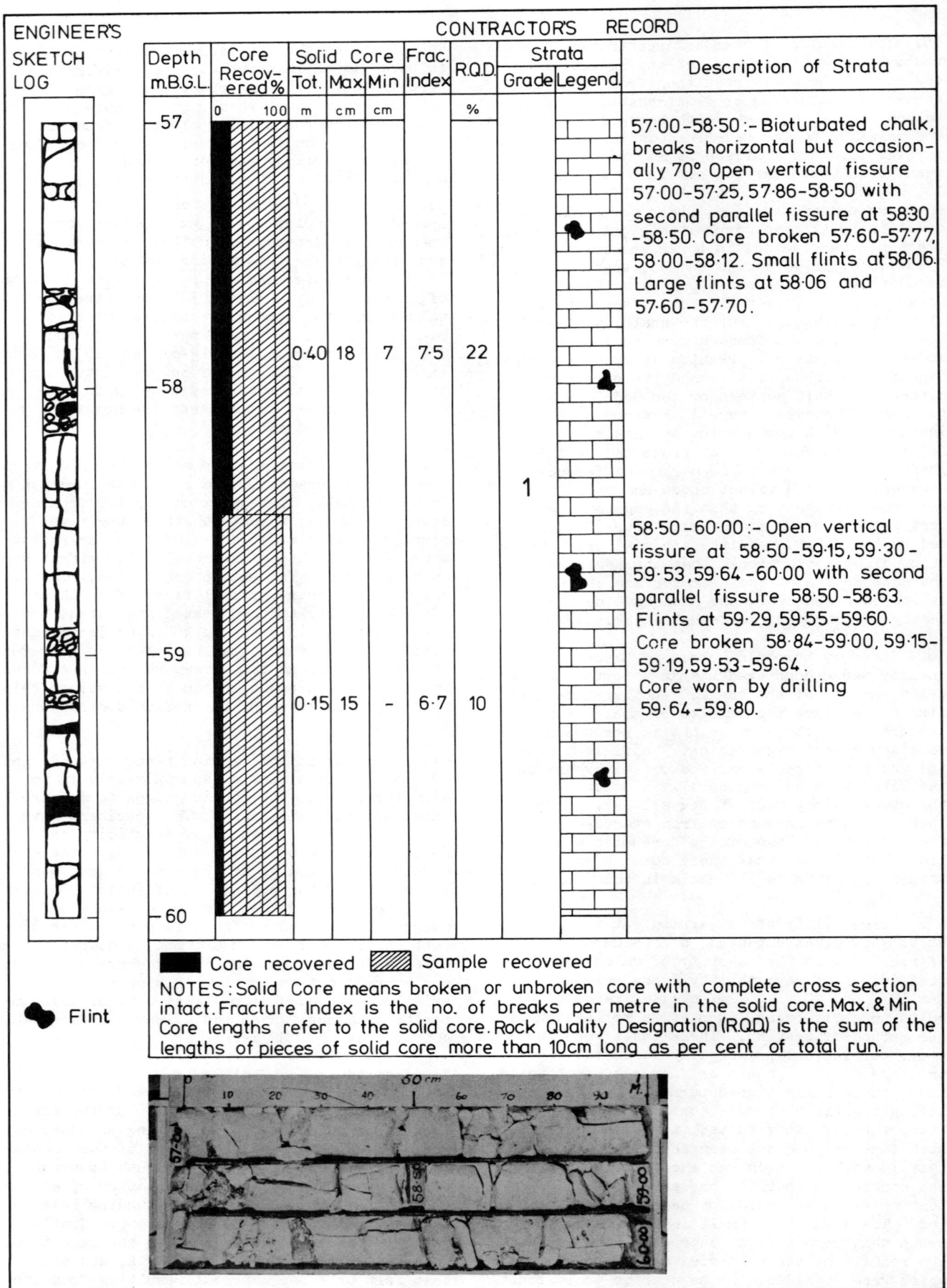

Fig.10. Engineer's sketch log and part of contractor's record for borehole 52 with photograph of core

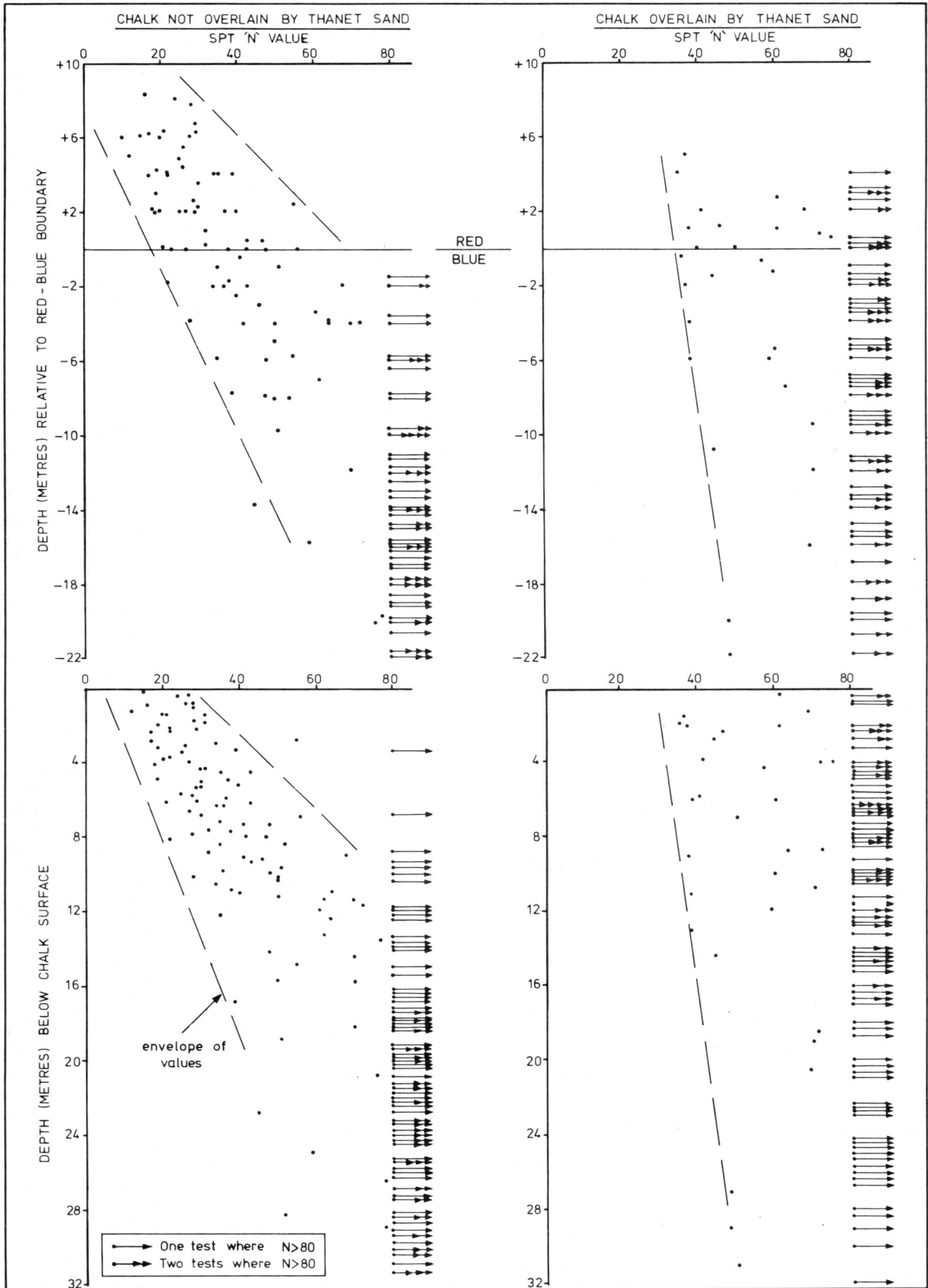

Fig.11. SPT results for chalk with and without Thanet Sand cover plotted in relation to the red-blue boundary and depth below chalk surface

jointed and fissured chalk were not measured,
and the deformation moduli obtained from the
results of laboratory testing are appropriate
only to intact chalk, the settlement
characteristics for the in-situ chalk were assessed from the consideration of the quality of the
in-situ chalk (as indicated by the S.P.T's and
from inspection of samples) in relation to
published information.

63. A variety of relationships between the
standard penetration test value, N, and deformation modulus, E, for chalk have been published,
e.g. Wakeling, 1966 (ref. 23), Kee and Clapham,
1973 (ref. 24), and Lake and Simmons, 1970,
(ref. 25), which are given in Fig. 12. These
relationships are principally based on the results of small diameter plate bearing tests in
boreholes, pile tests and unconfined compression
tests on open-drive samples, and it was considered that they would underestimate the deformation modulus of the chalk at the Site. The
results of the field loading tests on chalk at
Mundford (refs. 26, 27, 28), would appear to
support this view.

64. Although the chalk in which the Barrier is
to be founded is of a higher stratigraphical
horizon than the chalk tested at Mundford the
results of unconfined compression tests on intact specimens indicate that the deformation
modulus of the Grade I/II chalk at the two sites
was comparable (Fig. 12). The quality of the
in-situ chalk as indicated by the results of
standard penetration tests and the inspection
of open-drive samples and cores was also judged
to be similar. The relationship between E and N
suggested by Wakeling (ref. 22) based on the
Mundford results (Fig. 12) was considered,
therefore, to give a better indication of the
deformation characteristics of the chalk at the
Site. In design a slightly more conservative
assessment of E was adopted, the minimum value
of deformation modulus for the lower part of
Grade III chalk being taken as 1000 MN/m^2.

65. After consideration of all the information
obtained from the site investigation it was
decided that Grade III chalk with standard penetration test value N greater than 25 should provide a competent foundation stratum for the main
piers where the bearing pressure will not exceed
600 kN/m^2 (Piers 4 to 8) and that for Pier 9 and
the South Abutment, where the bearing pressure
will not exceed 450 kN/m^2, Grade III chalk (N
greater than 20) should provide an adequate
foundation. It was considered that these criteria would ensure that differential settlement
would be less than the limits considered tolerable for the structures. In Fig. 13 the results of standard penetration tests are given
for Piers 8 and 9 showing the minimum design
values of N of 25 and 20 respectively, and the
founding levels proposed.

66. Due to the possibility of local minor variations in the level of the upper surface of the
chalk the foundation levels were proposed at
least 2 m below the top of the chalk as indicated by the contour plan of the chalk surface

prepared from the site investigation information.
Additional investigation by probe or borehole at
each pier position, prior to excavation, was
recommended to enable a more precise assessment
to be made of surface level and further information to be obtained on the quality of chalk, to
confirm the foundation levels to be adopted.

67. In general, the differences in depth to the
competent foundation stratum on either side of a
fault, as indicated by the penetration tests,
were not considered large enough to justify a
stepped foundation. If investigation prior to
cofferdam construction showed, however, that
significant economy could be gained by incorporating a step in the foundation, it might be considered if it were evident that it would not
lead to construction complications.

CONCLUSIONS

68. The detailed investigation, which cost
£108,634, is considered to have been successful
in providing the geotechnical information required for the design of the foundations to the
Barrier and the final outcome will be judged
after construction.

69. The investigation showed that the Site is
underlain at relatively shallow depth by strata
of adequate bearing capacity and suitable settlement characteristics for the foundation of the
piers and abutments of the Barrier. The Thanet
Sand in the northern part of the Site overlies
the chalk in considerable depth and could in its
undisturbed dense state provide a foundation for
the North Abutment, the smaller inshore piers
and possibly Pier 3, provided that precautions
were taken against scour. The structures south
of Pier 3 can only be founded in the chalk.

70. Probably the most important conclusion
drawn from the site investigation was that there
is, at the Site, a general absence of soliflu-
tion material, deep frost-shattering and major
solution features which are often found in the
Chalk of South-east England. The possible presence of narrow pipes was not discounted and
allowance was made for their discovery during
excavation for the pier and abutment foundations.
Where Tertiary bedrock overlies the chalk it has
apparently acted as a protective capping and the
effects of the Quaternary Ice Ages on the chalk
at the northern end of the Site are not significant to the design of the Barrier structures.
At the southern end of the Site the chalk is
affected by frost-shattering to a shallow depth.
Deep buried channels were not found at the Site.

71. The existence of small scale faulting was
established by the micropalaeontological and
geophysical studies. The faults were not considered to have affected significantly the
engineering properties of the material in the
vicinity of the fault zones, and it was considered that the presence of faults would not
adversely affect the foundations to the Barrier
structures. From a study of seismic records obtained from the Global Seismology Unit of the
Institute of Geological Sciences, together with

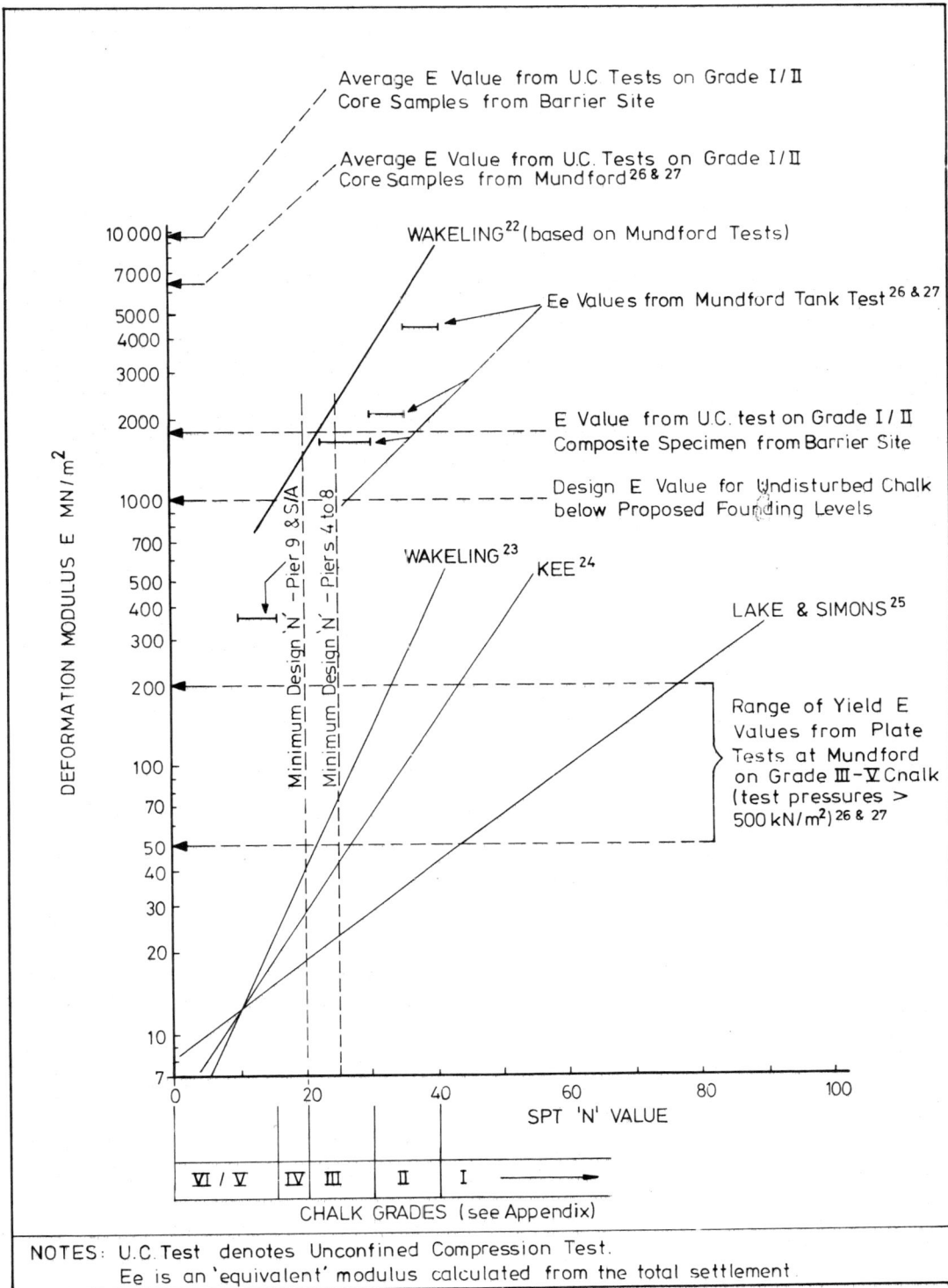

Fig.12. Deformation modulus in relation to SPT value and chalk grade, and criteria adopted in the determination of pier founding levels

assessments of these records by Dr. A.T.J. Dollar of Birkbeck College and Dr. N.N. Ambraseys of Imperial College, it was concluded that there was no more than an insignificant probability of risk of damaging movements or tremors resulting from these faults or from the larger faults elsewhere in the region.

72. Following the site investigation, the proposed founding levels for the piers and abutments were decided from consideration of the total and differential settlements likely to occur under operational loadings unless this resulted in a founding level higher than that required for the seating of the sills. In estimating the settlement of the structure it was appreciated that some settlement would occur due to the reconsolidation of material, at or below founding level, disturbed during excavation. It was considered, however, that most of the reconsolidation would occur during construction and would not, therefore, contribute significantly to differential settlement affecting the sills and gates as these would be installed at a late stage in the construction of the Barrier. Long term consolidation settlement of the chalk and Thanet Sand was considered likely to be negligible in comparison with the elastic deformation of these materials under the dead and live loads.

73. The construction of piers within sheet-piled cofferdams dewatered after the placing of several metres of tremie concrete, was considered feasible. The results of permeability tests carried out in the Thanet Sand and the chalk indicated that the dewatering of temporary works cofferdams could probably be achieved without undue difficulty. However, although open fissures or joints in the chalk were not encountered, the presence of zones having a much greater permeability than that measured, could not be discounted. Because of the very dense nature of the Thanet Sand and general hardness of the chalk it was anticipated that driving of sheet-piles for the cofferdams, some of which would be up to 35 m in length, would require the use of very heavy piling equipment.

ACKNOWLEDGEMENTS

74. In the preparation of the Paper, the Authors have naturally drawn heavily upon the information given in the factual and interpretive Reports on the detailed site investigation which were prepared by Mr T.R.M. Wakeling and Mr R.A.J. Jennings of Foundation Engineering Ltd. The Authors also wish to acknowledge, Mr. E.R.F. Lord and Mr. T.J. Kayes, both formerly of Rendel, Palmer & Tritton, who were, respectively, geotechnical engineer for the Barrier at the time of the investigation and Engineer's Representative for the investigation, and Mr. P. Shorten of Rendel, Palmer & Tritton who has assisted in the preparation of the Paper.

REFERENCES

1. WOOLDRIDGE, S.W. The minor structures of the London Basin. Proc. Geol. Assoc., 1923, 34, 175-193.

2. WOOLDRIDGE, S.W. and LINTON, D.L. Structure, Surface and Drainage in South-East England. 2nd Edition, George Philip and Son, Ltd., London 1955.

3. HIGGINBOTTOM, I.E. and FOOKES, P.G. Engineering aspects of periglacial features in Britain. Q. Jour. Engng. Geol., 1970, 3, No.2, 85-117.

4. WEST, G and DUMBLETON, M.H. Some observations on swallow-holes and mines in the Chalk. Q. Jour. Engng. Geol., 1972, 5, No.1 & 2, 170-177.

5. Thames Flood Prevention: Second Report of Studies Vol.3: Report on Subsoil Investigation at Woolwich Reach No.2 Site. 1970.

6. PITCHER, W.S., PEAKE, N.B., CARRECK, J.N., KIRKALDY, J., HESTER, S.W. and HANCOCK, J.M. The London Region. Geologists' Assoc. Guides, No.30, 1958.

7. CARTER, D.J. and HART, M.B. Micropalaeontological investigations for the site of the Thames Barrier, London. Q. Jour. Engng. Geol., 1977, 10, No. 4, 321-338.

8. KERENSKY, O.A. and LITTLE, G. Medway Bridge: Design. Proc. I.C.E., 1964, 29, Sept, 19-52.

9. KIER, M., HANSEN, F. and DUNSTER, J.A. Medway Bridge: Construction. Proc. I.C.E., 1964 29, Sept., 53-100.

10. Joint Discussion on Medway Bridge: Design and Construction. Proc. I.C.E., 1965, 31, June, 160-204.

11. WEST, K.H., HOLLEYWOOD, J., JACKMAN, S.V. and WOOD, D.J.D. Three ferry terminals. Proc. I.C.E., 1968, 39, March, 397-432.

12. Discussion on Three ferry terminals. Proc. I.C.E., 1968, 41, Nov, 597-612.

13. PEEL, C., CARMICHAEL, A.J. and SMEARDON, R.F.J. No.1 Berth, Tilbury Dock. Proc. I.C.E. 1957, 8, Dec., 331-362.

14. Discussion on No.1 Berth, Tilbury Dock. Proc. I.C.E., 1958, 10, June, 201-211.

15. SMEARDON, R.F.J., NEWTON, E. and PAGE, F.A. Engineering Works at Tilbury Docks, 1963-67. Proc. I.C.E., 1967, 38, Oct., 177-228.

16. Discussion on Engineering Works at Tilbury Docks 1963-67. Proc. I.C.E., 1968., 40, June, 223-247.

17. U.S. Department of the Interior, Bureau of Reclamation. Earth Manual, 1st Edition U.S. Govt. Printing Office, Washington, 1968, 467-474.

18. PECK, R.B., HANSON, W.E. and THORNBURN, T.H. Foundation Engineering. John Wiley, New York, 1953.

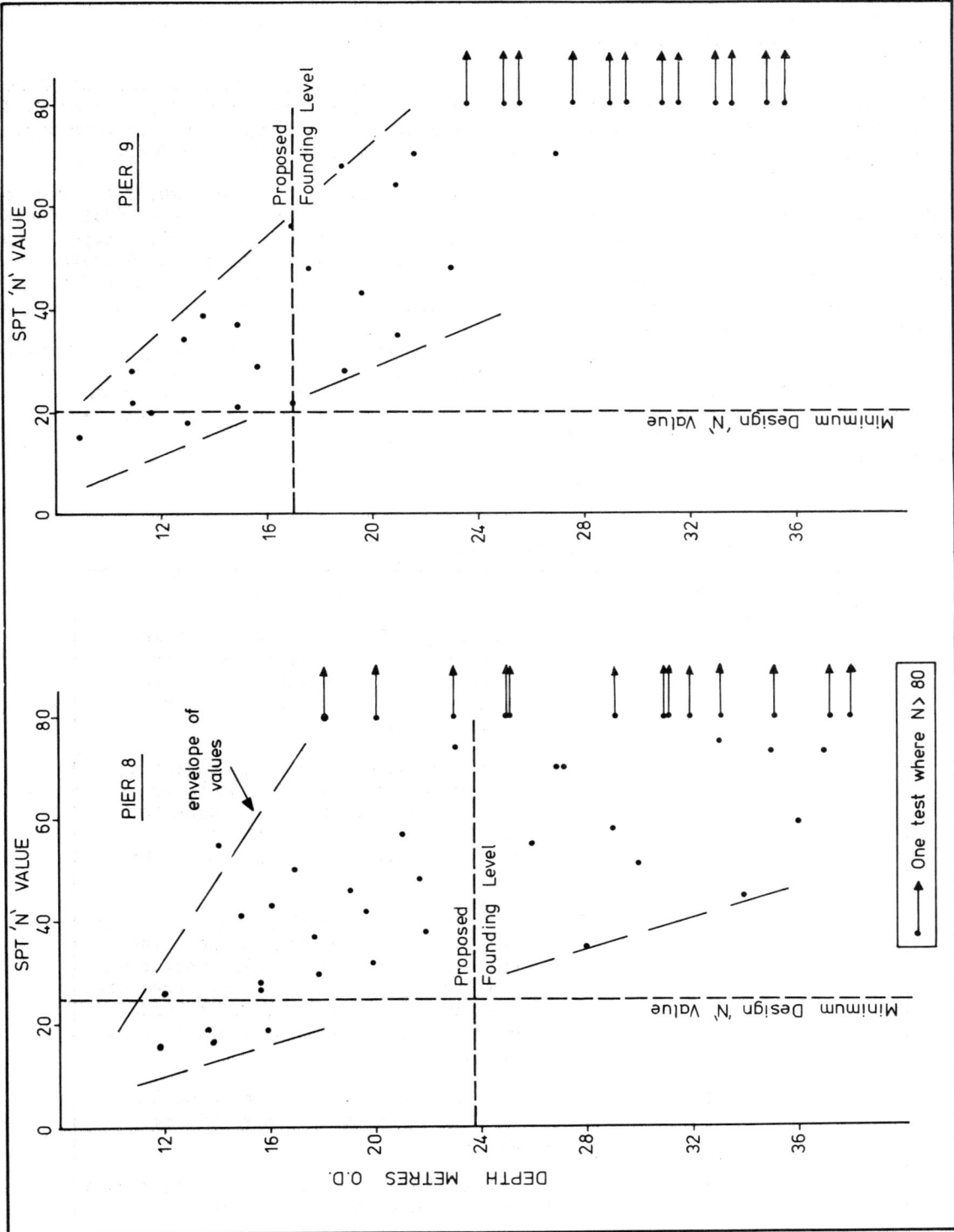

Fig.13. Standard Penetration Test results for piers 8 and 9

19. D'APPOLONIA, D.J., D'APPOLONIA, E.E. and BRISETTE, R. Settlement of spread footings on sand. A.S.C.E. Jour. of Soil Mechanics and Foundations Division, 1968, 94, No.SM3, May, 735-760.

20. WEBB, D.L. Settlement of structures on deep alluvial sandy sediments in Durban, South Africa. Proc. B.G.S. Conf. on In Situ Investigations in Soils and Rocks, 1969, Paper 16, May, London, 181-188.

21. MEYERHOF, G.G. Penetration tests and bearing capacity of cohesionless soils. A.S.C.E. Jour. of Soil Mechanics and Foundation Division, 1956, 82, No.SM1, Jan., pp 19.

22. WAKELING, T.R.M. A comparison of the results of standard site investigation methods against the results of a detailed geotechnical investigation in Middle Chalk at Mundford, Norfolk, Proc. B.G.S. Conf. on In Situ Investigations in Soils and Rocks, 1970, London, 17-22.

23. WAKELING, T.R.M. Foundations on chalk. Proc. I.C.E. Symp. on Chalk in Earthworks and Foundations, 1966, London, 15-23 and contribution to the discussion of Session 85-87.

24. KEE, R and CLAPHAM, H.G. An empirical method of foundation design in chalk. Civil Engineering and Public Works Review, 1971, 66, Sept., 981-986.

25. LAKE, L.M. and SIMONS, N.W. Investigations into the engineering properties of chalk at Welford Theale, Berkshire. Proc. B.G.S. Conf. on In Situ Investigations in Soils and Rocks 1970, London, 21-27.

26. WARD, W.H., BURLAND, J.B. and GALLOIS, R.W. Geotechnical assessment of a site at Mundford, Norfolk, for a large proton accelerator. Geotechnique, 1968, 18, 399-431.

27. BURLAND, J.B. and LORD, J.A. The load-deformation behaviour of the Middle Chalk at Mundford Norfolk; a comparison between fullscale performance and in situ and laboratory measurements. Proc. B.G.S. Conf. on In Situ Investigations in Soils and Rocks, 1970, London, 3-15.

28. Thames Barrier Project: Report on Site Investigation at Woolwich. Vol.3: Report on the geotechnical aspects of the design and construction of the foundations. Foundation Engineering Ltd., 1972.

Appendix

Chalk Classification (after Wakeling, 1972 (ref. 28)

Grade	S.P.T. Values	Description
VI	Less than 8	Extremely soft structureless chalk, consisting of unweathered and part weathered small angular chalk fragments, set in a matrix of deeply weathered remoulded chalk.
V	8-15	Soft structureless chalk, consisting of unweathered and part weathered angular chalk blocks and fragments, set in a matrix of weathered remoulded chalk; bedding and jointing absent.
IV	15-20	Soft, friable to rubbly chalk possibly weathered with bedding and jointing present. Joints and small fractures from 10 mm to 60 mm apart are present. The joints are commonly filled with remoulded chalk and small unweathered fragments.
III	20-30	Rubbly to blocky unweathered medium chalk with joints 60 mm to 200 mm apart.
II	30-40	Medium chalk with widely spaced, closed joints, more than 200 mm apart.
I	Over 40	Hard, brittle chalk with widely spaced joints similar to Grade II but harder chalk.

5. Effects of a barrier on existing estuary regime

Mary P. KENDRICK, BA, MWES*

B. V. DERBYSHIRE, MICE†

INTRODUCTION

1. This paper summarises the studies undertaken by the Hydraulics Research Station between 1968 and 1975, to determine the effect on the existing estuary regime of the construction and operation of a Thames Barrier. A separate paper by Mr. J.S. Burgess describes more recent model tests made by H.R.S. in 1975-76 to establish the bed protection necessary to prevent local scour at the Barrier.

2. The studies carried out during the first seven years involved extensive measurements in field and laboratory (ref. 1) and although the paper is primarily concerned with the results of physical and mathematical model tests, the field and laboratory measurements formed an essential part of the whole investigation. They provided the basic data from which the models were designed and operated and they gave a greater understanding of the way in which the Thames estuarine system works. A short summary of this work is therefore a necessary introduction.

FIELD AND LABORATORY MEASUREMENTS

3. The field measurements included four large-scale estuarine surveys, two smaller-scale local surveys of the reaches adjacent to the Barrier site, in-situ silt settling tests, and the establishment and operation of stations at six locations for continuously monitoring the suspended solids content of the river (Fig. 1). The data from the surveys demonstrated how current-velocity, suspended silt concentration, salinity and temperature at different depths along the estuary, change throughout a spring and a neap tide during periods of both high and low river flow (Fig. 2), (ref. 2). The results of the silt monitoring exercise supplemented these data by indicating how suspended sediment concentrations in the estuary vary - with tidal range, with point of time in the bi-monthly spring-to-neap cycle, with position in the estuary and with seasonal variations in fresh river discharge (ref. 3).

4. The object of the laboratory tests was to determine some of the more important properties of Thames silt - in particular those properties governing its behaviour during the cycle of silt movement; from the initiation of bed erosion, through the process of transport in suspension, to final deposition on the bed. The results of these tests using flume and settling tube, together with those from the field settling tests using a new instrument developed by Mr. M.W. Owen (ref. 4), provided values for the four parameters needed to develop and operate the silt mathematical model. These were: the "critical shear for erosion" (the force exerted on the bed by a particular current-velocity gradient immediately above it, at which silt starts to move); the "erosion constant" (the rate of increase of the erosion rate with shear force); the "limiting shear for deposition" (the value of shear force at which sediment starts to deposit out of suspension on to the bed); and the "settling velocity" of the silt.

MODEL TESTS - SURGE-EXCLUDING BARRIER

5. The tests to establish the effects on the distribution of a barrier closed only to exclude a surge-tide, employed two hydraulic models, investigated two different types of structure and examined four possible barrier sites.

6. The models were: the "Thames Estuary Model" which extended from Teddington to Southend to a horizontal scale of 1/600 and a vertical scale of 1/60, and the "Silvertown Barrier Model" which reproduced the tidal river between Teddington and Tilbury to horizontal and vertical scales of 1/300 and 1/60 respectively. They were built in a shed at Didcot, 10 miles from the Hydraulics Research Station, due to shortage of space in the Station's main experimental hall at the time the studies were commissioned. The type of structure tested represented either a vertical drop-gate barrier or a Rising Sector Gate barrier and the sites examined were at Crayfordness, Woolwich, Silvertown and Blackwall. It is with the studies at Silvertown, the selected site, that this paper is concerned.

Effects on Flow Distribution

7. The first indication of whether the presence of a new structure in the river is going to affect the estuary overall is obtained from a comparison of tide levels recorded before and after the installation of the structure on the model. If the rate of change of water level

*Principal Scientific Officer, Hydraulics Research Station
†Senior Scientific Officer, Hydraulics Research Station

Thames Barrier Design. Institution of Civil Engineers, London, 1978, 51-58

51

throughout a tide at a number of gauging stations along the estuary stays the same as before, (i.e. tidal propagation stays the same), then any changes in flow distribution will be confined to those reaches nearest the structure, and the estuarine system as a whole will remain un-affected.

8.   Several tests were conducted on the Thames Estuary Model to check the effect of a barrier on tidal propagation. The earliest series, completed in 1969 (ref. 5), used a barrier re-presenting a 137 m (450 ft)-wide central drop-gate flanked by two smaller drop-gates each 61 m (200 ft) wide, still smaller gates occupying the spaces between navigable openings and river banks. The main sill level was set at -10.97 m (-36 ft) ODN. Water levels were monitored on a spring tide, on the 1953 surge-tide and on three surge tides rising respectively 0.6 m (2 ft), 1.21 m (4 ft) and 1.82 m (6 ft) higher than the 1953 surge-tide at Southend. Each tide was tested under four different conditions of river discharge - zero flow, 70.79 m^3/s (2,500 ft^3/s), 283 m^3/s (10,000 ft^3/s) and 1132 m^3/s 40,000 ft^3/s).

9.   The next series of tests to check tidal pro-pagation was carried out in 1971 (ref. 6), follo-wing the GLC recommendation (ref. 7) to construct a Rising Sector Gate barrier at Silvertown. The model barrier had four openings 61 m (200 ft) wide flanked by two subsidiary openings 31.5 m (103 ft)-wide, the remaining space between openings and river banks consisting of 31.5 m (103 ft)-wide openings for falling gates- three to the north and one to the south. The tests were made on a spring tide under three different arrangements of main barrier sill level: all four main sills at -9.14 m (-30 ft)ODN; all four main sills at -10.06 m (-33 ft)ODN; the two inner main sills at -10.06 m (-33 ft) ODN, stepped up to the two outer main sills at -9.45 m (-31 ft) ODN. In all cases the two sills flanking the main spans were at a level of -4.60 m (-15 ft) ODN, and the invert level for the falling gates on the north and south sides of the barrier were at ODN.

10.   The results for the tests with both drop-gate barrier and Rising Sector Gate barrier demonstrated that neither structure will have any measurable effect on tidal propagation during the normal astronomical spring-to-neap tides at times of average river discharge. The drop-gate tests showed in addition that this statement is still valid for typical winter flood flows down the Thames, whilst the Rising Sector Gate tests established that it remains true irrespective of whether main barrier sill levels are set at -9.14 m (-30 ft) ODN or -10.06 m (-33 ft) ODN.

Effects on Sediment Distribution

11.   The first siltation experiments undertaken on the Thames Estuary Model (1969-70) (ref. 8) included an investigation of how the distribu-tion of bed sediment in the main navigation channel would be affected by the presence of the drop-gate barrier at Silvertown.

Since the flow tests had demonstrated that tidal propagation along the estuary was un-changed by the structure, the study area was limited to the stretch of river between the Isle of Dogs and Erith - an area which extended about seven miles above and below the barrier and in-cluded the main deposition zone known as the "Mud Reaches". Initially, tests were made with-out a barrier on the model, the three-fold objec-tive being to check its ability to reproduce natural conditions, to establish the repeata-bility of results, and to provide a standard with which to compare the later results obtained with the barrier.

12.   The model sediment of granulated Perspex (density 1.18 g/cm^3, median grain diameter 0.3 mm) was confined to the main channel, the vertical exaggeration of 10 precluding the possibility of extending the study over the steeply sloping riverside banks which became exposed at low tide level. Twenty-eight tides of a typical spring-to-neap cycle were continuously generated at the estuary mouth, where the introduction of salt water maintained a constant level of salinity: the fresh water discharge over Teddington Weir was pre-set to give the average river flow of 70.79 m^3/s (2500 ft^3/s). The natural salinity distribution in the estuary was thus reproduced in the study area. Sediment was introduced at the downstream limit at the rate necessary to provide constant conditions at the boundary where a progressive up-river drift of material occurs. Successive surveys, made at a given stage in the tidal cycle, indicated when the model bed had reached a state of dynamic equilibrium.

13.   This test procedure was repeated with the drop-gate barrier installed at Silvertown: the results are compared with those from the pre-vious test in the form of an accretion/erosion chart (Fig. 3). Clearly bed-level changes were negligible, being everywhere less than those normally experienced in the river due to typical seasonal variation.

14.   With the publication of the GLC's Second Report of Studies (1971) which recommended not only the location but also the form of the barrier, it became important to obtain informa-tion on its impact on the river in more detail than had been possible on the smaller-scale Thames Estuary Model. It was no longer enough to know what general pattern of scour and de-position could be expected in the navigation channel; it was necessary in addition to establish the likely changes both in the immediate vicinity of the structure and along the jetty frontages in Woolwich and adjacent reaches.

15.   The Silvertown Barrier Model was designed and built to satisfy these needs. The first siltation experiments undertaken on this model (1972-73) (ref. 9) included an investigation of how the distribution of bed sediment in the river would be affected by the presence of the Rising Sector Gate Barrier at Silvertown. The study area was confined to the 10 miles of river be-tween upper Bugsby's Reach and Cross Ness, which

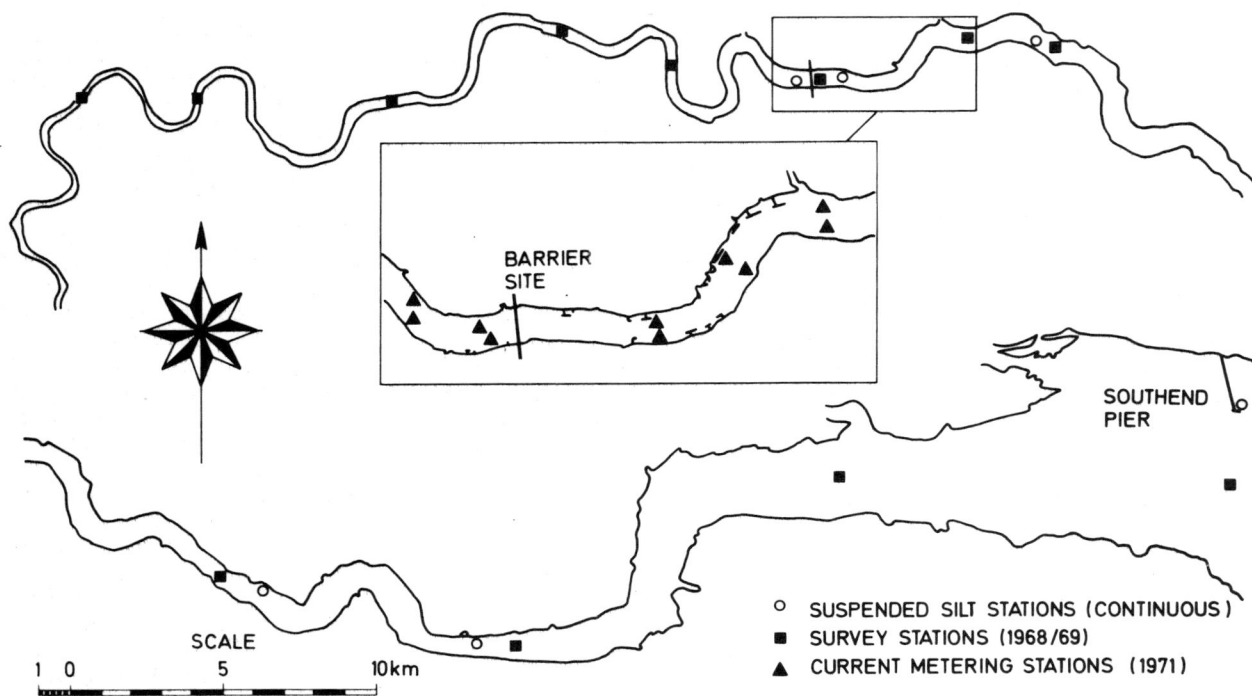

Fig.1. Location of field stations

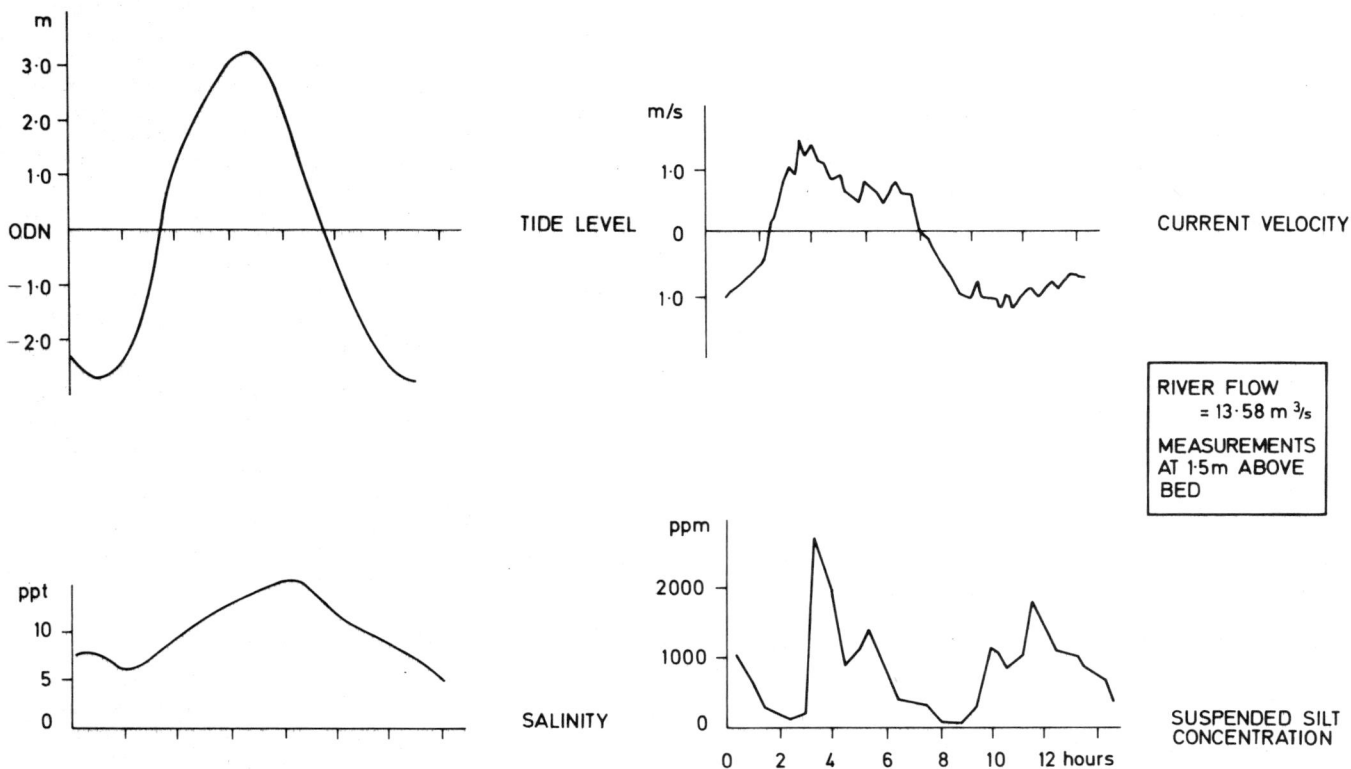

Fig.2. Typical spring tide survey data: Barking Reach

included the area bounded by the 'limits of compensation' as defined in the Thames Barrier Act - an area stretching roughly from the Western Entrance of the Royal Docks to Woolwich Power Station.

16. As with the drop-gate studies on the Thames Estuary model, and for the same reasons, the siltation experiments were carried out first without and then with a barrier. The model scales dictated a lighter-weight, larger-grained bed material than the Perspex of the 1969-70 tests, and so obeche wood grains (density 1.10 g/cm^3, median grain diameter 0.8 mm) were used. The sediment was no longer confined to the navigation channel but extended across the whole width of river from bank to bank, additional material being introduced to maintain constant conditions at the boundary of the study area as wood grains passed out of the working section due to the prevailing net drift of sediment. As before, typical, or average, conditions were reproduced on the model since the object was to establish the effects of the barrier structure in the middle-to-long term rather than the very short term (i.e. over a period of years rather than weeks).

17. Fig. 4, the accretion/erosion chart which compares conditions before and after barrier construction, demonstrates that although bed levels beyond about 1000 m of the structure were changed more than with the drop-gate barrier, the changes were relatively insignificant, still not exceeding seasonal variations and nowhere reducing the ruling depth for navigation. It also indicates, however, that near to the barrier, there was a much greater redistribution of sediment.

18. Fig. 5, showing river sections at the barrier site with the two designs of structure, goes some way towards explaining these quite large changes in bed level (approximately 2 m) which appear as an increase in depth on the northern side of the navigation channel and a decrease in depth in the centre and south.

19. Although design sill levels for the Rising Sector Gate Barrier accorded roughly with general bed levels in upper Woolwich Reach, the barrier ran through an area which was locally 1-2 m deeper in the centre and south (Fig. 5(b)). The main sill for the drop-gate structure, by contrast, was virtually coincident with the bed in the centre of the section (Fig. 5(a)), thus affording no obstruction to flow in that part of the channel which carried the bulk of the discharge. Another difference between the two structures as tested on the model was that the full length of the dredged Northern Division Channel was reproduced for the Rising Sector Gate experiments, but not for the earlier tests with the drop-gate. (The former involved the removal of material along a 1500 m length of river in line with Piers 3 and 5 (Fig. 5(b)) to provide adequate depths in the approaches to the northern navigable spans and a consequent lowering of the bed of between 3 m and 5 m).

20. Subsequent siltation tests (ref. 9), comprising a detailed examination of the effect on

bed levels of different barrier construction stages, established that the dredging of the Northern Diversion Channel was the factor primarily responsible for redistributing the flow near the structure and so modifying the topography of the river bed.

MODEL TESTS - TIDE CONTROL BARRIER

21. In addition to being operated solely as a flood protection device during surge-tides, a barrier could be used on a regular basis to control the flood and ebb of the tide, thus affording some of the benefits associated with a dam or barrage whilst at the same time preserving a degree of tidal flushing. Used in this way with the gates closed to impound the lower half of every tide and left shut in the event of a surge-tide, it was found to give rise to smaller changes in down-river water levels than full barrier closures made after the onset of the flood tide. The decision was therefore made to explore the effect of continuous half-tide control on the distribution of flow and sediment in the estuary. The studies were carried out in 1969 and 1970 employing two models - the Thames Estuary physical model with drop-gate barrier, and a specially-developed mathematical siltation model (ref. 10), which proved to be the first of its kind.

Effects on Flow Distribution

22. The model experiments already described showed that the presence of the barrier structure had no effect on normal tidal propagation in the estuary. The results of the measurements of (a) tidal level, (b) current-velocity, and (c) salinity, made above and below the barrier under these normal conditions, are compared on Fig. 6 with those obtained in tests in which the barrier gates were fully closed at every mid-ebb and not re-opened until the following mid-flood when water levels on either side of the structure had again equalized.

23. Below the barrier, the response to half-tide control was for the ebb period to be shorter, the flood period correspondingly longer, maximum flood current-velocities lower, and maximum ebb current-velocities higher. Above the barrier, once the initial fall in flood tide level following gate closure had occurred, levels increased slowly with incoming river flow until the gates were re-opened; current-velocities on both flood and ebb were lower. The main effect of half-tide-control on the longitudinal salinity distribution was to move the upstream limit of saline penetration about 5 km seawards.

Effects on Sediment Distribution

24. In the siltation tests on the Thames Estuary physical model, the same procedure and model operating conditions were adopted for the initial runs as those described for the drop-gate studies with the surge-excluding barrier. Once the bed had reached a state of dynamic equilibrium with the barrier gates permanently open, continuous half-tide control operations were begun and maintained until the new dynamic equilibrium was achieved.

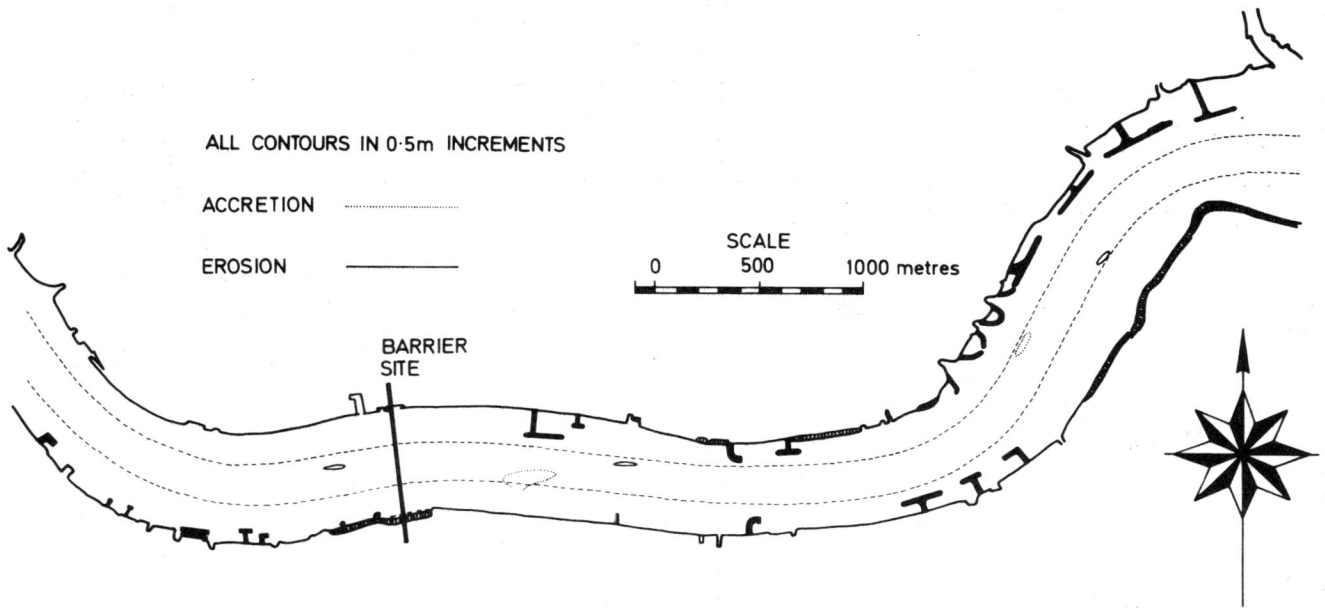

Fig.3. Effect of drop gate barrier on sediment distribution

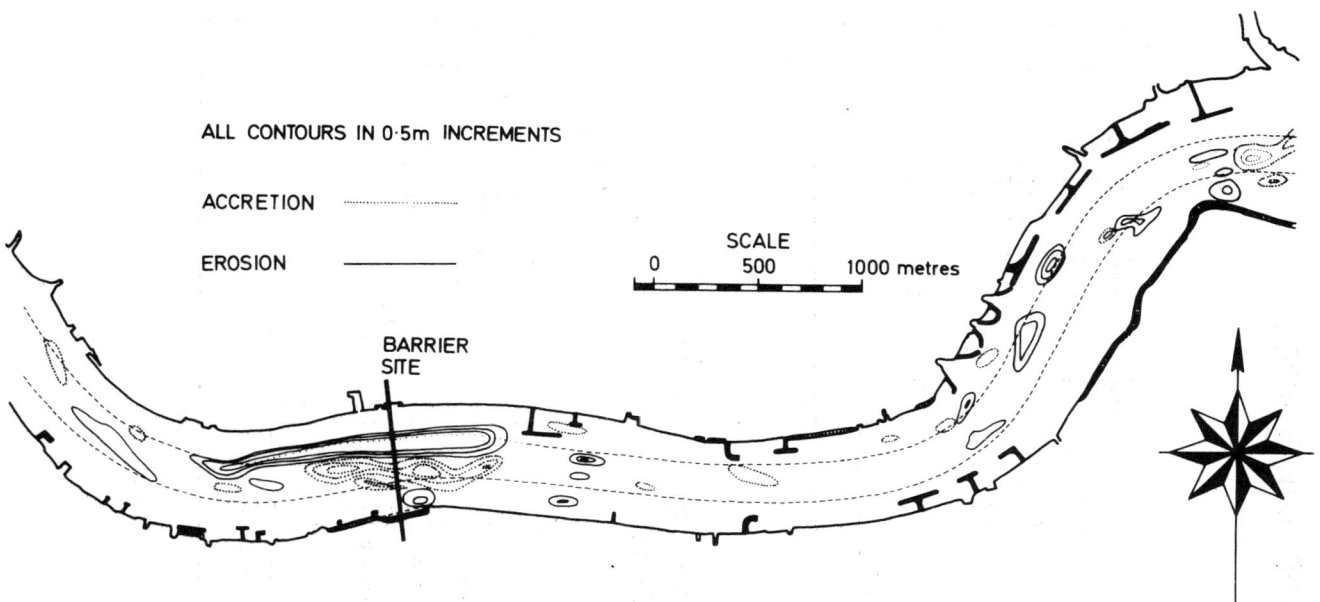

Fig.4. Effect of rising sector gate barrier on sediment distribution

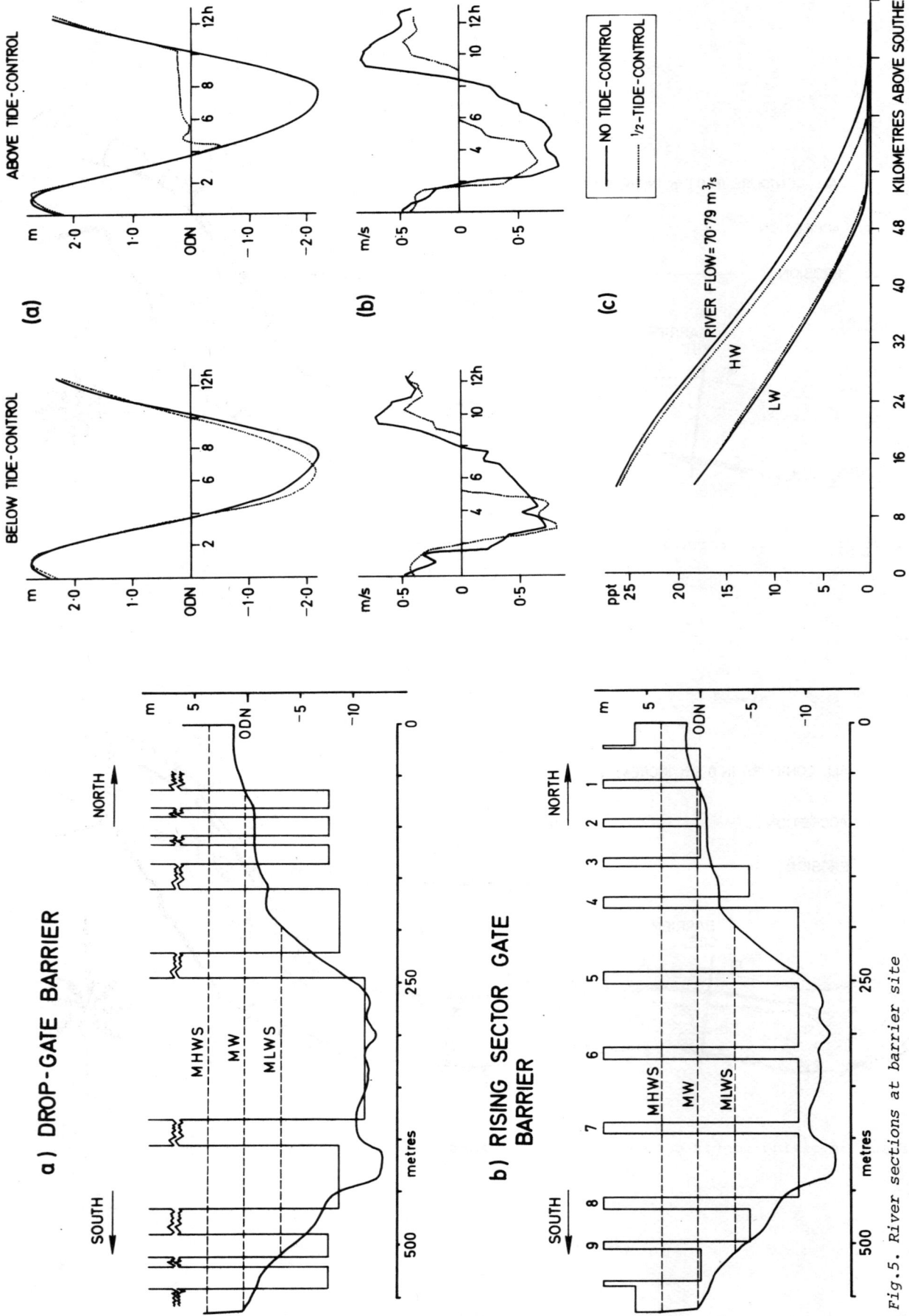

(a)

ABOVE TIDE-CONTROL

BELOW TIDE-CONTROL

(b)

(c)

— NO TIDE-CONTROL
········· ½-TIDE-CONTROL

RIVER FLOW=70·79 m³/s

HW

LW

KILOMETRES ABOVE SOUTHEND

Fig.6. Effect of tide control on water movement

a) DROP-GATE BARRIER

NORTH

SOUTH

MHWS
MW
MLWS

metres

b) RISING SECTOR GATE BARRIER

NORTH

SOUTH

MHWS
MW
MLWS

metres

Fig.5. River sections at barrier site

PHYSICAL MODEL

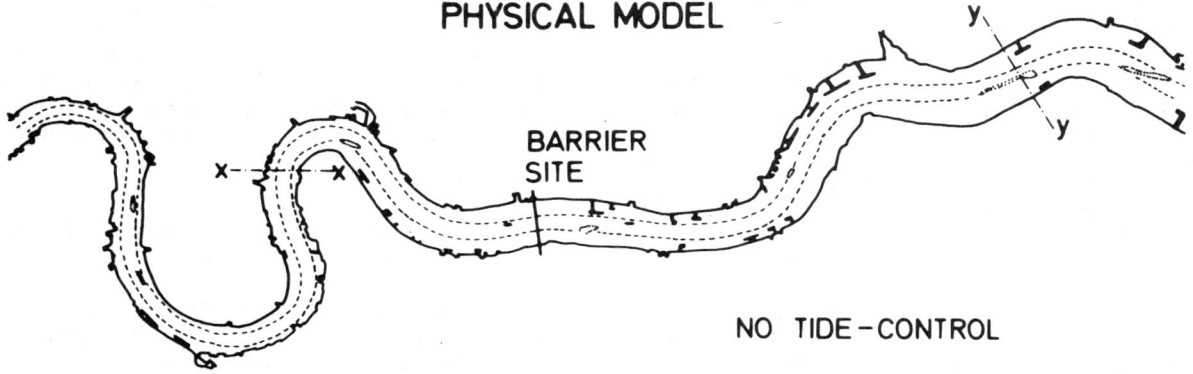

BARRIER
SITE

NO TIDE-CONTROL

ACCRETION ....

EROSION ⸺⸺⸺

ALL CONTOURS IN 0·5m INCREMENTS

½-TIDE-CONTROL

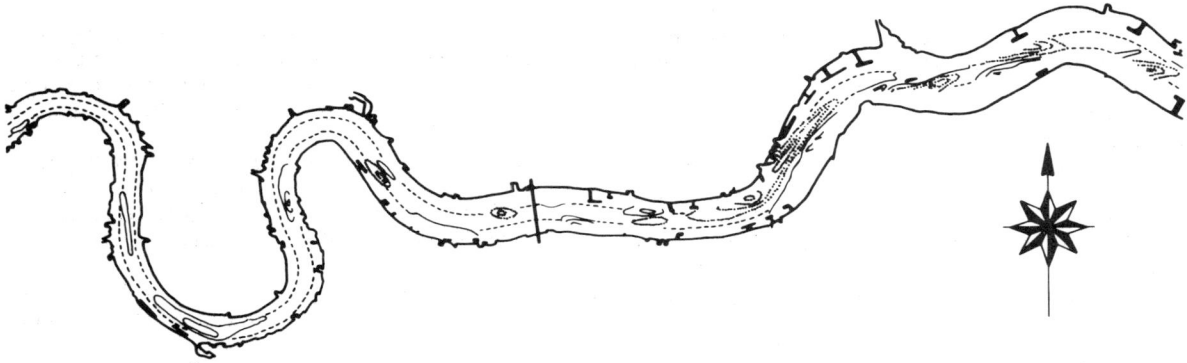

MATHEMATICAL MODEL

kg/m³

NO TIDE-CONTROL ⸺⸺⸺
½-TIDE-CONTROL ....

1·0

ACCRETION

0

EROSION

1·0

x   y

90   60   30   0

km ABOVE SOUTHEND

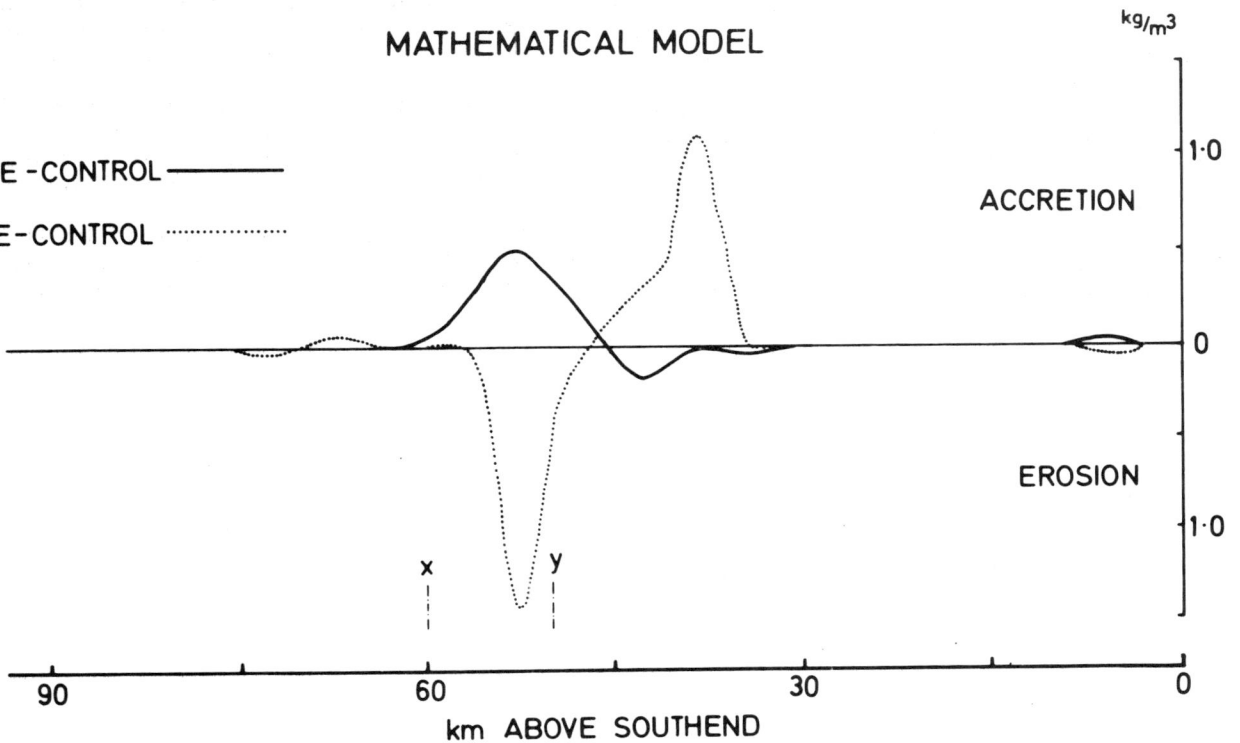

Fig.7. Effect of tide control on sediment distribution

25. For the tide-control tests however, one change in operating conditions was found to be necessary; the new flow distribution resulting from regular half-tide gate operation had reversed the direction of sediment drift from up-river to down-river, so the introduction of Perspex at the seaward limit of the study area was no longer necessary. Nor was material injected at the landward limit, since a plentiful supply existed upstream of the study area and the transport rate was relatively slow.

26. In comparison with the upper chart, the lower accretion/erosion chart on Fig. 7 demonstrates the large changes in sediment distribution resulting from continuous half-tide control. The main effect was for erosion in the upper reaches of the study area to be accompanied by deposition further down-river: siltation in the traditional major deposition zone of the estuary, – the Mud Reaches of Gallions, Barking and Halfway, – was increased and its seaward limit extended downstream to Erith.

27. Whereas the physical model reproduced the behaviour of sediment which in nature moves on, or just above, the bed, the mathematical model provided additional information on the behaviour of finer material which is transported in suspension throughout the depth of flow. Having initially calculated tidal propagation in the estuary between Southend and Teddington from the field survey data, the model used the results of the field and laboratory measurements of the movements and properties of silt to reproduce firstly the variation of the average silt concentration in an upper and a lower layer of flow throughout the tide with distance along the estuary, and secondly the variation of silt concentration in each layer with time at twelve river sections. Then continuous half-tide control was introduced, Fig. 7 illustrating longitudinal changes in the location of zones of accretion and erosion. The results support the findings of the physical model tests in that they demonstrate a seaward migration of the major deposition zone in the estuary.

ACKNOWLEDGEMENTS

28. This paper is presented by permission of the Director of Hydraulics Research, Wallingford.

The Authors gratefully acknowledge the co-operation of those members of the Station who, at one stage or another between 1968 and 1976, were engaged in the investigation, particularly Mr. M.A. Littlewood whose involvement with the work on the Didcot models was unbroken.

REFERENCES

1. KENDRICK, M.P., 1972, Phil Trans. Roy. Soc London A., 272, 223-243.

2. Hydraulics Research Station Report EX542-555, 1971. Thames Estuary Flood Prevention Investigation Field Survey Data.

3. Hydraulics Research Station Report in preparation, Transport of suspended sediment in the tidal River Thames.

4. OWEN, M.W., 1971, The effect of turbulence on silt floc settling velocities, International Association for Hydraulics Research, 14th Congress, Paris, August 1971. Societe Hydrotechnique de France.

5. Hydraulics Research Station Report EX479A, 1969. Thames Estuary Flood Prevention Investigation – First Report on Hydraulic model studies, January 1969 – September 1969.

6. Hydraulics Research Station Report EX613, 1972, Thames Estuary Flood Prevention Investigation – Sixth Report on hydraulic model studies, July 1971 – December 1971.

7. Greater London Council, 1971. Thames Flood Prevention, Second Report of Studies.

8. Hydraulics Research Station Report EX491, 1970. Thames Estuary Flood Prevention Investigation – Second Report on hydraulic model studies, October 1969 – January 1970.

9. Hydraulics Research Station Report EX679, 1975. Thames Estuary Flood Prevention Investigation – Tenth Report on hydraulic model studies, November 1972 – March 1975.

10. Hydraulics Research Station Report EX479, 1970. Thames Estuary Flood Prevention Investigation – Mathematical silt model studies.

Discussion on Papers 4 and 5

DR FOOKES and MR MARTIN

1. The modifications made to the local landscape by the last major geological events in the area, in particular the Ice Age, which caused for example frost-shattering of the near-surface chalk, the development of solifluction soils and the development of certain solution and erosion features in the Chalk, are important. The significance of solifluction soils, solution and erosion features in relation to the engineering, their presence or absence, and their size and form, was examined at the time of the site investigation.

2. At the Barrier site it was considered that pipes, i.e. approximately circular, narrow and more or less vertical infilled channels (see Fig.7 of Paper), could be present at the positions of the piers and abutments as they could be small enough to remain undetected by the boreholes; allowance was therefore made for their discovery during construction. However, deep depressions as, for example, those found during the construction of Battersea Power Station (Fig.1),[1] which would have seriously affected the foundations, were shown by the site investigation to be absent. In Fig.2 a contoured plan of the top of the Thanet Sand shows that, although there are small undulations in its surface, there is no evidence of a major depression. Figure 3 shows contours of the chalk surface, and although there are also undulations, which may be related to the fault pattern, they do not amount to major depressions of the size encountered at Battersea.

3. Figure 4 illustrates a series of narrow pipes within the Chalk and Fig.5 shows a larger feature of similar size to the depression found at Battersea. Figure 6 shows an exposure of weathered chalk overlain by soliflucted material grading upwards into top soil. The classification of chalk grades is in accordance with that given in the appendix to the Paper.

MR F.J.T. KESTNER (formerly Hydraulics Research Station)

4. The Eider discharges into the North Sea north of the Elbe. The problem there of safeguarding the low-lying areas in the estuary is the same as in the Thames, although on a much smaller scale. The solution was to build a barrage and this was done in 1935.

5. The predictions of hydraulic conditions immediately after the structure was built were fulfilled. Essentially, these were that high water of mean spring tides would be slightly higher (a few inches) and that low water at mean springs would be substantially lower (about 2 ft). These changes did indeed take place immediately after the barrage was completed, and as the lowering of low water improved drainage conditions, it appeared at first that the scheme was very successful. However, the barrage failed because the predictions had not taken into account the movement of the very fine mobile sands (60-120 microns) which caused heavy siltation seaward of the barrage and hence disastrous flooding landward.

6. Following the failure of the barrage in the Eider the Germans put forward a number of solutions which included a barrier, similar to the one for the Thames, which would turn only the surges at high spring tides if they occurred. It was hoped that such a barrier could be used as a control structure, for instance to hold back the ebb, and so influence the relationship between flood and ebb tide velocities.

7. As this solution and other alternatives were not recommended strongly enough an expert committee was set up, headed by Professor Walter Hensen, who had been the Director of the Tidal Institute at Hannover for twenty years. This committee recommended[2] that no new barrier should be built. They felt that it would be impracticable to maintain the essential cross-sectional area at the proposed new site. (This was much further seaward than the Thames Barrier.) The effect of the siltation had already spread from the upper reaches into the seaward stretches where the new barrier would have to be built.

8. The Eider local authorities did not take the committee's advice and have now built a barrier which turns only a surge tide and very high spring tides. The heavy siltation that this seaward barrier is causing is already giving rise to anxiety, and german engineers do not feel they have solved their problem.

9. Unlike the one in the Eider, the Thames Barrier fulfills the essential criterion that cross-sectional flow under normal conditions is not obstructed. If this pioneer venture succeeds we should not forget to acknowledge our debt to the german experience.

MR J. WATERSTONE (Polytechnic of South Bank)

10. One of the factors which causes a surge in London is the existence of river banks downstream which funnel the surge of the high tide up into the London area. Before there were such works the surge tide simply spilled on to the marshes and the upsteam effect was less. During the period of construction of the Thames Barrier the downstream embankments are being raised, which will put London in greater danger than it was before.

MR H.L. WAKELING (Rendel Palmer and Tritton)

11. One of the problems arsing from closing the barrier is its effect on water levels down river due to the reflected wave which is formed. This had shown up on the hydraulic models but as in these models the tide generator was situated just below Southend and the effect of the wave was suppressed as soon as it reached the generator.

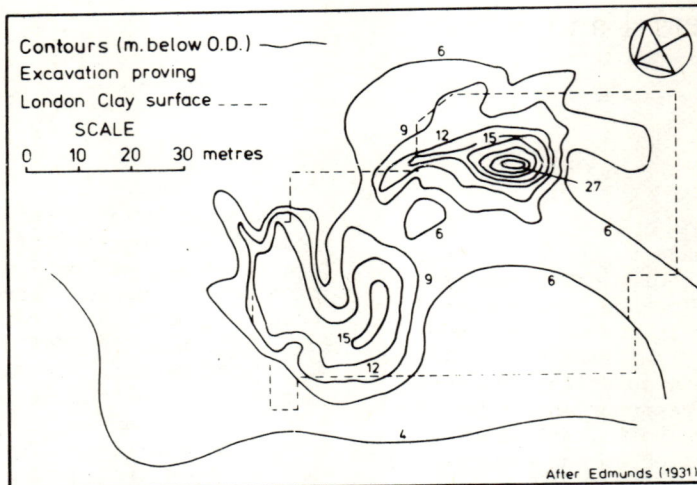

Fig.1. Form of gravel filled depression in the London Clay at Battersea Power Station

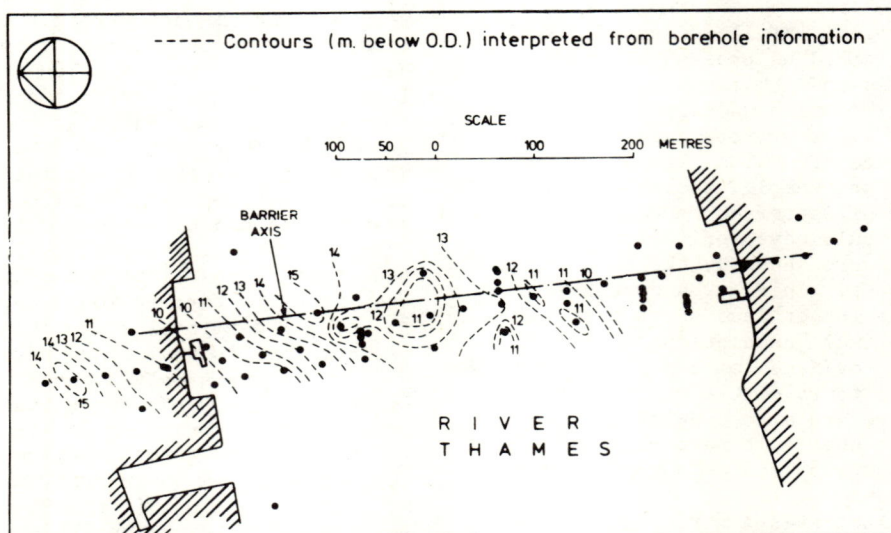

Fig.2. Plan of barrier site showing contours of Thanet Sand surface

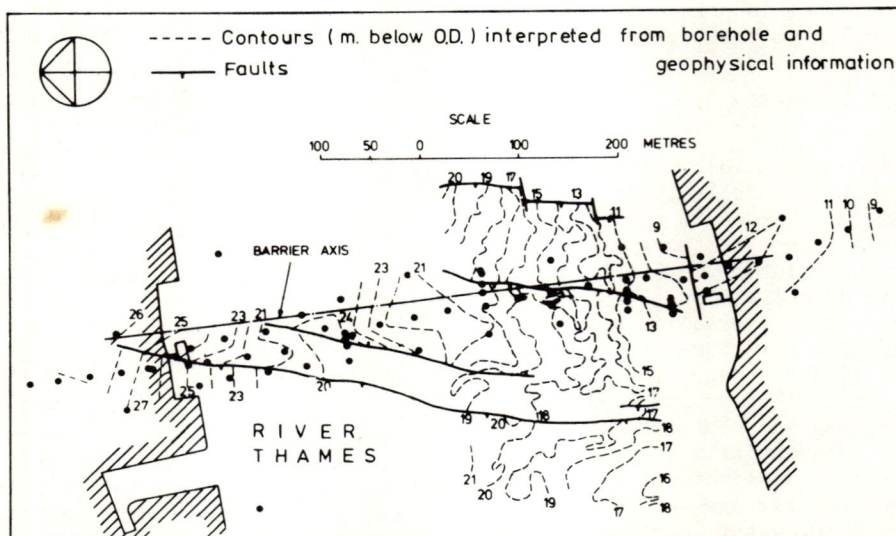

Fig.3. Plan of barrier site showing contours of chalk surface

Fig.4. Narrow pipe solution features exposed in a chalk quarry, Kent

Fig.5. Large scale erosion feature in sea cliffs, Kent

Fig.6. Weathered chalk overlain by soliflucted material exposed in a chalk quarry, Kent

12. During early studies Rendel Palmer and Tritton developed the first mathematical model in the United Kingdom, possibly in the world, of a tidal river using a computer.[3] With this computer model it was possible to overcome the limitations of the physical model by projecting the river out into the Thames Estuary in an exponential form and producing a tide at the new boundary which produced the correct tide at Southend. In this way it was possible to eliminate the effects of the hydraulic model tide generator and assess the effect of the reflected wave.

13. During the studies a lot of attention was paid to conditions during construction because the Port of London Authority was concerned that river passage should not be interrupted. In the first stage of construction the navigation channel was diverted towards the north bank; this condition was investigated extensively on the Silvertown Barrier model. The results were pessimistic and indicated that heavy siltation could be expected in the diverted channel. In fact, this has not happened. Do the Authors consider that this is due to the recent cleaning up of the Thames or are there other reasons for the difference between model predictions and what is actually happening?

MR S.S.D. FOSTER and MR A.C. CRIPPS (Institute of Geological Sciences)

14. The results of the investigations[4-7] referred to in paragraph 17 of Paper 4 pertain primarily to the operation of the tidal barrier rather than its design, but they may also have some implications for construction. (Details of the location and construction of the boreholes used for the investigations in Woolwich Reach are given in Fig.7 and Table 1.)

15. Along Woolwich Reach tidally-induced fluctuations in pore-water pressure are observed over substantial distances from the river in the more permeable formations: the Flood Plain Terrace Gravel and the Chalk (Figs 8 and 9). The Thanet Sand whose permeability is 1-2 orders of magnitude lower, partially confines the Chalk groundwater on the northern bank and probably only exhibits tidal fluctuations in pore-water pressure close to the river bank and to its contacts with the more permeable formations both above and below (Fig.5 of Paper 4).

16. The transmission of diurnal changes in head in a tidal river to adjacent groundwater systems is a function of two essentially distinct processes: transfer through the river bed and river bank to the groundwater system and propagation away from the river, laterally through the groundwater system.

17. In the ideal case of perfect hydraulic communication between a tidal river and any permeable strata into which it has channelled, the tidal ratio (TR*) beneath the river bed would be 1.00; but the presence of any low-permeability, compressible deposit in the river bed would tend to reduce this ratio substantially. The observation of considerable damping (TR = 0.39) and measurable time-lag (3-10 mins) in the response of piezometer 24B (Fig.8), which is sealed through the river bed into the Flood Plain Terrace Gravel of 24 m offshore from the southern bank of the river, is an illustration of the considerable 'impedance' of the river bed deposits.

18. Propagation away from the river will be a function of the hydraulic properties of the groundwater system and of any confining strata.[6, 8, 9] The observation of a sizeable tidal fluctuation (TR = 0.11) in observation borehole W7 (Fig.8), 110 m from the southern bank of the Thames, is a reflection of the high permeability coefficient and low specific storage of the confined groundwater system in the Flood Plain Terrace Gravel.

* The ratio between the response of the groundwater system and the corresponding diurnal change of tidal head in the river.

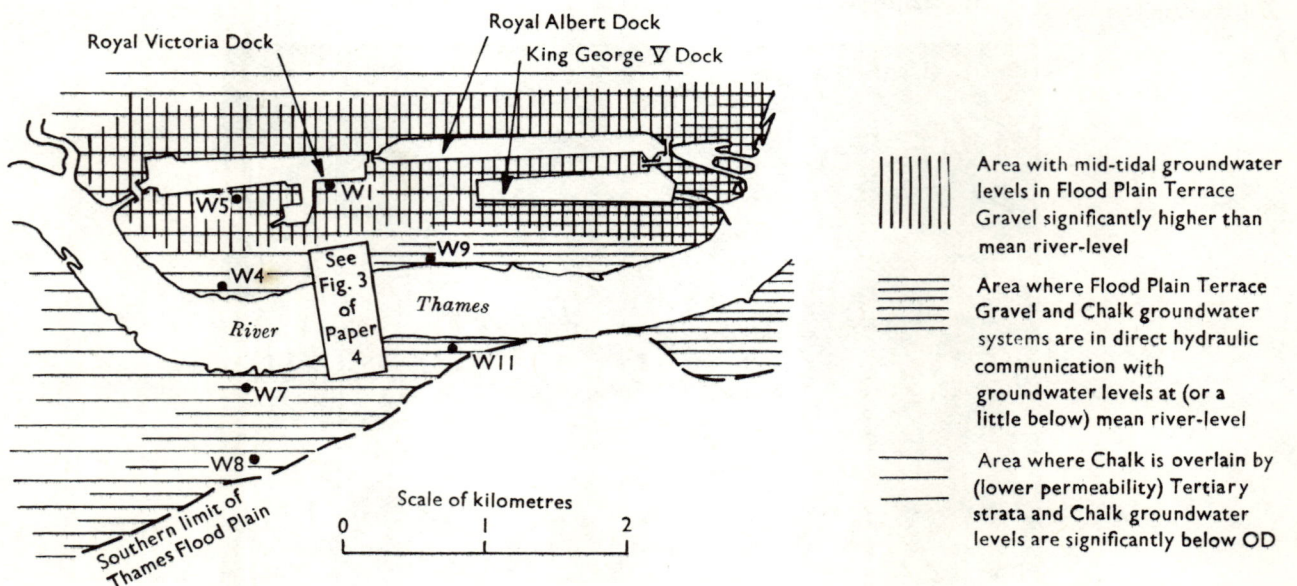

Fig.7. *Thames Flood Plain in Woolwich Reach - general location map with groundwater conditions*

Table 1. *Summary of construction, location and tidal response of observation boreholes and piezometers*

Borehole No.	Type≠	Geological formation*	Distance from river wall	Distance from edge of dredged channel	TR**	Time lag (mins)** at low tide	Time lag (mins)** at high tide
North Bank							
W4	O	FPTG	60	250	0.17	c 80	10
W5	O	FPTG	660	880	0.08	c130	c120
W12A	O	FPTG	12	260	0.19	70	5
W12B	O	sCK			0.15	c130	c 60
W13	O	FPTG	45	290	0.15	c 90	20
48	P	sCk	110	360	0.13	?	45
52A	P	sCk	19	270	0.18	c110	20
52C	P	dCk			0.09	c160	c100
South Bank							
W7	O	FPTG	110	300	0.11	30	c130
W8	O	FPTG	650	840	0.01	c 80	c100
W11	O	FPTG	75	250	0.25	c 50	25
W14A	O	FPTG	9	130	0.25	25	10
W14B	O	sCk			0.26	20	5
W15	O	FPTG	52	170	0.21	30	20
24B	P	FPTG	-24	95	0.39	10	3
30	P	FPTG	165	285	0.14	40	30
53	P	sCk	16	80	0.49	15	10
54	P	FPTG	54	120	0.42	25	10

≠   O Observation borehole completion; P piezometer completion

*   FPTG Flood Plain Terrace Gravel; sCk(dCk) shallow (deep) Chalk

**   data apply to a single moderate spring tide in 1972; some variations have been observed with fortnightly tidal cycle and with season

19. In analysing the TR data for Woolwich Reach, it was considered pertinent to express distances of observation boreholes and piezometers from the nearer edge of the main dredged channel (or dredged anchorage) and not from the river frontage. The data are then sufficiently ordered (Fig.10) to permit confident use of the analytical theory[8] for predictive purposes. The fact that the log TR-distance data can be extrapolated to the origin suggests that the 'impedance' to the transfer of tidal fluctuations from the river to the groundwater system is more or less zero in areas which are regularly dredged.

20. It was thus concluded that when the river bed sediments were removed from the foreshore areas (by dredging of diversion channels or excavation for pier foundations, for example) there would be substantial increases in the magnitude of the groundwater tides transmitted through the Flood Plain Terrace Gravel and the Chalk. This was, in fact, confirmed to be the case during the initial diversion dredging, which started in September 1974 and diverted the main navigation channel about 100 m to the north by removing the river bed sediments from the Flood Plain Terrace Gravel. The increases in the TR observed in the gravel strata on the northern bank during the succeeding months are indicated in Table 2, and compared with those predicted.

21. A similar phenomenon affecting a more extensive area on the southern bank is expected to occur following the planned diversion of the main navigation channel to the south, so as to pass between the piers currently under construction. The side-effects of the transient increases in pore-water pressure during high tides could include:

(a) increased peak flow to the drains protecting non-waterproofed buried structures
(b) increased maximum hydrostatic uplift on tanked sub-surface structures
(c) decreased overall stability of the river embankment and the flood plain cover.

Although no major problems appear likely to develop, it would be prudent to keep the groundwater situation under continuous observation.

22. The values for permeability coefficient of the Chalk given in the Paper (paragraph 57) could, from hydro-geological knowledge, be easily exceeded locally by one order of magnitude or more. In strata such as Chalk, rising and falling head tests in single boreholes are prone to give unreliable results. Have the experiences to date from the operation of wellpoints in the Chalk for relief of uplift pressures on the cofferdams at piers 8 and 9 borne out the quoted permeability values?

23. The basal Thanet Sands will be subject to large tidally-induced fluctuations in pore-water pressure from the underlying Chalk. Will it therefore be necessary to relieve pressure, by pumping from wellpoints in the Chalk around the sites of piers 1, 2 and 3 during the excavation in the Thanet Sand for formation of their foundations and emplacement of the cofferdams?

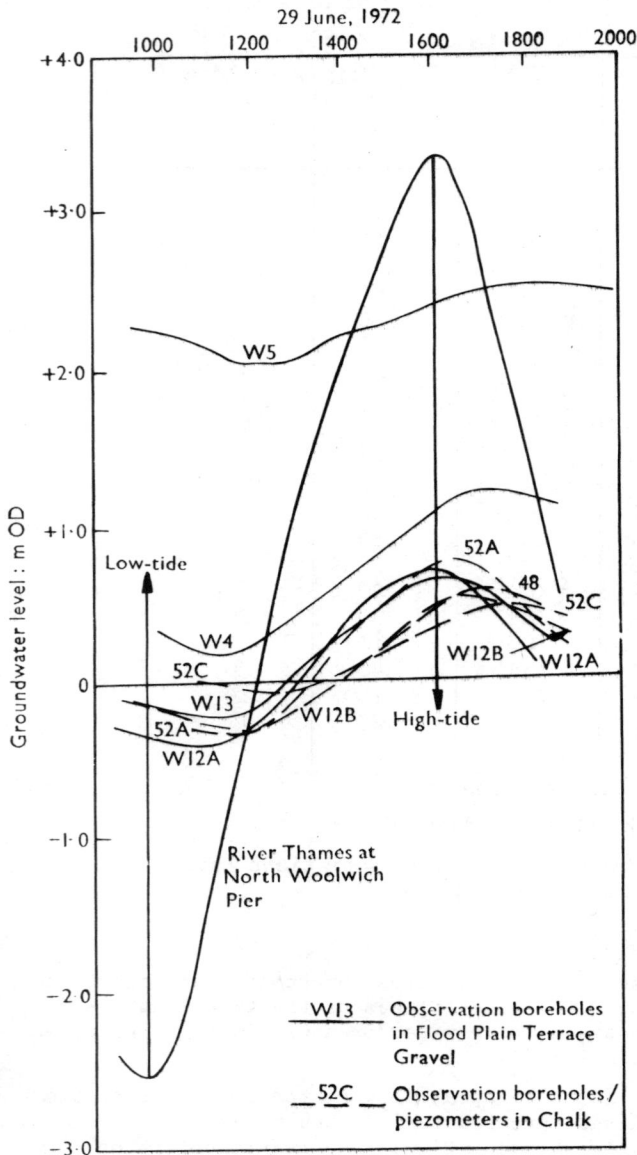

Fig.8. Groundwater response to diurnal tidal fluctuation along Woolwich Reach (North Bank)

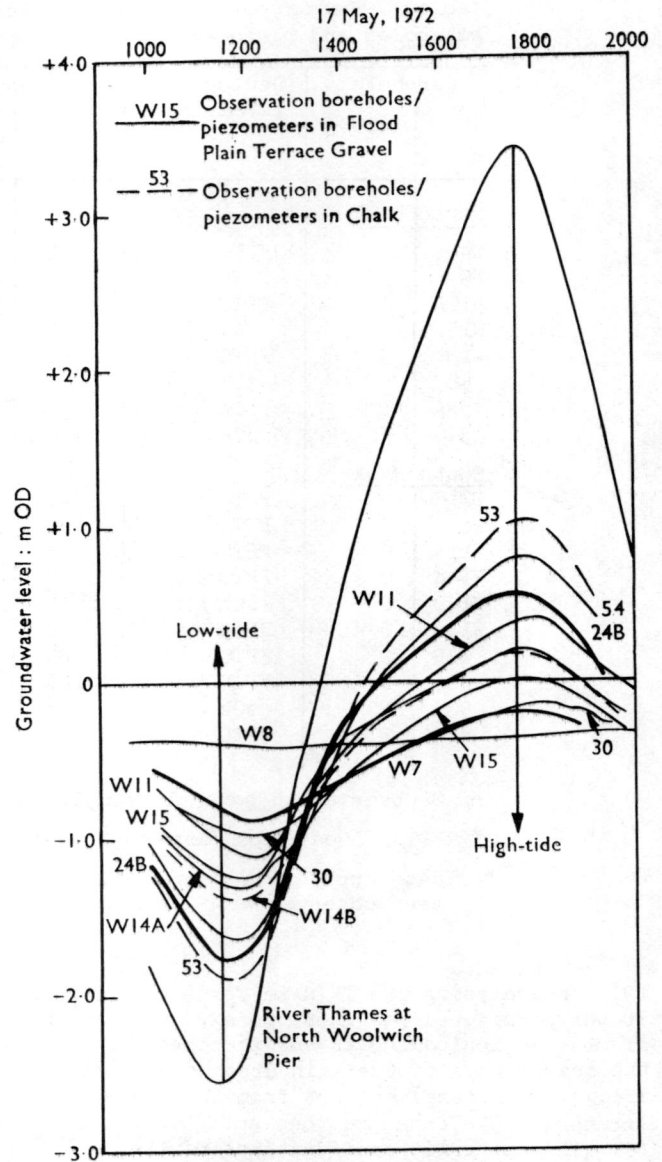

Fig.9. Groundwater response to diurnal tidal fluctuation along Woolwich Reach (South Bank)

24. A survey of the chemistry of groundwaters along Woolwich Reach, based only on pumped samples, was also undertaken by the Institute of Geological Sciences.[6] Such samples may not reflect fully the chemistry of the most aggressive groundwaters present in situ, as a result of dilution from other inflow levels into the borehole concerned. Most of the groundwaters appeared to have fairly high chloride and sulphate concentrations (2000-3500 and 100-750 ppm respectively) but were not markedly acidic. However, a sample from borehole Wl3 (on the northern bank of the barrier site) had about 2500 ppm SO_4^{2-} with a pH of 4, and lenses of a similar groundwater could be present elsewhere locally. The existence of groundwater of this quality is attributed to previous disposal of industrial effluents to the ground near the site - this was also the case at a site in West Bromwich where spectacular attack of the concrete linings of a new main drainage tunnel was reported.[10]

25. On the subject of costs (paragraph 68), what proportion of the estimated total cost of the structure (at 1972 prices) was represented by the site investigation? How does this proportion compare with that for other major river and marine engineering works? Does the figure quoted include the staff costs of the consulting engineers for supervision and interpretation of the site investigation?

MR D.P. SHIPMAN (Anglian Water Authority)

26. I should like to comment on the question of whether there would be any need for a barrier at Silvertown if the downstream tidal walls had not been raised to prevent a surge from spilling over on to the marshes. Over 40 000 people live on the north bank marshes at Canvey and Tilbury and, among the many installations built on the north bank, there are two oil refineries, three generating stations, Tilbury Docks and Container

Terminal, petro-chemical storage facilities and
many factories producing paper, cement, soap,
detergents, foodstuffs, vehicles, and so on, not
to mention group facilities for the export/import
trade at deep water berths in the river. The
north bank marshes are thus effectively ruled
out as a storage area for surge tides. As
regards the south bank, the question of using
certain marshes in agricultural use as an overspill area was investigated.[11] Under most conditions little benefit would accrue from such a
scheme and indeed under certain conditions, an
adverse effect could well be produced.

DR FOOKES and MR MARTIN

27. The foundation work for piers 8 and 9 has
been completed. In reply to the question by Mr
Foster and Mr Cripps on the operation of the
pressure relief systems installed for these
piers, the method of relieving the uplift pressures beneath the pier foundations did not consist of wellpoints but of large diameter wells
sunk up to 10 m deep into the Chalk below foundation level. Before dewatering, a plug of tremie
concrete several metres thick was placed on the
prepared formation, holes being left through the
plug so that during the subsequent dewatering of
the cofferdam water flowing into the wells was
discharged at the surface of the concrete. The
uplift pressures were, therefore, independent of
the permeability of the Chalk as the wells and
discharge tubes did not restrict the flow. It
was not practical to measure the discharges from
the wells during the dewatering of piers 8 and 9
but the flows were judged to be of an order commensurate with the coefficients of permeability

Table 2. Increases in Tidal Ratio (TR) following initial diversion dredging

Observation borehole	TR		
	Pre-dredging mean	Post-dredging observed	predicted
W12A	0.16	0.36-0.40	0.40
W13	0.15	0.30-0.34	0.34
W9	0.18	0.26-0.30	0.34
W1*	0.04	0.08	?

* Influenced by dock water-levels

of the Chalk determined during the site investigation.
28. With reference to their second question concerning the need for relieving uplift pressures
beneath piers 1, 2 and 3, a pressure relief system consisting of a number of pumped wells outside the cofferdam has already been installed
for pier 2, where foundation work is currently
in progress, to reduce the uplift pressure on
the Thanet Sand. A similar installation is envisaged for pier 1. For pier 3 the means of
ensuring an adequate factor of safety against
uplift has yet to be decided.
29. The original contract price for the main
civil works was approximately £38 million but
we do not consider that a meaningful breakdown
of costs can be given in view of the scope of
work included in that sum, very little of which
is directly controlled by the foundation conditions. We do not believe that representative
figures can be quoted for other river or marine
structures because there is so much variation
between sites both in ground conditions and in
structure configuration.
30. The figure quoted for the investigation
(£108 634) did not include the Consulting
Engineers' supervision costs. It also did not
include the cost of several boreholes made in
Woolwich Reach during the preliminary studies
or those made during the construction to confirm
detailed foundation design.

MRS KENDRICK and MR DERBYSHIRE

31. Mr Kestner's history of german experience
on the Eider has some relevance in that the
Eider barrage provides engineers with a very
good example of what not to do in a sandy or
silty estuary if flood protection is to be
achieved without loss in channel depth. It
demonstrates clearly a basic concept in tidal
hydraulics, namely that in a sediment-laden
estuary, a significant reduction in tidal volume,
and hence tidal discharge, results in sediment
deposition.
32. This was one of the considerations which
prompted the Hydraulics Research Station to seek
wider terms of reference when, in early 1968,
the Government asked us to be prepared to design
and build an hydraulic model to examine the
effects of a barrage across the Thames to protect central London from flooding by surge tides.
The problem was much too urgent to allow the
time needed to make a thorough appraisal of a
scheme which would virtually have reduced the

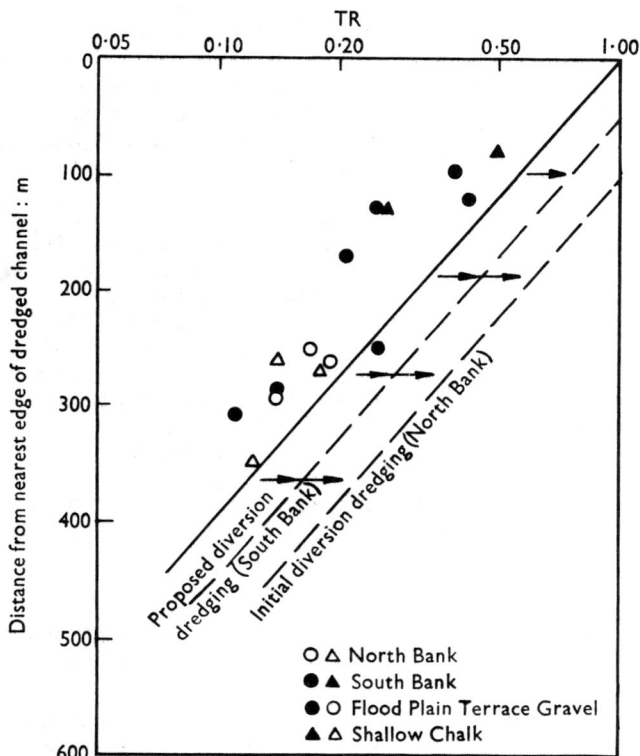

Fig.10. Analysis of propagation of groundwater tides

length of the tidal Thames by about 40%.
Accordingly we recommended the less drastic solution of a barrier with movable gates, closed in the first instance only to exclude surge tides. When open such a structure, if carefully designed, need have no significant effect on existing tidal propagation and so the overall regime of the estuary would be unchanged. Any problems of siltation or scour would be confined to those reaches nearest the barrier and, with the help of model experiments, could be kept to a minimum.

33. Mr Wakeling's reminder of Otter's early mathematical model of the Thames is timely because at the Hydraulics Research Station we have recently completed a series of tests for the Greater London Council in which two existing physical models were used to determine to what extent increases in water level, due to reflections from a barrier closed against an in-coming surge tide, are affected by the proximity of the tide generator. The first model had its seaward boundary at Southend, the second extended for a further 50 km into the southern North Sea.

34. Comparison of the results of a variety of barrier closures made on the two models demonstrated that the influence of the tide generator varied according to the type and time of closure, being least with gates partly closed to allow undershot flow, and greatest with gates closed fully at approximately mid-flood.

35. Mr Wakeling's second point concerns the prediction of siltation in the northern diversion channel which was dredged before the start of barrier construction. He suggests that a possible reason for the discrepancy between the predicted and actual siltation in the channel is the cleaning up of the river in recent years. Certainly records from a number of silt monitoring stations on the Thames indicate that at the present time concentrations of suspended solids tend to be lower than in the past and so less material is available to deposit from the flow than before. There are, however, other factors which shoud also be taken into account when seeking an explanation for the lack of channel siltation.

36. Although, as Mr Wakeling says, the northern diversion channel was extensively investigated on the Silvertown Barrier model, there was in fact insufficient time to examine all the possible, or even likely, combinations of tide and river flow as well as the possible barrier construction layouts, since each model test with movable sediment took several weeks to complete. In such circumstances a decision has to be made on the test conditions to apply and, following consultation with the GLC and their consultants, it was agreed to opt for average or 'typical' conditions.

37. In the event, when the northern diversion channel came to be dredged in the winter of 1974-75, high freshwater flow down the Thames was sustained for an unusually long time. This resulted in fine sediments being carried down river from Woolwich and the adjacent mud reaches - not by virtue of any significant increase in current velocity, but more probably because silt particles which flocculate and deposit when river salinities exceed a certain value, appear to experience deflocculation and become easier to erode and transport when the environment remains fresh for several weeks. As a consequence, the northern diversion channel fortunately remained free from siltation during that very vulnerable period which came after dredging but before the Stage 1 barrier construction works had advanced sufficiently far to confine the bulk of the tidal flow to the northern side of Woolwich Reach, thereby assisting in the maintenance of the channel.

38. A second factor which probably contributed to the well-being of the northern diversion channel was the dredging of material for reclamation which was being carried out further seaward on the Thames. Fine material which might otherwise have been carried upstream by the flood tide, particularly during the unusually long dry season (18 months) which followed the winter freshets, would be more limited in supply than usual due to having deposited in the excavated areas down river.

39. We agree with Mr Waterstone that the historic canalisation of the lower reaches of the estuary has served to funnel surge tides further upriver. Some reasons for embanking to protect these down-river areas are stated by Mr Shipman.

40. Under the terms of the 1972 Thames Barrier and Flood prevention Act, the entire estuary is to be protected against a 1:1000 year storm surge. To achieve this standard in the shortest time, barrier construction and the raising of banks down river must proceed concurrently. It was appreciated that during barrier construction central London would be at risk, and interim measures were taken which raised riverside defences throughout the area by up to 0.45 m. These works offset the relative deterioration in London's defence standards caused by the bank raising and strengthening in the lower reaches following the 1953 storm surge disaster, and made an allowance for the secular increase in surge tide heights.

41. The effects of the present bank raising activities below the barrier might constitute a much smaller risk element in central London than envisaged by Mr Waterstone. Indeed it is not until the works are well advanced that any change in water level will be apparent in the capital. As Mr Shipman points out, overspilling in the lower reaches does not always lead to a reduction in high river levels. Certainly hydraulic model studies of the 1953 surge tide demonstrated that overspilling and bank breaching kept levels in central London 0.08-0.11 m lower than they would otherwise have been, 0.03 m of which was attributable to overspilling, the remainder to bank breaching. However, model tests also showed that when overspill fills areas of limited storage capacity, it can result in increases in peak river levels.

REFERENCES

1. EDMUNDS F.H. Some gravel filled pipes in the London Clay at Battersea. Summ, Prog. Geol. Surv., 1930-31, 12-24.

2. KÜSTENAUSSCHUSS NORD- UND OSTSEE, GUTACHTER-GRUPPE EIDER. Gutachten über die Vorschläge zur Behebung der Schwierigkeiten in der Eider. Die Küste, 1964, 12, 30-60.

3. OTTER J.R. and DAY A.S. Tidal flow computations. Engineer, Lond., 1960, 29 Jan.

4. FOSTER S.S.D. Groundwater conditions in the Thames Flood Plain and their relation to the Thames Barrier. Thames Flood Prevention. Greater London Council, London, 1971, 2nd report of studies, Appendix 6(1).

5. GRAY D.A. and FOSTER S.S.D. Urban influences upon groundwater conditions in the Thames Flood Plain Deposits of Central London. Phil. Trans. R. Soc., Series A, 1972, 272, 245-257.

6. CHILTON P.J. and FOSTER S.S.D. Groundwater conditions in the Woolwich Reach of the Thames with reference to the effects of a tidal-surge and tidal-barrier. Institute of Geological Sciences, London, 1973, unpublished report WD/73/5.

7. FOSTER S.S.D. and CRIPPS A.C. Groundwater conditions in the vicinity of the Rotherhithe Underground Railway Tunnel and their relation to the River Thames. Institute of Geological Sciences, London, 1974, unpulished report WD/74/1.

8. FERRIS J.G. Cyclic water-level fluctuations as a basis for determining acquifer transmissibility. Wat.-Supply Irrig. Pap., Wash., 1963, No. 1536, I, 305-318.

9. VAN DER KAMP G.S.J.P. Periodic flow of groundwater - a systematic study of wave propagation under confined, semi-confined and unconfined flow conditions. Editions Rodopi, Amsterdam, 1973.

10. Tunnel faces acid test - and disintegrates. New Civ. Engr, 1975, 30 Jan., 21-22.

11. GREATER LONDON COUNCIL. Thames Flood prevention barrier/barrage project. First report of studies, 1969, section 5.4.5., 24.

6. Basic concept of the rising sector gate

B. G. R. HOLLOWAY, FICE, FIStructE*

G. MILLER RICHARDS, BSc(Eng), FICE†

R. C. DRAPER, FIED‡

INTRODUCTION

1.   Paper 6 describes the basic principles of the Rising Sector Gate concept finally chosen for the navigation spans of the Barrier. Fundamental to the adoption of the Woolwich Reach as the Site for the Barrier was the decision by the Greater London Council that there should be four main navigation openings, each 61 metres (200 ft) clear. The ability to adopt this width of main opening rather than the greater widths necessary further downriver allowed the use of the Rising Sector Gate.

GENERAL REQUIREMENTS

2.   The decision to select the Rising Sector Gate to span the main navigation openings satisfied the design criteria given in Paper 3, the navigational requirements being repeated here for convenient reference:-

(i) Four 61 m (200 ft) clear width main navigational openings.

(ii) Two subsidiary navigational openings of 31.5 m (103 ft) clear width.

(iii)Piers to be as narrow as possible consistant with their function.

(iv) Sills of the main openings to be at a level which conforms to the "ruling" depth of the present channel and hence which did not restrict navigation in the Reach, or lead to excessive siltation.

(v) Adequate overhead clearance or no obstruction with the gate open.

3.   The criteria of unrestricted navigation within the main openings and the capability of Barrier operation in flowing water at any state of the tide, required a gate which, in the open position, would not impede the river flow or river traffic. To be capable of operation under all conditions of river flow, the gate should resist the thrust due to the water pressure without creating major out-of-balance forces about

the line of action of the pivots. Consideration of how best to fulfil these basic requirements has led to the concept and development of the Rising Sector Gate.

GENERAL DESCRIPTION OF THE BARRIER GATES

4.   The complete Barrier comprises the four main Rising Sector Gates of 61 m clear span for the main navigation openings, with similar Rising Sector Gates of 31.5 m clear span at either side of the main openings providing a waterway for smaller vessels, the remaining river width being protected by four conventional falling radial gates of 31.5 m span each.

5.   The gates are supported by a series of nine piers and two abutments, with concrete sill units spanning between the piers at river bed level. In the open position the Rising Gates are housed in shaped recesses in the sill units, so that the flat upper surface of the gate is flush with the sill surface and does not intrude on the full navigational depth of the opening between the piers.

6.   To close these gates to form a Barrier they are rotated through approximately 90° until the flat skin plate is near to the vertical and the curved skin plate is facing seawards.

7.   To facilitate periodic maintenance and inspection, these gates can be rotated through a further 90° until the gate is lying horizontally in the inverted position with the flat face above the normal high water level.

8.   The four spans of falling radial gates, of similar structural form to the Rising Gates, are stored in the inverted position above high water level and permit normal river flow. When closure takes place, these gates are lowered to seal on a flat concrete sill, the curved surface of the gate facing seawards, in a similar manner to the Rising Sector Gates.

9.   The gates are operated by a hydraulically powered mechanical system described in Paper 8, giving a positive means of control for all gate movements.

DEVELOPMENT OF THE RISING SECTOR GATE

The Gate Span

10. The segmental cross-section of the gates is

---
*Consultant, formerly Partner, Rendel,Palmer and Tritton
†Managing Engineer, Rendel, Palmer and Tritton
‡Senior Design Engineer, Rendel, Palmer and Tritton

formed by steel skins stiffened by internal webs and diaphragms, serving to distribute the water pressure and transmit the load to the hollow steel discs at each end (the gate arms) which are centrally pivoted about stub shafts projecting from the piers.

11. In selecting a radius of curvature for the gate it was necessary to fix the required chord length compatible with the sill level and surge water level and the depth of the segment consistent with acceptable skin plate thicknesses.

12. A satisfactory solution was arrived at which positioned the pivot level at about mean low water and a radius of curvature of 12.2 m (40 ft). This gave a chord width of approximately 20 m (66 ft) and a maximum section depth of 5.3 m (17 ft 6 in) at the centre of the segment. The length of the chord is slightly more than that required as a theoretical minimum, so that when the gate is rotated into the closed position, the trailing edge of the gate span remains well within the sill recess and prevents a significant flow past the gate.

13. In order to eliminate the out-of-balance loading effects from differential water levels acting about the pivots, it was decided that the gate should be a free-flooding structure with the curved skin forming the separation between upstream and downstream water levels. This ensured that the resultant of all hydrostatic forces, acting in either direction on the gate, passed through the centre of curvature of the gate skin and the centre of rotation of the gate structure.

14. Preliminary calculations were made to determine approximate plate thicknesses for the curved and flat skins of the gate section. The gate span was assumed to be simply supported and was to be designed as a stressed skin structure. A multiple system of longitudinal webs and transverse diaphragms was positioned inside the span, suitably perforated to allow the free flow of air and water through the structure during the operation cycle. This early concept of the gate structure is illustrated in Fig. 1 which shows the structure stiffened by fifteen longitudinal web plates composed entirely of mild steel.

15. A check was made on the weight of a 61 m span of this cross-section and it was found to be of the order of 1750 tonnes. Even taking into account the reduction in the dead weight of the span due to its displacement in the water, this would be a formidable load for a gate-operating mechanism to deal with. It was concluded that the moment of the dead weight of the span about the centre of rotation, should be counterbalanced by using an equivalent ballast weight suitably positioned on the opposite side of the axis of rotation. The accommodation of the counterweights dictated the form of the gate arms.

The Gate Arms

16. The optimum form for the gate arms was seen to be circular and of cellular construction. This shape facilitated the connection of the operating mechanism of the gate, and provided space for incorporating the counterweight ballast. With a gate arm radius equal to that of the gate span surface and a width of 1.5 m, the required volume of cast iron could be accommodated.

17. The cellular construction of the gate arms permits a solution to the problem of how to provide for the gate span to flood freely, whilst preventing as far as possible the entry of silt and debris by providing a one-way system for water flowing in and out of the structure during gate rotation.

18. For water entry, a row of ports are positioned in the gate arms in both skin plates below the pivot bearing, the ports on the river side of the gate arms being fitted with grills to prevent large floating debris entering the structure.

19. Where the chord of the gate span connects with the chord of the gate arm, water passages are provided into the span. Along the trailing edge of the flat skin plate of the span are positioned a series of flap valves, which open away from the outer face of the flat skin. Hinged at their upper edges, they are restrained from opening more than a fixed amount.

20. When a gate is rotated upwards from its open position and the leading edge breaks surface, air will be drawn into the gates through a suitably positioned port above water level in each gate arm, and water within the gate span will open the flap valves by means of the calculated low differential head between the carried water level and that of the external water. Hence the water contained in the span will discharge through the flap valves until the level within the span has equalised with the up-river level.

21. Conversely, when a gate is lowered from its closed position, the flap valves will remain closed under the depressed head of water within the gate during lowering, and water will be drawn in to the gate arms and gate, as the gate arm ports enter the water. The flap valves thus permit equalisation of the water level in the gate and arms and the upriver water level whilst at the same time diminishing the extent that silt and debris can enter the gate interior. While the gate is lying in its recess in the open position, the water within the structure does not change since a no-flow condition exists.

22. Silt drawn into the structure during lowering will tend to be discharged during the raising cycle; any silt remaining in the gate can be manually flushed out during maintenance periods when the gate span can be brought wholly out of the water into its inverted position.

Fig. 2 illustrates the raising and lowering cycles of the gate structure.

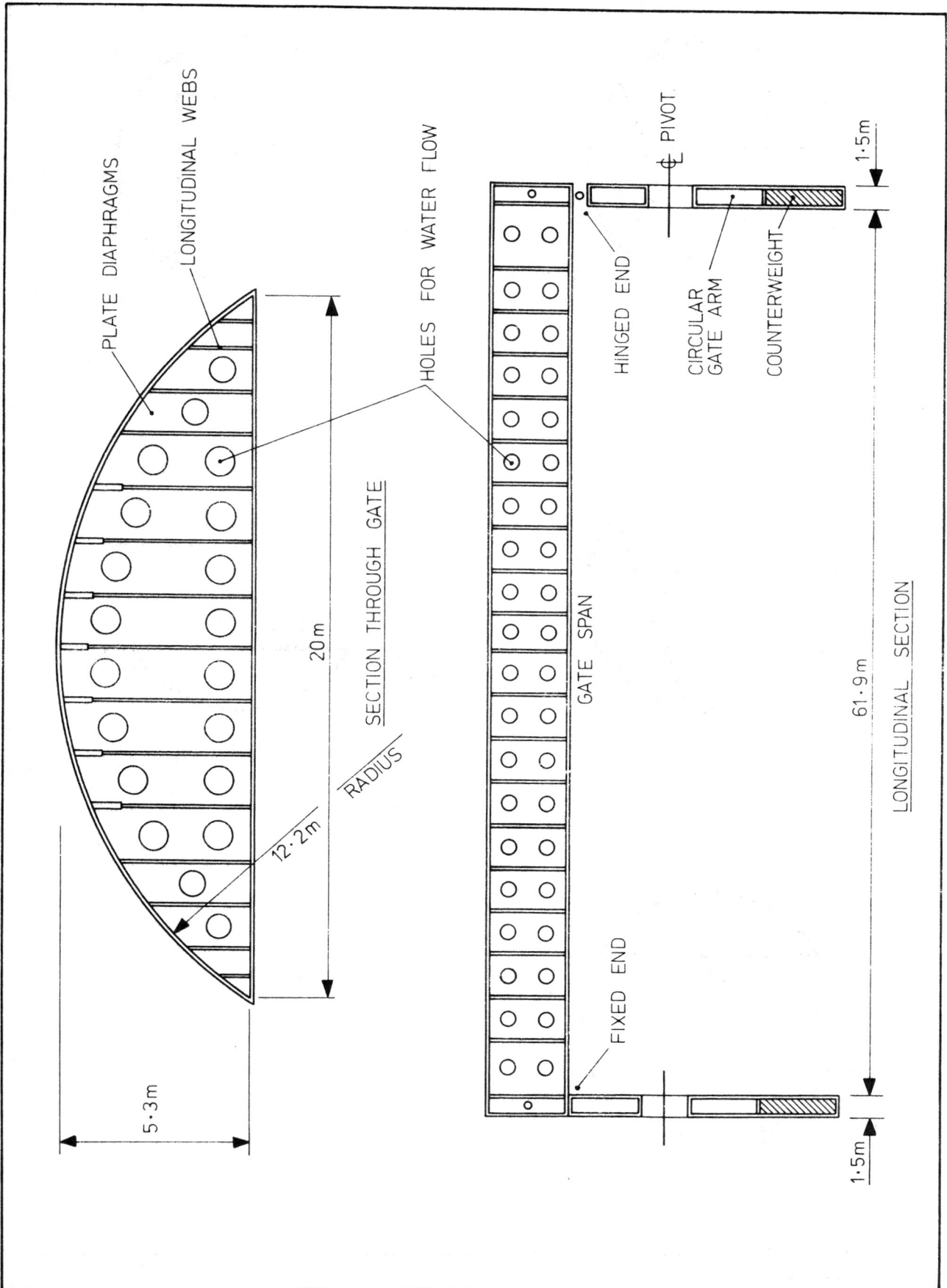

Fig.1. Early form of gate structure

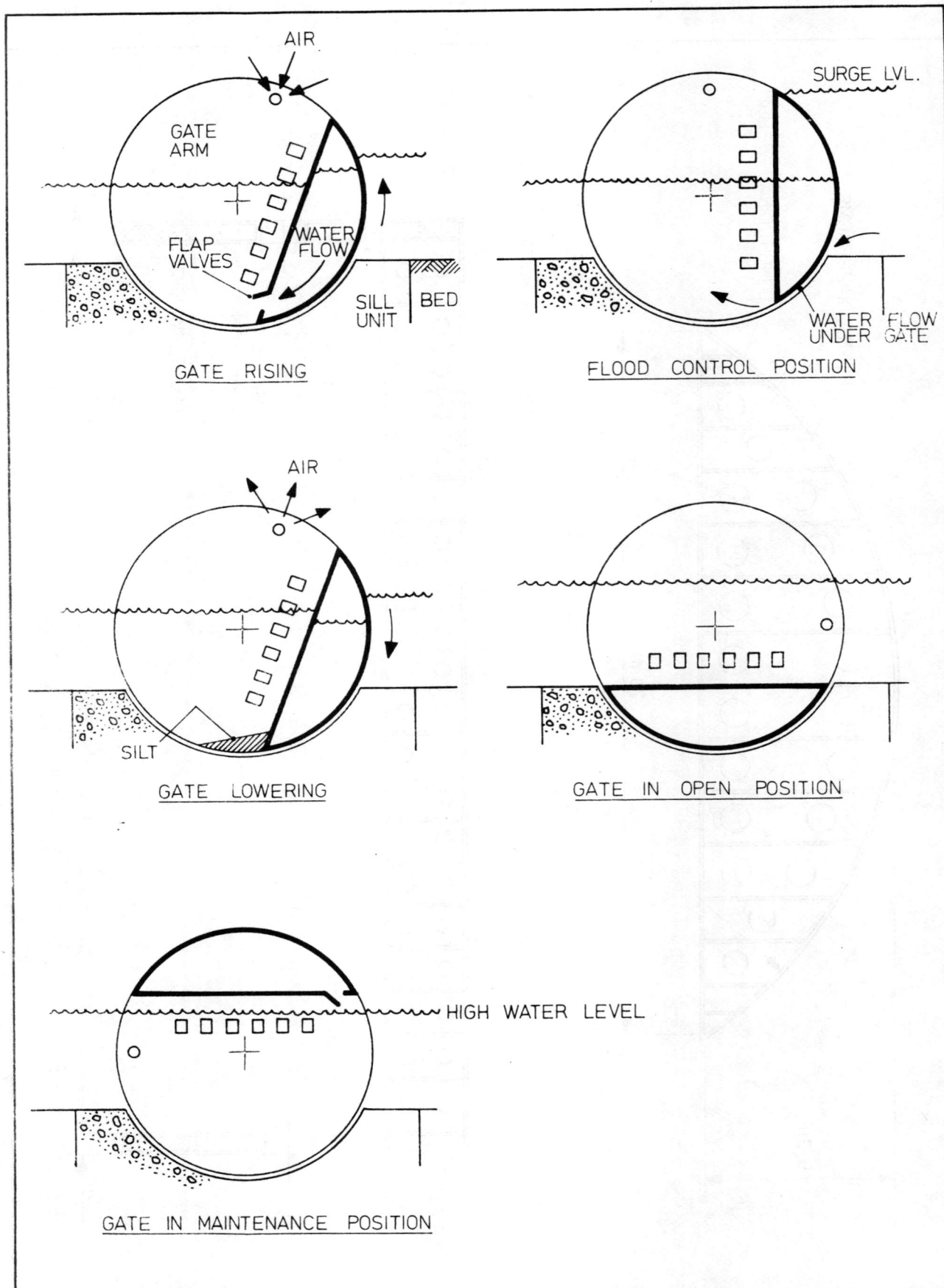

Fig.2.   Basic concept of gate operation

The Articulation of the Gate Structure

23. Earlier calculations to determine a possible gate section, had assumed the gate span to be simply supported at its connection to the gate arms. It was recognised that some form of lateral restraint would be required to centralise the span and consideration was given to the provision of full moment connections between the arms and the span, thus forming a portal structure. A further possibility considered was the use of side rollers bearing against the piers. The problem of high lateral thrusts to be absorbed by the pivots in the case of the former and the complications of a multiple roller system for the latter, were felt to make these solutions not worth pursuing further.

24. Finally a scheme was developed whereby the gate span was fixed at one end to a gate arm by a full moment-carrying connection but would be free to rotate about a hinge-connection at the other end.

25. In order to eliminate large thrusts being transmitted to the main pivots, spherical bearings would be employed to mount the gate arms on the pivots. This arrangement, shown in Fig. 3, allows the span to deflect freely in the closed position under hydrostatic loading, with the fixed gate span/arm connection providing the stabilising element against transverse collapse of the mechanism.

Pivot Bearings

26. Each spherical bearing is housed in the gate arm and located on a static stub shaft, tapered to facilitate initial erection of the gate arm, and its subsequent removal in the event of severe damage necessitating replacement. Rotational motion of the gate rising to any position within its operating sector, together with the lateral movement of the gate arms as the span deflects under load, occurs at the spherical interfaces of the bearings, which are permanently lubricated by a proprietary dry lubricant system.

27. Since the lower portion of the spherical bearings is operating permanently under water, a system of seals is provided to prevent the ingress of silt to the spherical surfaces and the shaft/bearing interface. Eventual deterioration of the seals, is of course recognised and arrangements have been made in the bearing housing to allow periodic flushing of the bearing with clean water.

Pivot Support System

28. The maximum loading on a stub shaft under design surge loading is of the order of 5,500 tonnes, and therefore a basic requirement for the design of the shaft support system, was that it should sustain the design loads with a safety margin in keeping with those used for the remainder of the project, and to distribute such loading into the piers in an efficient manner. A design concept was evolved whereby the cantilever shaft was supported on a massive casting,

housed in the walls of the hollow pier. It was not found practical to absorb the resultant very large loads into the pier walls, and a solution was finally adopted, consisting of a 3 metre diameter heavily stiffened tube incorporating steel castings at each end, which form the face plates on to which the stub shafts are bolted, the whole structure being cast into a solid portion of the pier. The details of the finalised gate support system are given in Paper 7 following.

The Gate/Gate Arm Hinge

29. The chosen articulation of the gate and gate arm structures, depended on the formation of a line hinge at one end of the gate along the chord length of its junction with the gate arm, in order to allow the gate end to take up its simply supported end slope under hydrostatic loading.

30. An early scheme indicated a continuous rocker bearing, but this was rejected; the possibility of some form of preloaded rubber hinge was investigated, but this too was eventually discarded in favour of individual rockers of simple but robust construction. The selection of a system of individual rockers arranged along the 20 m length of the chord, gave rise to the problem of the distribution of the end reaction of the gate span under load; the manner in which an even distribution of load is ensured is described in Paper 7.

Seals for Rising Gates

31. No attempt is made to seal the gap between the gate and sill, since the amount of water passing through would not be sufficient to significantly affect upstream levels in the time scale of a tide cycle.

32. When the gates are holding back a differential head of water, high velocity flow will occur through the gap, which will effectively scour the sill recess of any accumulated silt or debris that may have been lying on the back faces of the gate prior to closure. In the event of single gates being raised for maintenance or inspection, a differential head of water is not built up and a significant flow of water under the gate would not occur to scour out deposited debris in the sill recess. To cater for this condition the trailing edge of the gate is provided with a robust edge casting so that as the gate is lowered back into the recess, silt and debris is "bull-dozed" back on to the flat face of the gate by the shearing action of the reinforced gate edge.

33. The castings to both the leading and trailing edges of the gate, are shaped to restrict the particle size of any debris collecting in the gap between gate and sill when the gate is lying in its recess.

34. In order to avoid the build-up of high level water between the gate arm and the pier recess in which the arms revolve, double rubber seals are provided to separate the downriver and upriver water levels at this point. Should

severe leakage occur through deterioration of these seals, then a severe build up of hydrostatic pressure behind the gate arms is avoided by flow through the open ports in the gate arm skin plates.

GATE PRELIMINARY DESIGN

35. Subsequent to the development of the gate concept and its articulation and method of support, basic calculations were carried out for the gate span, to determine approximate plate thicknesses and to check its torsional rigidity and its probable vibration characteristics.

Design Load Cases

36. The three basic load cases for which the gates were to be designed are as follows:-

(all levels are related to Ordnance Datum Newlyn)

(a) Design Surge -

Downriver water level	+ 6.9 m
Gate crest level	+ 6.9 m
Upriver water level	- 1.5 m
Differential head	8.4 m

(b) Extreme Surge -

Downriver water level	+ 7.2 m
Gate crest level	+ 6.9 m
Upriver water level	- 2.7 m
Differential head	9.9 m

(c) Design Reverse Head -

Downriver water level	+ 1.8 m
Upriver water level	+ 4.3 m
Differential head	6.1 m

In considering the case of the extreme surge condition, a lower load factor was considered to be acceptable.

37. These three load cases represent the basic hydrostatic loading on the gates, but a further requirement was to consider the effect of a controlled flow of water upstream of the Barrier, to alleviate the problem of reflected water artificially raising surge water levels downriver.

38. At the early design stages, it was considered that a flow over the crest of the gates could be permitted by lowering the gates an appropriate amount, but later model investigations to determine the dynamic loads on the gates and sills resulting from such a method of operation, indicated a condition of mass oscillation of a gate about its pivots and severe vertical loading on the sills. These effects were acceptable only if the volume of water passing over the gates was kept below a maximum limit, and the gates themselves were held rigid in the inclined position by a locking mechanism separate from the main hydraulic operating mechanism.

39. The model tests, to substantiate and supplement the theoretical vibration analysis, is described in Paper 11. Subsequent model tests indicated that a more acceptable method of passing a volume of water upstream of the Barrier would be by undershot flow operation, whereby the Rising Gates are raised a limited amount beyond their vertical position, so that the trailing edges of the gates pass out of the sill recesses. This condition is further described in Paper 12.

40. The hydrostatic load cases listed above, represent the most severe loading on the gates for any operating conditions within the stated ranges. The effect of impact loading, arising from the improbable event of a vessel colliding with a raised gate, could not be determined with any degree of assurance, but the final design of the gate, as described in Paper 7, makes some provision against impact loading.

Preliminary Analysis of the Gate

41. As described earlier in this Paper, the gate was considered as a stressed skin structure composed of the curved and flat skin plates supported by a system of multiple webs and diaphragms. For the preliminary analysis, the unstiffened plates of the webs and outer skins were analysed by simple beam theory based on the cross-sectional area and moment of inertia of the overall section. Stresses in the curved plate due to local water pressure were ignored at this stage.

42. It was found that with the use of Grade 43 mild steel of a suitable notch ductile quality and permissible stresses within the requirements of B.S. 153 "Girder Bridges", the maximum thickness of plate required for the curved skin plate was 50 mm. However, in order to reduce skin plate welding costs, it was decided to reduce the maximum thickness of plate to 40 mm and where necessary employ Grade 50C steel to accommodate the resulting higher stresses within the permissible limits.

43. This simple approach was adequate to establish the feasibility of the gate cross-section during the development stage of the gate concept. The subsequent refinement of the analytical approach and the consequent economy achieved in the overall weight of the gate structure is described in Paper 7 following.

44. A comparison of approximate weights of the gate span at the various stages of design is of interest:-

Preliminary design

M.S. plate only (max. thickness 50 mm) - 1750 tonnes

M.S. plate plus some H.T.S. (max. thickness 40 mm) - 1600 tonnes

Final design

M.S. plate plus some H.T.S. - 1300 tonnes

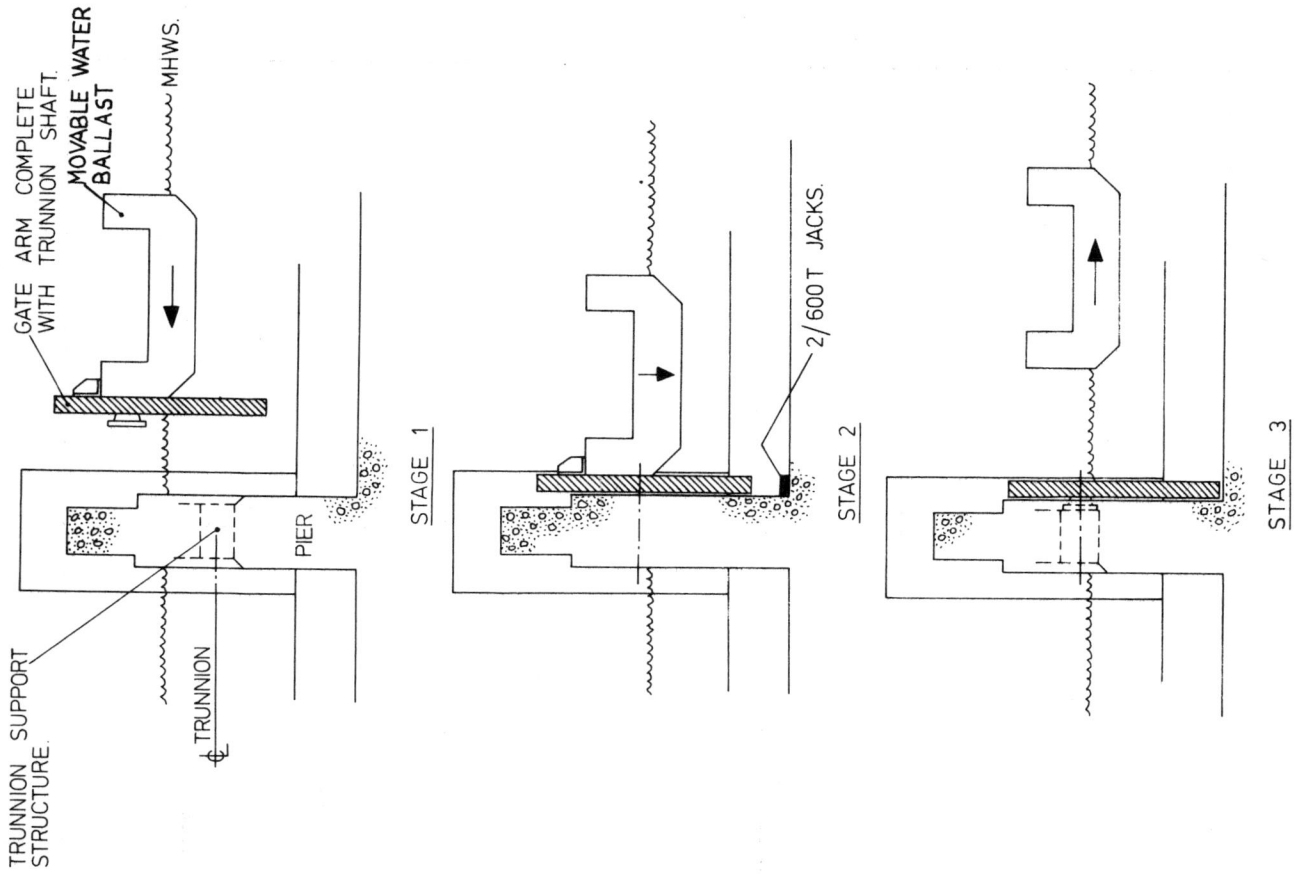

Fig.4. Gate arm installation

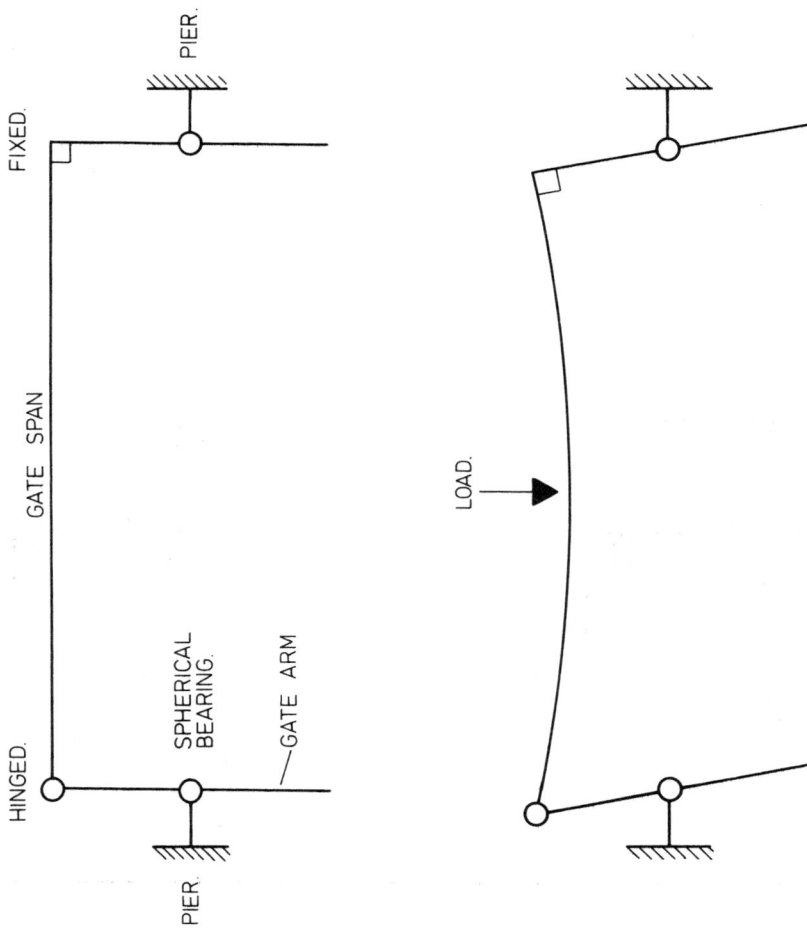

Fig.3. Articulation of gate structure

MHWS.

GATE SPAN

PONTOON

GATE ARM

STAGE 1

℄ TRUNNION

GATE SPAN

STAGE 2

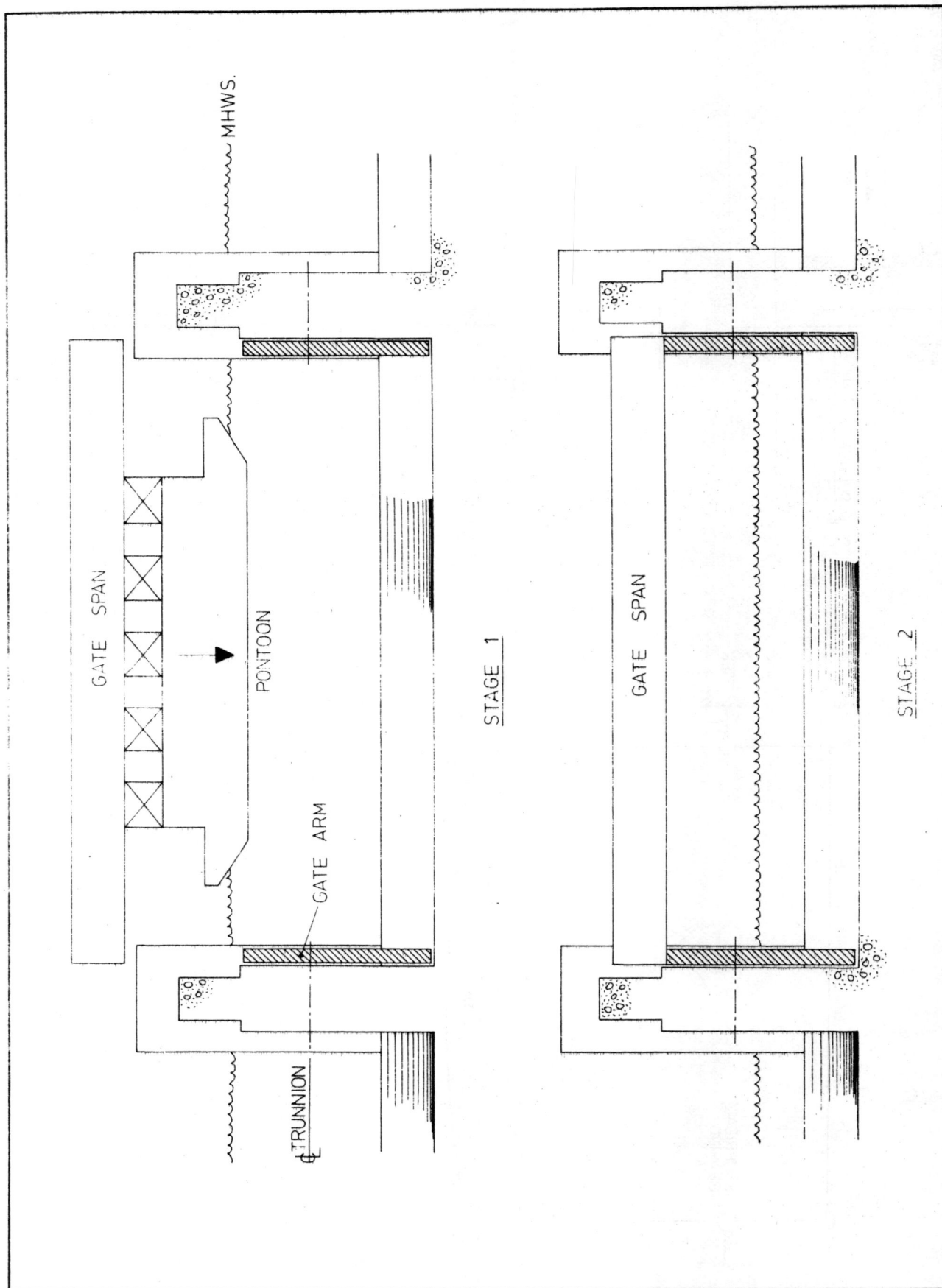

Fig.5. Gate span installation

45. It had been established at an early stage that the shape of the gate provided high torsional rigidity enabling the gate to resist the considerable torsional loading resulting from the hydrostatic loads. This torsional strength is also essential to the concept of operation of a gate from one end only, in the unlikely event of machinery failure.

PRACTICABILITY OF GATE FABRICATION AND ERECTION

Fabrication

46. It was appreciated early in the design that in view of the shape of the gate, the fabrication procedure could have a considerable influence on residual stresses. One of the first considerations was whether the curved skin plate should be made from plates running transversely on a true curve, or whether longitudinal plates would be used to form the curved skin of a series of chords. General drawings were prepared, and the concept discussed with leading fabricators to obtain their views and to ensure that the design was within their capabilities. As a result of this investigation, a longitudinal plate arrangement was adopted for the gates and in the case of the 61 m span gates, the curve was approximated by a series of flats.

47. Views of fabricators were also obtained on the likely method of the fabrication of the gates, with either the curved face uppermost, or the curved face formed on a shaped stallage. These early discussions proved very beneficial when formulating the final design of the gates, and gave assurance that the gate could be fabricated using well-established techniques.

Erection of Gates

48. Before finalisation of the gate concept, it was considered necessary to examine in detail a possible method of installation of the gate structure. The final documents, prepared for the Gate Contract, required the Tenderers to submit their outline erection scheme and accept the responsibility of installation, but it was incumbent upon the designers to prove the practicability of their design by formulating a possible method by which the gate structures could be installed.

49. The following is a description of the envisaged method of installing the Rising Sector Gate arms and spans and is illustrated in Figs. 4 & 5.

(a) Equipment assured necessary for installation

One Large Pontoon (for supporting a gate span).
Two Heavy duty jacks (600 tonne capacity).
Jacks for use between gate arms and gate spans.
Temporary brackets on pier side face of gate arms at fixed end connection of span.
Bottle screw, double-acting jacks and links for positioning gate arms relative to piers.
Large 'U' shaped inflatable seals (to be positioned between pier and gate arms around the trunnion zone).

One special limpet box on each gate arm.
Temporary timber guides and various ropes and tackles as necessary.
A lifting gantry on the pontoon for holding a gate arm in the vertical position and lowering.

(b) Method of Installation

(i) The Gate Arms

50. At the fabrication site the gate arms are fabricated complete with ballast, and loaded onto a special pontoon for transporting to site. All necessary lifting brackets and the limpet box will have been attached to the gate arm. The side of the pier to which a particular gate arm is to be fitted is prepared by arranging guide timbers, anchorages for tackles, hydraulic pumps and jacks. The gate arm complete with counterweight, stub shaft and bearing is brought alongside the pier and lowered, counterweight section at the bottom, onto two 600 tonne jacks positioned in the segmental recess of the pier foundation. The jacks will have been specially adapted to allow for rotational and translation movements. The gate arm is secured against the pier by means of a number of double-acting jacks, and the gantry pontoon is released and removed.

51. The gate arm is jacked up so that the centre of the pivot shaft is 80 mm above its final required level, lateral alignment being attained initially during off-loading the arm onto its jacks, by means of pre-set guides. The guide timbers between the gate arm and the pier are removed and an inflatable seal, in the form of a large 'U' around the shaft, is installed and inflated.

52. A limpet box is fitted to the outer face of the gate arm, covering the outer end of the shaft bearing zone and extending above the highest water level. The complete zones within the limpet box and 'U' seal are pumped dry and the top of the gate arm is moved towards the pier by the double-acting jacks, so that the whole arm pivots about the 600 tonne jacks on which it is sitting, until the complete face of the trunnion shaft is in contact with the face of the trunnion support structure.

53. The bottom jacks are lowered until the holes in the shaft and support structure are in line, when the bolted face-plate connection can be made. After all bolts have been fitted, and final cover plates installed, the limpet box is flooded, the seals deflated, and all temporary erection fittings removed.
The gate arm to the pier at the opposite end of the span is installed in a similar manner.

(ii) The Gate Span

54. The gate span is brought into position with the flat chord face of the gate downwards, mounted on stillages on the pontoon, so that the span will clear the ends of the piers. The pontoon is accurately positioned by means of guide ropes etc., and flooded down so that the span lands on pre-positioned support jacks on the flat chord surface of both gate arms.

The position of the fixed-joint gate arm is adjusted to line up bolt holes in both span and arm, and the end of the span is lowered and the joint bolted up.

55. At the hinged joint end the span and the arm are adjusted in line and level until the bolt holes in the span are in line with the previously mounted hinges positioned on the gate arm. The bearings are checked for line and level and any necessary packings inserted before the span is lowered and the final connections made.

56. The machinery links are attached to the gate arms, all temporary erection equipment is removed, the gate arm side seals fitted, and the gate lowered to its operational position, and finally to its open position in the sill recess.

CONCLUSION

57. It will be seen that a stage in the development of the gate design had been reached where the basic shape of the gate and its sections had been established by conservative design methods. It had been shown that the resulting proportions, dimensions and weights were viable and manageable for both fabrication and installation of the completed structural components.

58. Paper 7 describes the refinement of the design, and later Papers describe the various model tests commissioned to prove and to supplement the theoretical analysis.

ACKNOWLEDGEMENTS

59. The Authors wish to thank the Department of Public Health Engineering of the Greater London Council, and Messrs Rendel, Palmer & Tritton for permission to write this Paper. Also to give thanks to the many people involved during the early development of the design of the gate structures, and particularly to Mr. H. Scrutton, former Senior Partner of Rendel, Palmer & Tritton, and to Mr. A.R. Young, former Project Manager of the Consultants' Barrier team.

7. Final design of Thames Barrier gate structures

P. J. CLARK, ACGI, BSc, FICE*

R. G. TAPPIN, MA, MICE†

INTRODUCTION

1.   This paper describes the design of the
Thames Barrier Gates from the establishment of
the practicability of the Rising Sector Gate
concept to the finalisation of the detailed design of these Gates and the Falling Radial Gate.
Reference is made to design details, to the
various structural and vibration analyses
carried out and briefly to the model work which
was done in parallel with the design.

LOADING

2.   Mention has been made in earlier papers of
the factors determining the design water levels.
In addition to the upstream and downstream water
levels, the load on the gate is a function of
the operating position of the gate. For design purposes, characteristic loadings were
based on maximum calculated forces and their
probability of occurrence. The key values are
stated below and relate to water levels and
gate crest levels given in Paper 6. N is the
force in the direction normal to the flat face
of the gate, and P that parallel to it.

61 m Rising Sector Gate

(i) Design Surge N = 7,600 t, P = 1,340 t.

(ii) Reverse Head N = 7,700 t, P = 2,000 t.

31 m Rising Sector Gate

(i) Design Surge N = 2,400 t, P = 200 t.

(ii) Reverse Head N = 2,230 t, P = 620 t.

31 m Falling Radial Gate

(i) Design Surge N = 805 t.

(ii) Reverse Head N = 270 t.

BASIC STRUCTURING AND DESIGN DETAILS

61 m Rising Sector Gate and Gate Arms

3.   The overall shape of the gate was unaltered
during the detailed design but the internal
structuring was modified as shown on Figs. 1 & 2.

This resulted in a reduction in the weight of
the gate span to approximately 1,300 t. The
skin panels stiffened with flats or bulb flats,
have panel dimensions which are typically 3.0 m
by 3.4 m and have a maximum plate thickness of
40 mm. Each internal diaphragm or cross-frame
consists of transverse members integral with the
skin plate and the webs, and a diagonal universal column. This arrangement permits the flow
of water longitudinally within the gate and
forms a stiff framework with which to distribute
the local water loads to the structure and to
limit the distortion of the gate cross-section.
The multiplicity of braced cross-frames give
extensive capacity for redistribution and will
also help to minimise any damage which could
occur from ship collision.

4.   The gate webs, which form the longitudinal
boundaries of the skin panels, are holed in
each panel in the central portion of the gate
to allow the flow of water transversely. These
holes are bounded by the stiffening arrangement necessary to stabilize the web. The
thicker portions of the webs adjacent to the
curved skin contribute significantly to the
bending resistance of the gate and give some
measure of resistance to local impact damage.
The location of the changes in web thickness
were chosen to permit maximum width of rolled
plate to be used for the main part of these
webs, thus minimising welding. At an early
stage in the design, consideration was given
to the fabrication methods for the gate. The
splitting up of the skin plates into part panels
over the main webs, together with infill panels
between, can be seen on the initial detail
design drawings,(see Fig. 2). This system
allows for the fabrication of this multi-cell
structure with the minimum of locked-in
stresses, and has been adopted by the fabricator.

5.   Due to the articulation of one of the gate/
gate arm connections and the presence of the
spherical support bearings, the gate load will
be transmitted into the two outer skins of each
gate arm approximately equally. Two diaphragms
are located at each end of the gate, in line
with the gate arm discs. One is a full plate
diaphragm covering the segmental cross-section
at the end, and the other a deep beam supported
on the webs of the gate.

*Partner, Rendel, Palmer and Tritton
†Senior Engineer, Rendel, Palmer and Tritton

SECTION THROUGH RISING SECTOR GATE.

UPRIVER.

DOWNRIVER.

∇ +6·900m SURGE.
∇ +3·690m M.H.W.S.
∇ −2·830m M.L.W.S.

LEVELS RELATE TO ORDNANCE DATUM NEWLYN.

10 5 0 10m

∇ −2·375m
∇ −9·250m

GATE ARM.
TRUNNION ASSEMBLY.
HINGED END.

FIXED END.
GATE SPAN.

64900

SECTION ON ₵ OF GATE LOOKING DOWNRIVER.

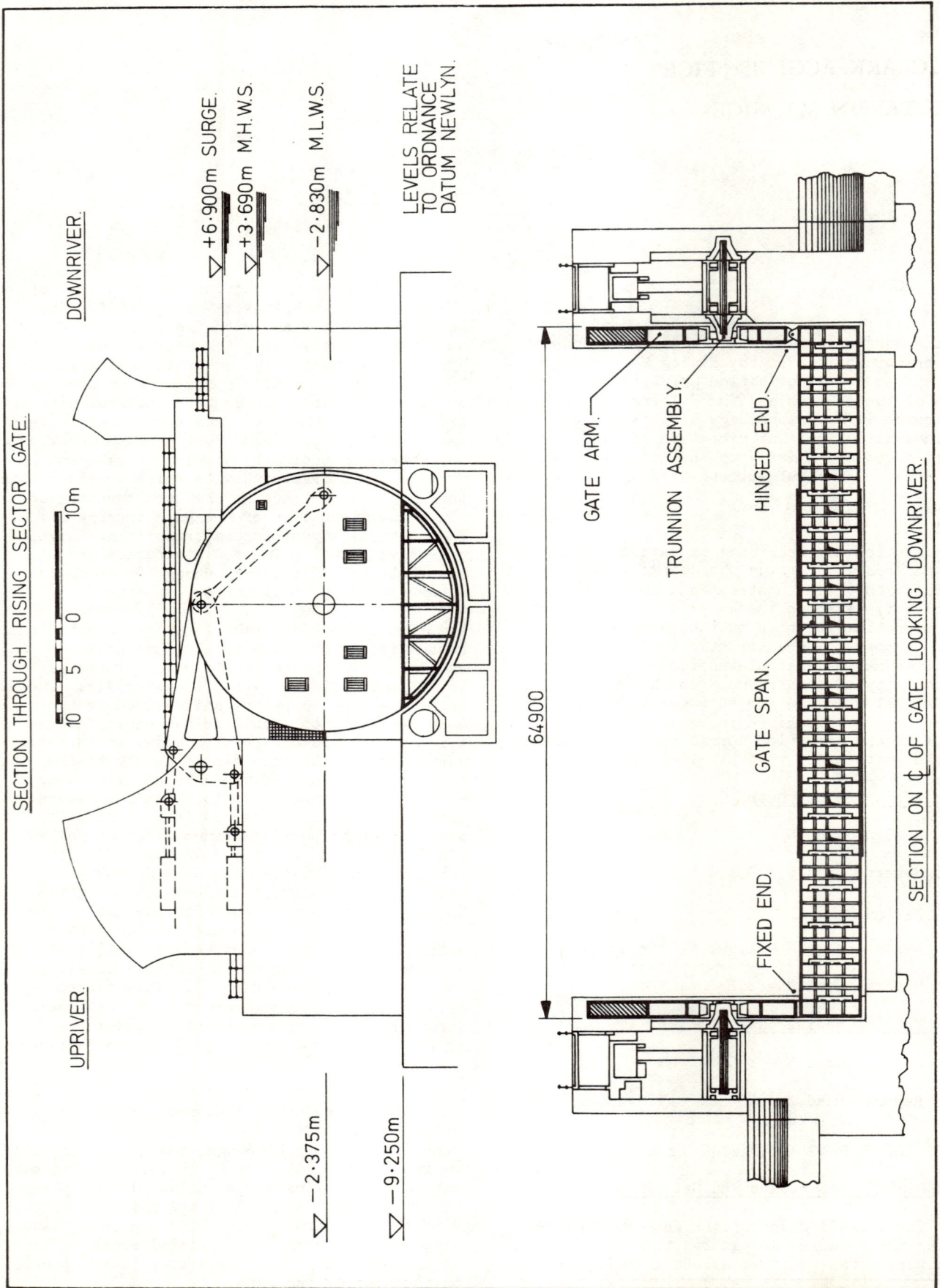

Fig.1. 61.0 m rising sector gate

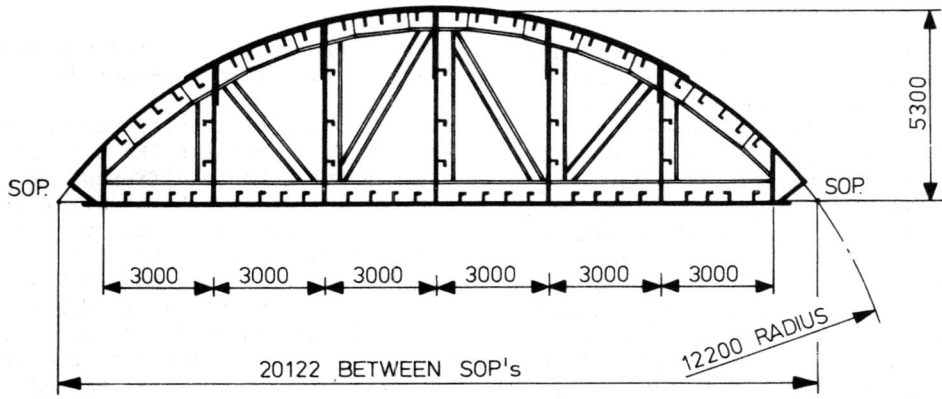

5300

SOP.   SOP.

3000   3000   3000   3000   3000   3000

20122 BETWEEN SOP's

12200 RADIUS

Fig.2. 61.0 m rising sector gate cross section

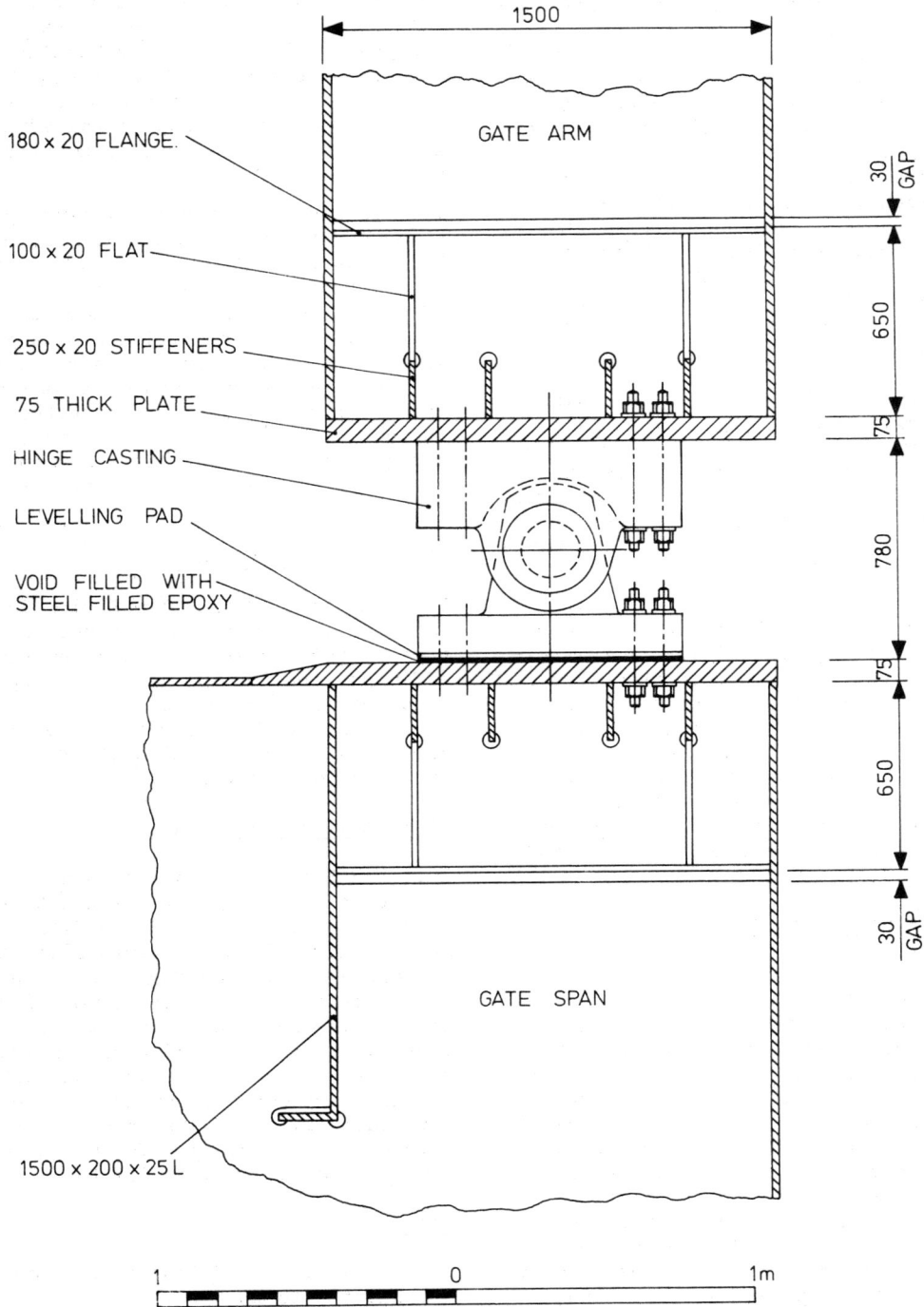

1500

GATE ARM

180 x 20 FLANGE.

30 GAP

100 x 20 FLAT

650

250 x 20 STIFFENERS

75 THICK PLATE

75

HINGE CASTING

LEVELLING PAD

780

VOID FILLED WITH
STEEL FILLED EPOXY

75

650

30 GAP

GATE SPAN

1500 x 200 x 25 L

1   0   1m

Fig.3. 61.0 m rising sector gate hinge

6. The hinged connection between the gate and the gate arm is shown in Fig. 3. Each hinge is tensioned onto the gate and the gate arm with sufficient force to transmit the shear in friction under the reverse head load. To ensure that the load is transmitted as intended along the joint, each hinge is designed to carry shear in one direction only. Of the thirteen hinges, nine are orientated to carry shear in one direction, while the remainder will be effective in the other, this arrangement satisfying the design requirement. Each hinge is designed to carry part of the direct force across the joint. The transverse rigidity of the gate, in particular the end diaphragms, renders the hinge loads extremely sensitive to fit up. This point was demonstrated very clearly in the structural model tests. Also the alignment of these hinges is critical, in spite of the very small rotational requirement. To this end, considerable attention is being given to obtaining a very high degree of flatness on the adjoining areas of the gate and gate arm, and a steel filled epoxy is to be used to seat the hinges on the gate side. As an added precaution against excessive direct load on an individual hinge, the support steelwork between the gate arm discs and between the two end diaphragms in the gate has been designed to yield safely in shear while maintaining its load, before any damage could be done to the hinge.

7. The gate arms consist of two outer skins of stiffened plating, separated by an internal cellular arrangement of webs and bracing. The gate loads will be carried through these skins to the large castings at the centre of the arms. These in turn are supported on the spherical bearings. The weight of each gate arm excluding the counterweight is 210 t.

8. While it is not possible to design the gate against ship collision, recognition of this extremely unlikely but possible occurrence has been made in the detail design, as has been noted above. In the event of a major collision, the gate would tend to jam within the pier recess developing considerable resistance to the colliding vessel. In the extreme case, where one 61 m gate becomes inoperable, the Barrier would, in some measure, continue to fulfil its function.

Trunnion Support Structure for 61 m Rising Sector Gate

9. The large forces for which the gates are designed are transmitted to the concrete pier through forged trunnions, post-tensioned onto 3 m diameter steel finned tubes embedded in the concrete, (see Fig. 4). The rigidity of these large tube structures enables the gate forces to be spread effectively into the narrow piers. The tubes are formed of 45 t castings at either end, joined by a fabrication of 75 mm steel plate. For the piers located between two 61 m gates, a trunnion is connected to either end of the tube. For the piers located between a 61 m gate and a 31 m gate, separate support structures are used for the adjacent gates, each structure therefore supporting only one trun-

nion. Descriptions of the trunnions and gate bearings are given in Paper 8.

10. In order to develop the required post-tension force between the trunnion and the support tube, three different systems of tensioning are used. Thirty-six 75 mm diameter rods are connected through the end casting only. These will be the only tensioning members necessary during the initial installation of the gate arm and gate. Twenty-four additional 75 mm diameter rods will be connected through the tube, either to the trunnion at the opposite end, or, in the case of the single trunnion support structures, to the end casting of this structure. The 75 mm diameter rods will be tensioned by means of a system of hydraulically linked nuts. The third system consists of sixty-seven 40 mm diameter Macalloy bars, stressed from the centre of the trunnion through the tube to the opposite trunnion or end casting. This system has the additional advantage that it reduces tensile stress concentrations in the trunnion itself.

11. In designing the tensioning system, it was necessary to ensure that under normal operation, the joint between the trunnion and the tube end plate cannot open. Also, in the unlikely event of a failure of all the through tensioning system due to ship collision on a gate or gate arm, it was necessary to ensure that the support to the adjacent gate would be maintained. Thus the 75 mm rods which are connected only on to the end casting, must be able to resist the full design force from the gate, although a nominal opening of the joint is acceptable in extreme cases.

31 m Rising Sector Gate

12. The form of the 31 m Rising Sector Gate, gate arm and support structure, is similar to that of the larger gates and demonstrates the efficient application of the concept to a gate required to carry a much smaller load over a smaller span. The weight of the gate span is approximately 307 t and that of each gate arm, excluding counterweight, is 80 t.

31 m Falling Radial Gate

13. The Falling Radial Gates are different in concept to the larger gates due to the smaller loads to be carried, and the differing requirements, but it is still appropriate to use the floodable box girder form. As such, the gate span section has much in common with the other gates. The horizontal thrusts on the gate bearings due to surge loading or temperature movements are acceptable without the articulation of one gate arm connection, and counterweights are not required. The gate arms are thus much simpler, (see Fig. 5). The weight of each gate, including gate arms, is 137 t.

Materials

14. Gate spans and Gate arms – Steel grades 43D and 50C to BS 4360.

Castings – BS 2456 Grade A to BS 3100

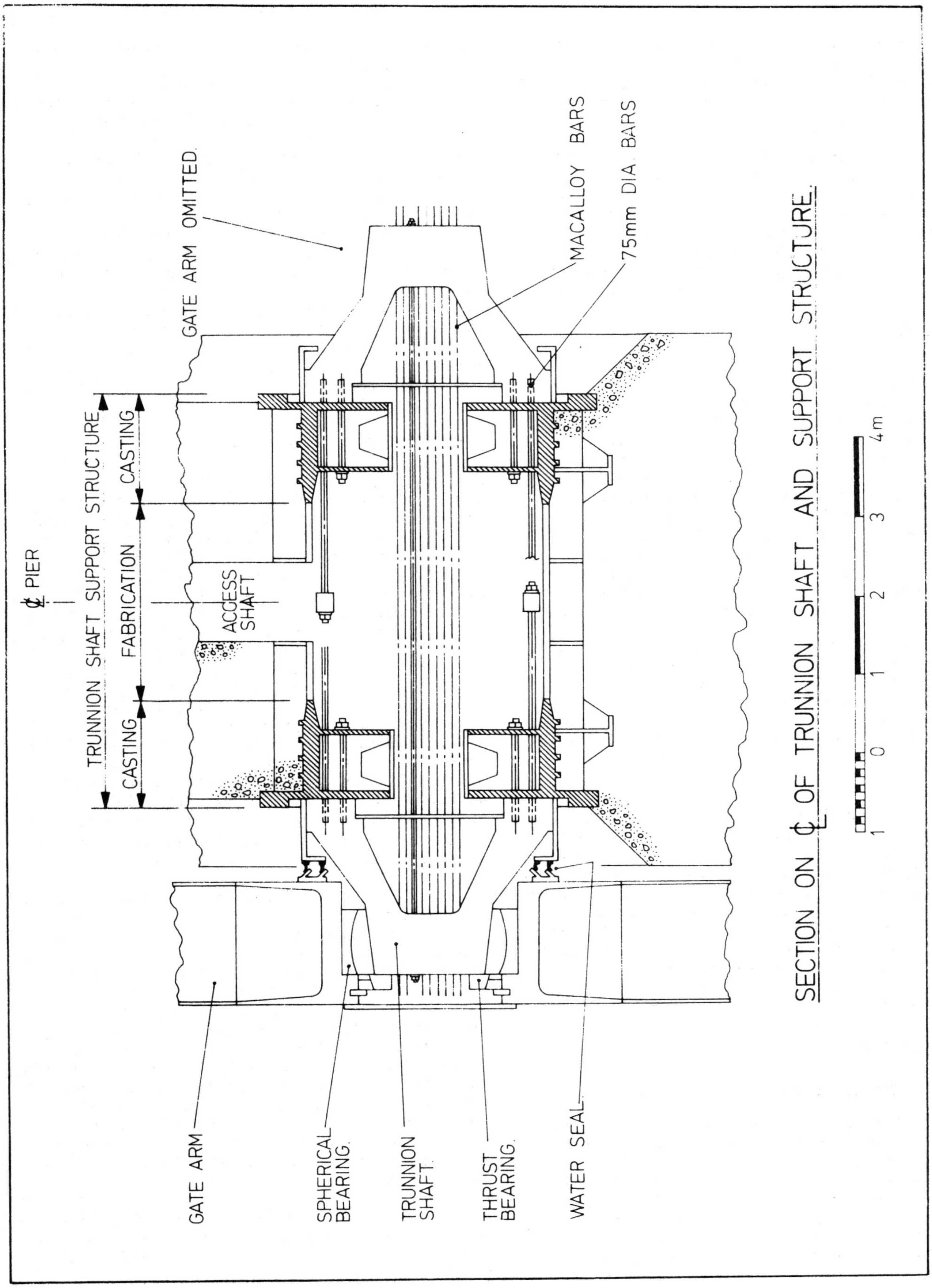

SECTION ON ℄ OF TRUNNION SHAFT AND SUPPORT STRUCTURE.

Fig.4. 61.0 m trunnion assembly

ANALYSIS

General

15. Preliminary calculations used to develop
the modified section of the gate were based on
the assumption that simple beam theory and
torsion theory apply, and that these are sufficient to describe the primary behaviour of all
gates. The distribution of shear between the
webs and the curved skin was evaluated using
shear flow analysis (ref. 1). This basic understanding of the structure was broadly confirmed
by subsequent finite element work and model
testing. The final design check was based on
panel loads derived from elastic analysis, with
characteristic strengths of the panels calculated in accordance with the Merrison Rules.

Load Factors

16. For this structure assurance against failure must be very high. On the other hand the
probability of the characteristic loads being
exceeded is very low. The design load is
evaluated from a one thousand year return
flood combined with a very low upstream water
level. To accommodate these considerations,
load factors of 1.6 for compression flanges,
and 1.5 for tension flanges, diaphragms and
webs were used for first panel collapse. The
factor on structure collapse will therefore be
in excess of 1.5. The different load factors
were chosen for this particular structure in
recognition of the capacity of the gate to re-
distribute loads, the greater predictability
of the tension flange behaviour, and the lower
sensitivity of the tension elements to local
damage. In relation to redistribution, the
model tests demonstrated clearly the ability
of the curved skin to carry considerable additional shear after the buckling of the webs.

Tolerances

17. It was considered appropriate to permit
tolerances greater than those specified in the
Merrison documents, but where this was done,
stability calculations were modified accordingly.
The additional costs expected from the fabricators for tolerances smaller than those adopted,
could not be justified on the basis of the consequent improved strength for these structures.
It should be mentioned that the gates are in
general stockier than the majority of bridges
for which the Merrison Rules were primarily designed, and as such are less sensitive to
imperfections.

Finite Element Analysis

18. The Imperial College Structural Analyses
System (ICSAS) program, was used for the global
analysis of the spans of the three different
gates and for the structural models. The part
of the program used was a 3-dimensional finite
element beam and plate assembly analysis, with
'biased' plate elements. Good idealisation of
a multi-cellular structure can be obtained using
the program with comparatively few elements.
The sizes of elements chosen was justified both

by previous experience of the program and by
comparisons made when varying the element sizes
on a smaller related finite element model.

19. The Rising Sector Gates are almost doubly
symmetric, and thus by applying symmetric and
antisymmetric loading, it was possible to
analyse the structure by modelling only $\frac{1}{4}$ of the
gate. The support conditions were assumed to
be as for the hinged end, as the moment at the
fixed end is comparatively small. Transverse
members were modelled as concentric beams to
ease the interpretation of the global analysis,
and detailed behaviour of these members was
assessed from separate plane frame analysis using the deflections from the three dimensional
runs. Additional plane stress analyses were
made in particular areas of the gate, notably
the webs, and gate arm skin plates. In addition to the determination of elastic stresses,
the plane stress analyses on the webs permitted
a realistic modification of their stiffnesses
due to the holes in the webs, to be used in the
3-dimensional work.

STRUCTURAL MODEL WORK

20. The objectives of the structural model testing were to prove, by reference to the largest
of the Barrier gates, the safety of the design
method to be used for all three gates, and to
determine the load factor to collapse of the
gate structure. As mentioned above, the final
appraisal of the design was to be based on
elastic analysis of the structures with characteristic panel strengths derived by reference
to design codes. Paper 10 gives a more detailed
description of the model testing. At the time
of the design, finite element work was still in
the stage of rapid development and it was considered advisable to prove the analysis of this
highly redundant box girder structure by comparison with the behaviour of a linear elastic
model. To this end a 1:25 scale Araldite model
was built and tested. Using this model it was
possible also to improve the computer idealisation of the structure, and to determine some of
the more complex aspects of the gates.

21. During the earlier stages of design there
was no comprehensive design code applicable to
the Barrier gates, and it was therefore considered expedient to test a 1:6 scale steel model
which would reproduce as faithfully as possible
the gate behaviour up to the collapse of the
structure. By the time of the model testing,
the Merrison Appraisal Rules had been issued,
to be followed later by the Merrison Design
Rules. From these, and modified BS 153 criteria,
characteristic strengths of the model panels
were evaluated and compared with the model
behaviour. This led to the adoption of the
Merrison Design Rules (Part II) with reference
as necessary to the Appraisal Rules (Part III)
in the final check of the structures.

22. The Merrison documents equate first panel
collapse with the failure of the structure. It
was possible from the tests on the steel model
to investigate, in some measure, the reserve of
strength in the structure after first panel

33500

FALLING RADIAL GATE

GATE SUPPORT ARMS

SECTION ON ℄ OF GATE LOOKING DOWNRIVER

LOWER EDGE SEAL

DOWNRIVER

UPRIVER

SIDE SEAL RUBBING PLATE

▽ + 6·900m SURGE

▽ + 3·690m M.H.W.S.

▽ − 2·830m M.L.W.S.

▽ + 3·650m

▽ 0·000 m

GATE PIVOT SUPPORT CASTING

ACCESS TUNNEL

SECTION THROUGH FALLING RADIAL GATE

10m

5

0

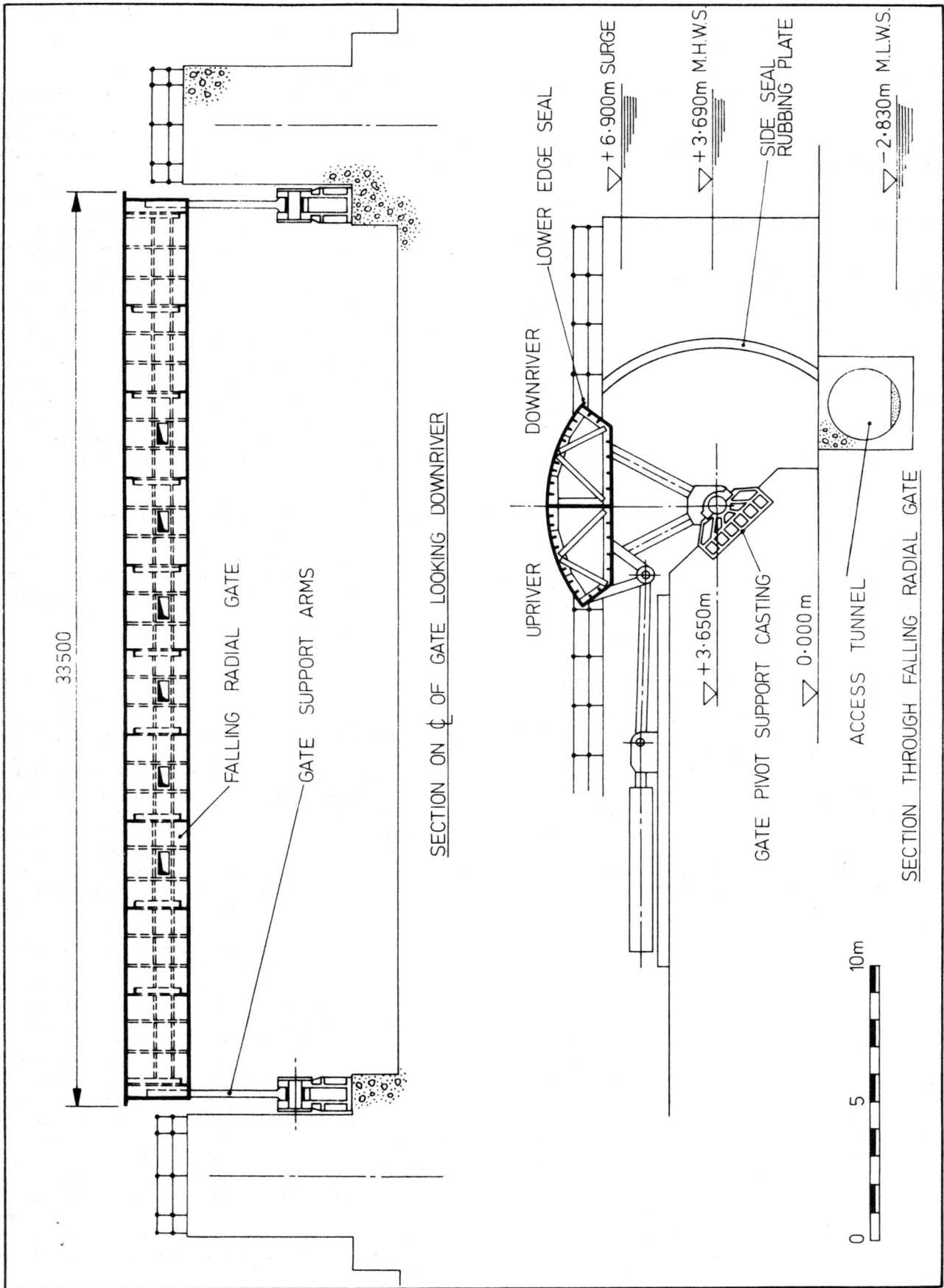

Fig.5. 31.5 m falling radial gate

collapse and the interaction between collapse of various panels. The load factor at which first panel collapse was predicted to occur, considering model thicknesses and yield stresses, is 1.89 to the Merrison Appraisal Rules and 1.74 to the Merrison Design Rules. (It should be noted that model yield stresses were in excess of those specified for the prototype). It was not possible to determine the point of first panel collapse for this particular structure, as a failed panel will move under the strain control of the remainder of the structure until a more general collapse. This occurred, at a load factor of 2.17 when over 90% of the plastic capacity of the section was developed. Failure occurred by compression plate failure at the quarter span position where the section changed, and was initiated by a horizontal shear failure of the webs. The most significant change to the design of the gate structures as a result of the model tests, was the avoidance of the previously proposed water passage holes in the webs of the regions of highest shear.

RESIDUAL STRESSES

23. While it has been established that the shrinkage forces due to welding procedures affect the strength of axially loaded compression members, information on these forces is sparse and has been based on relatively thin plating. Measurements were made on a full-scale panel of the 61 m Rising Sector Gate by Dwight (ref. 6) to verify the assumptions made in the design analysis and to determine the adequacy of these stress levels in the steel model. Residual stresses in the plate were similar to those calculated by the Merrison document although much of the stress came from flame cutting and little from the longitudinal butt welds. Compressive residual stress was also present in the outstand of the stiffeners, a result which had not been expected. The stresses measured in the 1:6 scale steel model were found to be in excess of those in the full-scale test and thus the main model conservatively accounted for this effect. Final calculations on the stability of the gate compression panels included the effect of compressive stress on the stiffener outstands.

VIBRATION ANALYSIS

General

24. During the early development of the Rising Sector Gate, preliminary hydraulic tests were conducted at the British Hydromechanics Research Association on a wooden model of the design current at that time (see Paper 11 and ref. 7). These investigated: the basic flow patterns to be expected during the operation of the gate; the hydrodynamic forces on the gate, in particular those causing torque about the gate supports; the vibration of the gate against the operating machinery. This work established that for flow over the gate, two distinct flow patterns existed, and a sudden change in torque on the gate was associated with change from one pattern to the other. It was also shown for certain overflow conditions that divergent oscillations could occur for the gate rotating about its supports against the simulated stiffness of the operating machinery.

25. It was therefore decided to study in greater detail the vibration characteristics of the gate with the aid of a full width hydroelastic model together with smaller sectional models (see Paper 11 and refs. 8 and 9). Thus, a 1:20 scale hydroelastic model of the 61 m Rising Sector Gate was constructed and mounted in a ten foot flume. This model was designed to reproduce the three dimensional interaction between hydraulic flow and the dynamic behaviour of the gate. In addition a 1:40 scale model of a section of the gate span was mounted in a flume, with supports designed to simulate the remainder of the gate and the gate machinery with regard to flexural and rotational motion. This model could only examine gate behaviour to two dimensional flow but enabled the flexural stiffness of the model to be varied. A range of spring stiffness representing the gate machinery was used for both models. A sectional model of the Falling Radial Gate was also tested.

26. Natural frequencies and modes of vibration were calculated for prototype behaviour in air and scaled as appropriate for the models. By equating certain non-dimensional numbers for model and prototype, it was possible to investigate the dynamic interaction between structure and hydraulic flow, and thus the anticipated behaviour of the prototype in service.

27. The models described above were initially used to investigate prototype gate response during normal gate operation from the open to the closed position for various surge conditions and ranges of differential head, and the situation which could occur should a gate be operated late, or become stuck. Conditions which could occur should the Barrier be used to impound the upriver level were also examined. During the development of the Project it became a requirement to consider partial closure of the gate during surge protection, to pass controlled discharges over the gate crests. This partial closure technique was initially investigated using the 1:20 scale hydroelastic model. In order to examine the hydrodynamic forces on the sill for this condition, a further sectional model to a 1:80 scale was tested at Imperial College (see Paper 12 and ref. 10). The results of this testing indicated that this method of operation would cause high loadings on the sills and as a consequence a further partial closure method, that of allowing flow under the gate, was considered. The vibrational stability of this condition was investigated using the 1:80 scale model. This was done by making a comparison between the vibrating behaviour of the gate oscillating against its machinery for the undershot flow condition and for the overflow condition and relating this to the results obtained on the larger model operating with overflow.

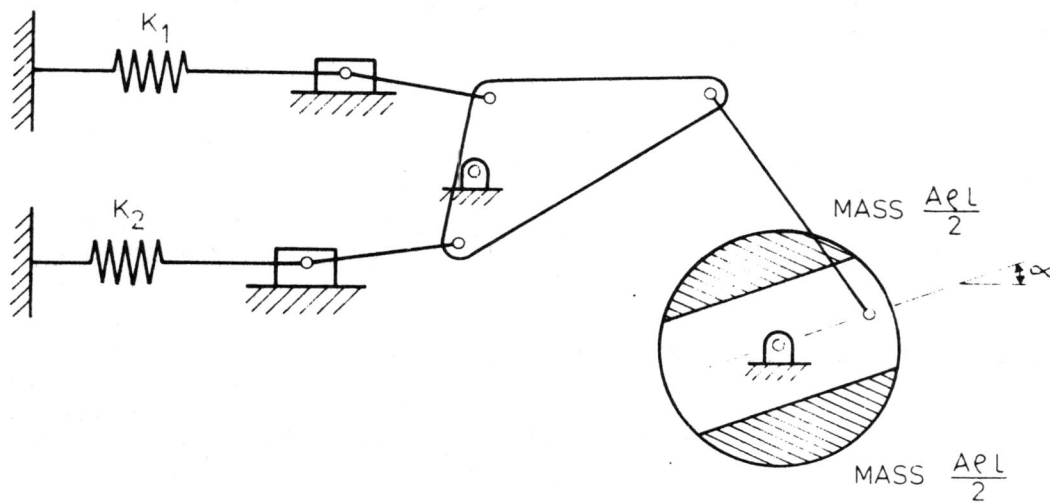

i) VIBRATION OF THE GATE AGAINST THE OPERATING MACHINERY

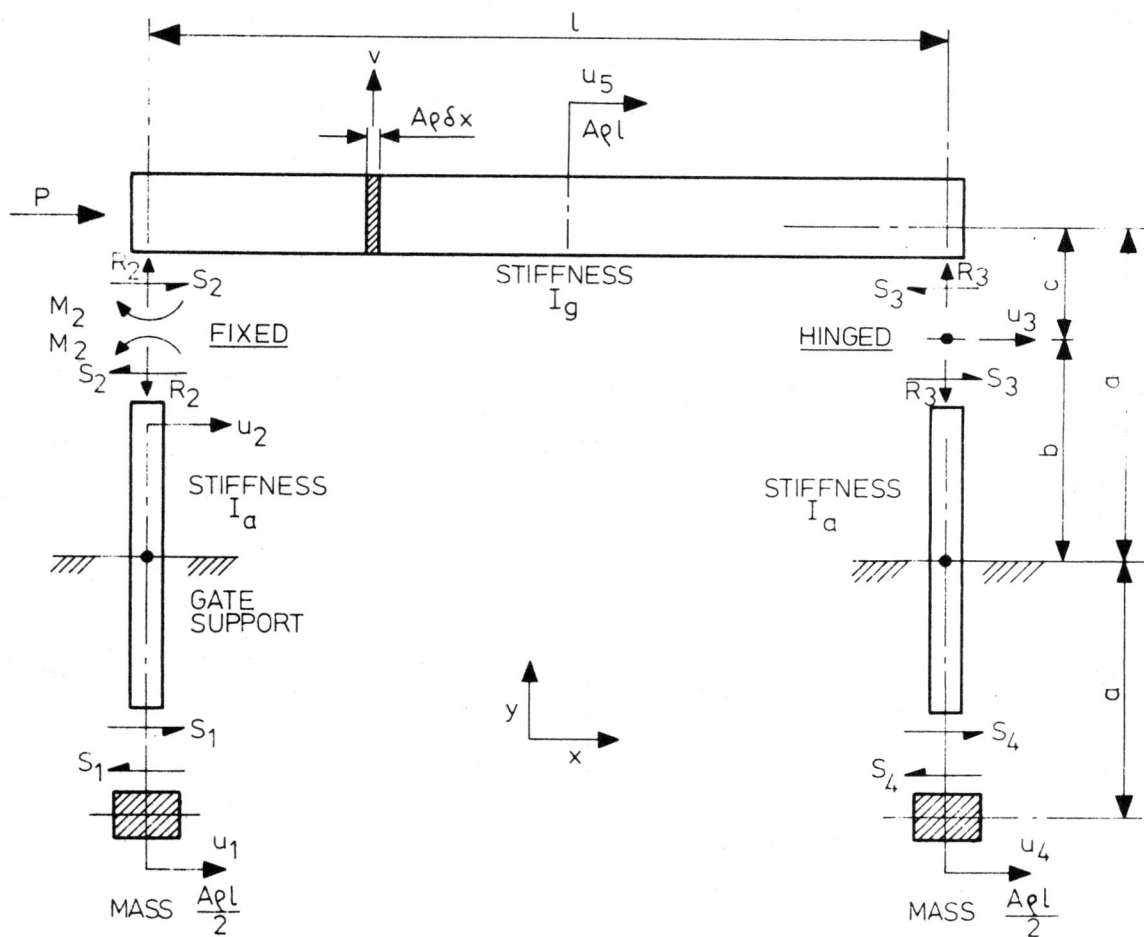

ii) STRUCTURAL VIBRATION OF THE GATE

Fig.6. Two-dimensional mathematical idealisations for vibration analysis

BASED ON COMPRESSIBILITY OF OIL AND FLEXIBILITIES OF HYDRAULIC CYLINDERS AND LATCH MECHANISMS. FREQUENCIES SHOWN ARE CALCULATED ON THE BASIS OF MACHINERY AT ONE END OF THE SPAN ONLY BEING CONNECTED. IF MACHINERY AT BOTH ENDS IS CONNECTED FREQUENCIES ARE INCREASED BY A FACTOR OF $\sqrt{2}$

HYDRAULIC CYLINDER CONFIGURATIONS
HELD——A B C D SEALED
RAISING—B C PRESSURISED A D OPEN
LOWERING—A D PRESSURISED B C OPEN

THE OPEN SIDES OF THE CYLINDERS ARE ASSUMED TO MAKE NO CONTRIBUTION TO THE STIFFNESSES.

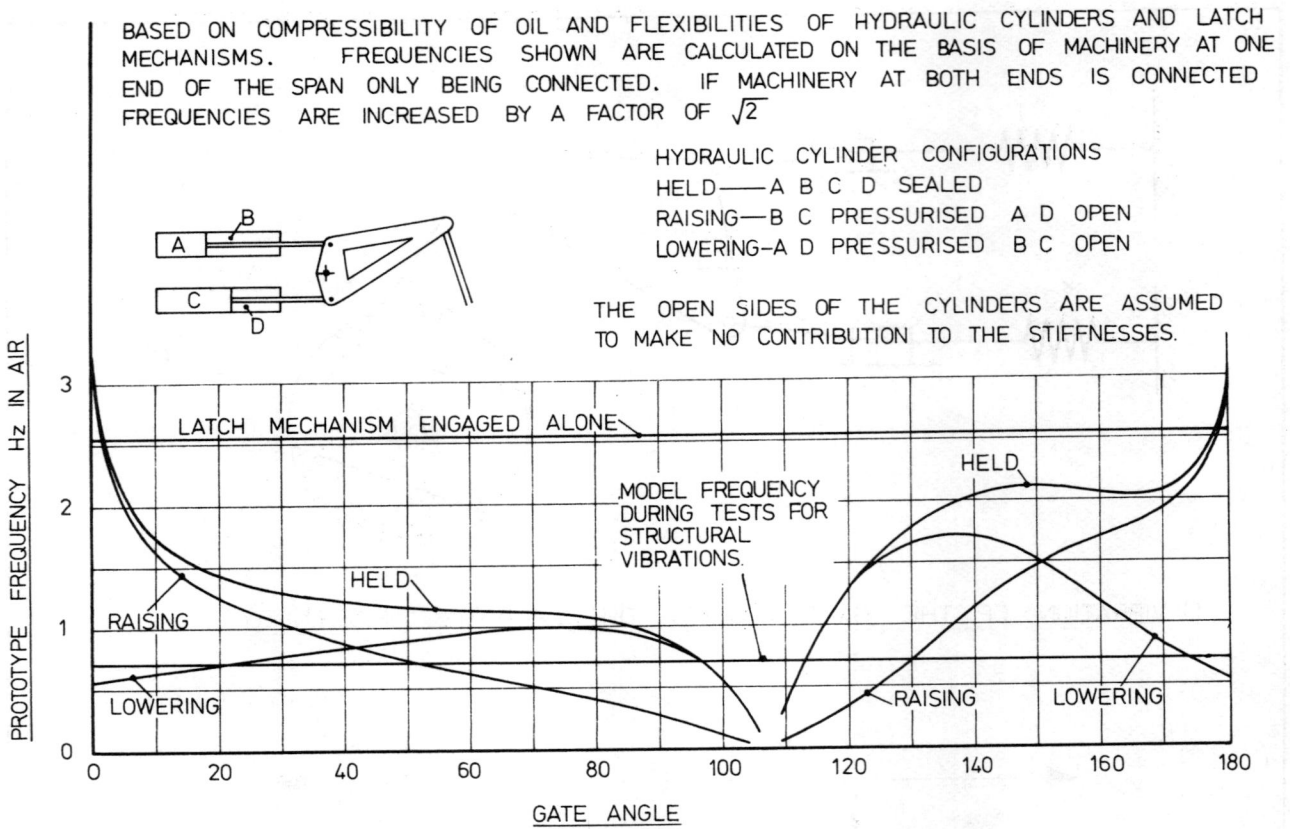

Fig.7. Natural frequencies of gate vibrating against operating machinery

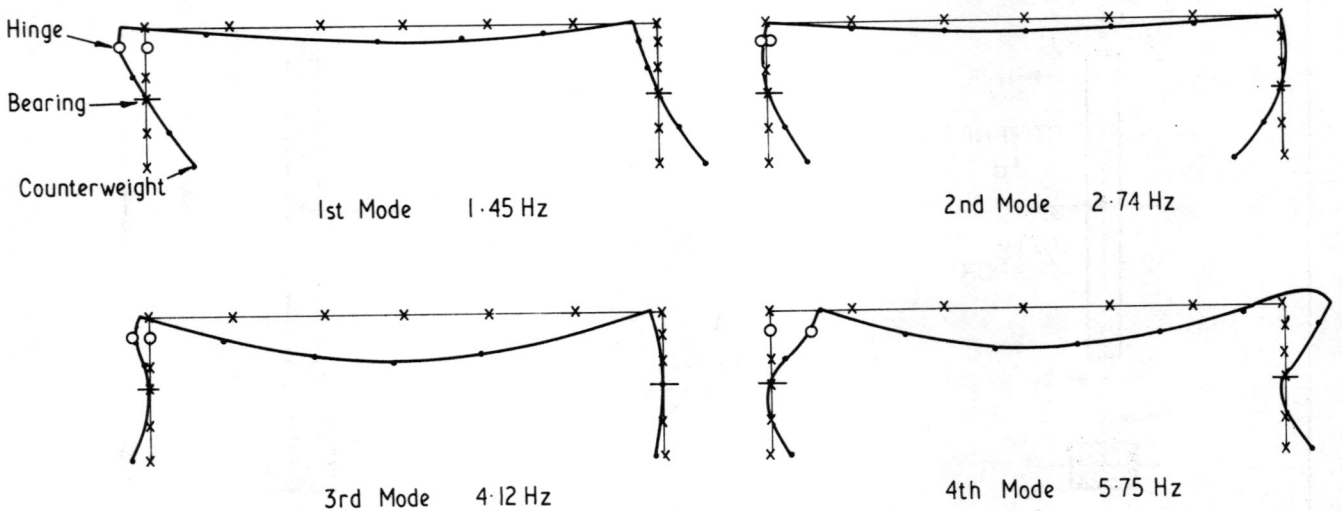

Fig.8. Natural frequencies and modal shapes of gate structural vibrations

Design of the Hydroelastic Model

28. The hydroelastic model was designed by dimensional analysis, equating non dimensional numbers for model and prototype. Reynolds scaling is not essential as viscosity is of secondary importance. As fluid dynamic damping predominates, no attempt was made to scale structural damping. (see para 34 for Notation).

(i) $l_p = n l_m$ where l is a typical length. This is the statement that the overall model dimensions are $1/n$th of those of the prototype.

(ii) Froude: $\left[\dfrac{V}{\sqrt{gl}}\right]_p = \left[\dfrac{V}{\sqrt{gl}}\right]_m$

For gravity effects to be modelled for free surface flow, scaling must be in accordance with Froude's law.

Thus: $V_p^2 = n V_m^2$

(iii) Strouhal: $\left[\dfrac{fl}{V}\right]_p = \left[\dfrac{fl}{V}\right]_m$ where $f = C\sqrt{\dfrac{EI}{A\rho l^4}}$

This relates fluid velocity with structural dynamics.

Thus, $\left[\dfrac{A\rho l}{EI}\right]_m = n^2 \left[\dfrac{A\rho l}{EI}\right]_p$ and $f_m^2 = n f_p^2$

(iv) Cauchy: $\left[\dfrac{E}{\rho' V^2}\right]_p = \left[\dfrac{E}{\rho' V^2}\right]_m$

This may also be written as
$$\left[\dfrac{EI}{\rho' A^2 V^2}\right]_p = \left[\dfrac{EI}{\rho' A^2 V^2}\right]_m$$
This relates structural stiffness to hydrodynamic force.

Hence, $[EI]_p = n^5 [EI]_m$, and $[A\rho l]_p = n^3 [A\rho l]_m$

These equations also determine that model amplitudes are $1/n$ th of those of the prototype.

29. Thus for testing a 1/20th scale of the prototype, prototype fluid velocities must be divided by $\sqrt{20}$ and structural frequencies multiplied by the same amount. Model and prototype stiffness must be related by the equation
$$[EI]_p = n^5 [EI]_m$$
As overall dimensions are fixed by the scale, and Youngs modulus selected with the material, it was necessary to vary the model thickness for correct scaling. The material chosen for the model was Darvic, a rigid P.V.C. This has a Young's Modulus approximately 1/60th of that of steel, and thus the model plate thicknesses were a little over three times that of scaled thickness. The model was internally structured similarly to the earlier gate designs referred to in Paper 6, but was of the mass and stiffness of the final design. Finally, the structural model having been defined, it was necessary to adjust its weight in order that the model weight should scale that of the prototype. For this purpose brass sheet was added to the diaphragms and lead shot placed in tubes at the neutral axis of the model.

Evaluation of the modes of vibration

a. The vibration of the Gate against the Operating Machinery

30. The gates are designed to be operated by hydraulic cylinders connected through a series of links to each gate arm. These are supplemented by latch mechanisms which are used to hold the gate in various positions and also to move the gate through the top dead centre position when it is necessary to move the gate to the maintenance position. Details of the Machinery are given in Paper 8. It may be seen, with reference to Figs. 6(i) and 7, that the spring stiffnesses which depend on the compressibility of the oil and the flexibility of the cylinders are also a function of the gate angle, and dependent on the operating mode of the machinery. The equation of motion can be written by relating the movements of the masses of the gate, gate arm and counterweights, to that of the hydraulic cylinder pistons, this relationship also being a function of the gate position. The structural flexibility and distributed mass of the gate was included but was shown to have negligible effect for the first mode of vibration.

b. The Structural Vibration of the Gates

31. The natural frequencies and modal shapes of the dynamic response of the gate structure were evaluated mathematically by different methods from two-dimensional idealisations of the gate structure. Structural damping was ignored.

(i) Rayleigh's Method. By considering the static deflection shape of the structure the first mode of vibration was calculated, using Rayleigh's method. The variation in gate stiffness and mass along its length was included and the effect of choosing equivalent parameters constant along the span was evaluated.

(ii) Method using equations of Motion. A two-dimensional idealisation, shown in Fig. 6(ii), was used to determine the equations of motion of the distributed mass of the gate span (having both u and v displacements) and the concentrated mass of the counterweights (u displacements only). The mass of the gate arms was ignored, although an allowance could have been included by an increase to the counterweight mass and to the gate mass in the x direction.

Table 1

	Frequencies Hz		
	Calculated		Observed
	$\sqrt{20}$ x Prototype	Model	Model
1st mode	6.4	6.0	5.5
2nd mode	12.2	11.6	9.2
3rd mode	18.4	16.9	15.4

The equations of motion, in matrix form, were solved by computer, giving the first four natural frequencies of the system.

(iii) <u>Finite Element Method</u>. A finite element program was also used to determine the first four natural frequencies and modal shapes. Again a two-dimensional idealisation was used, but in this case the variation of gate mass and stiffness along the gate span and the mass of the gate arms were included.

32. The three methods gave good agreement and the resulting natural frequencies and modal shapes are shown in Table 1 and Fig. 8. Also included are the observed natural frequencies of the model. It will be noted that agreement between observed and calculated values is reasonable, although the second harmonic shows a 20% difference. A modification to the mathematical idealisation was made, allowing for the three-dimensional behaviour of the gate arm by adding a spring in the x direction to the gate supports. This improved the agreement between observed and calculated frequencies and it could be concluded that the model did accurately reproduce the dynamic behaviour of the gate in air.

<u>Conclusion to the Vibration Studies</u>

33. The conclusions regarding vibrational stability of the Barrier Gates based on the experimental work conducted at the British Hydromechanics Research Association Laboratories and at Imperial College, London are summarised as follows:-

(a) The 61 m Rising Sector Gate is insensitive to flow-induced structural vibrations for all operating conditions anticipated. Measurements from the model tests indicated that only random excitation occurred, and that this was of insignificant amplitude.

(b) The higher natural frequencies for structural vibration of the smaller Rising Sector Gate and the Falling Radial Gate will make them less sensitive to flow-induced vibration than the 61 m Rising Sector Gate. It was therefore concluded that for these gates structural vibrations of significance are improbable.

(c) Under normal movement of the Rising Sector Gates from the open position to the closed position, the gates will not oscillate against their machinery.

(d) Should it be necessary to move or hold the Rising Sector Gate with a significant flow of water over the crest of the gate in the upriver direction, it will be necessary to attach the latches to prevent the possible build up of oscillation of the gate against the machinery. The attaching of the gate latches which have been designed to ensure the engagement of at least one latch throughout any movement of the gate to the closed position, increases the stiffness of the operating machinery and hence the natural frequency of this mode of vibration. The conditions to which this is relevant will not normally occur as partial closure will be accomplished by allowing flow under the gate and not over it.

(e) Under conditions of partial closure by allowing flow under the gate at least one gate latch must be connected to ensure the vibrational stability of the gate. Tests on the 1:80 scale sectional model showed that levels of excitation of the gate oscillating against the machinery for this condition were less than for equivalent conditions with overflow. In both cases the response was found to be essentially random.

(f) The torsional stiffness of the Rising Sector Gates is sufficiently high to preclude the possibility of significant torsional vibration.

(g) The Rising Sector Gates are vibrationally stable for conditions of overflow in the downriver direction. This relates to the possible future operation of the gate as a half-tide barrage.

(h) The natural frequencies of gate panels of all gates are sufficiently high to be insensitive to flow-induced vibration.

(j) The Falling Radial Gates are vibrationally stable during operation with flow in either direction. It may be noted that in practice the Falling Radial Gates will be closed well in advance of operation of the other gates, and hence during periods of very small flow only.

34. <u>Notation</u>

A Cross-sectional area of the gate.

E Modulus of elasticity of the gate material.

f Natural frequency of vibration of the gate.

g Acceleration due to gravity.

K Operating Machinery cylinder stiffness.

I Second moment of area of the gate span section.

L Length of gate span or characteristic length.

m,p Suffixes denoting model or prototype parameters.

n Scaling factor.

M) Internal forces on the gate structure
R) due to dynamic loads.
S)

u,v Gate element velocities in the x and y directions.

V Fluid velocity

ρ Density of gate material.

ρ' Density of fluid (water).

ACKNOWLEDGEMENTS

35. The Authors wish to express their grati-
tude to the Department of Public Health
Engineering of the Greater London Council and
Messrs. Rendel, Palmer & Tritton for permission
to write this paper and to the many Engineers
and those in Research Organisations who have
assisted in their work.

REFERENCES

1. KUHN, P., Stresses in Aircraft and Shell
Structures, New York, McGraw Hill, 1956.

2. Merrison Committee, Steel box girder bridge
design rules, Department of the Environment,
HMSO (London) 1973.

3. FAULKNER, D., Welded connections used in
Warship Structures. The Transactions of the
Royal Institute of Naval Architects, October,
1963.

4. DOWLING, P.J. and CARR, P., Thames Barrier.
Tests on an Araldite model of a Rising Sector
Gate. Engineering Structures Laboratories,
Civil Engineering Department, Imperial College,
London, CESLIC Report TB1, April 1974.

5. DOWLING, P.J., OWENS, G., et al. Thames
Barrier. Tests on a steel model of a Rising
Sector Gate. Engineering Structures Laborator-
ies, Civil Engineering Department, Imperial
College, London CESLIC Report TB2, April 1974.

6. KAMTEKER, A.G. and DWIGHT, J.D., Thames
Barrier Project – Welding stresses in Rising
Sector Gates. Cambridge University Engineering
Department, CUED/C-Struct/TR.36(1973).

7. KING, R. and PROSSER, M.J., Thames Barrier
Hydraulic Model of the Rising Sector Gate, BHRA
Report RR 1195, 1973.

8. KING, R., Thames Barrier Vibration Studies,
BHRA Report, RR 1193, 1973.

9. CROW, D., The Thames Barrier Rising Sector
Gate. Vibration Studies on a 1:20 scale hydro-
lastic model. BHRA Report RR 1225, 1974.

10. HARDWICK, J.D., Thames Barrier Rising
Sector Gate. Studies of Flow-induced Vibration
due to underflow and overflow made on a 1:80
scale sectional model. Civil Engineering
Department, Imperial College, London, 1974.

8. Operating machinery

D. M. S. FAIRWEATHER, BSc, FIMechE*

R. R. H. KIRTON, BSc(Eng), MIMechE†

INTRODUCTION

1. The Thames Flood Barrier is formed by ten movable gates spanning the river, supported by nine piers and two abutments. Four large Rising Sector Gates span the main navigation channels and are flanked by two smaller Rising Sector Gates and four falling radial gates adjacent to the abutments.

The gates are described in detail in Papers 6 and 7 but the following brief resumé will assist in understanding the operation of machinery.

2. The Rising Sector Gates comprise a gate segment supported at its ends by two disc shaped arms. At the centre of each arm is a shaft and bearing assembly, the shafts being bolted to and supported by an embedment in the adjacent pier while the gate is free to rotate on the bearings. When the gate is open to allow shipping to move up and down the river, the gate segment lies in the curved recess of the sill with its flat surface level with the bed of the river. The primary function of the Rising Sector Gate machinery is to rotate the gate until the gate segment is nearly vertical, with the curved surface facing downstream, to form a barrier against a tidal surge. Secondary functions of the machinery are to (a) rotate the gate beyond the vertical position, until a gap is opened up between the lower edge of the gate segment and the top of the sill, to allow water to undershoot the gate during a surge and (b) to rotate the gate through 180° until the gate segment is vertically above the support bearings and clear of the water, for the purposes of inspection and maintenance of the segment and adjacent parts of the arms. When the danger from a tidal surge is past, or maintenance or testing completed, the machinery lowers the gate segment to the open position again.

3. The falling radial gates carry the gate segment on non-circular arms which house the support bearings which, in turn, are mounted on stationary shafts carried on pedestals bolted to the outside of the adjacent piers. In the open position the gate segment is parked above the bearings, and the operating machinery rotates the gate until the segment is in a vertical position with its lower edge resting on the

top of the sill, thus completing the Barrier across the river.

4. Failure of one or more of the gates to close during a large tidal surge could cause flooding and damage upriver of the Barrier, and could also cause severe erosion of the river bed close to the piers. Reliability of operation was therefore a major factor in the choice and design of the machinery.

DEVELOPMENT OF THE MACHINERY

5. **Factors considered in deciding the type of machinery to be adopted**

(a) Reliability of Operation

(b) Infrequency of Operation

(c) Maintenance Requirements and Accessibility

(d) Constraint of Pier Size and Shape

(e) Misalignment Capability

(f) Gate Oscillation

(g) Feasibility of Manufacture

(h) Cost

6. Of the factors listed above, although all had to be satisfied to a large degree, reliability of operation was the most important. Tidal surges can be predicted many hours before the surge reaches London, but the actual height of the surge cannot yet be predicted with sufficient accuracy to give more than three or four hours warning that the Barrier is to be closed.

7. In certain emergency conditions the warning may be considerably less, therefore when the decision is made to close the Barrier there will be little time for fault finding and correction. When the buttons are pressed to close the Barrier, it must close; there will be no time to take alternative measures even if these were feasible. A variety of machinery arrangements were investigated and some typical schemes are illustrated in the Appendix to this paper.

8. The machinery scheme adopted for the Rising Sector Gates is shown in Fig. 2 and is a simple

---

* Partner, Rendel, Palmer and Tritton
†Managing Engineer, Rendel, Palmer and Tritton

Thames Barrier Design. Institution of Civil Engineers, London, 1978, 93-110

93

arrangement of links and levers, albeit of very large proportions, powered by large double-acting hydraulic cylinders or rams. Wear will be confined to the hydraulic cylinders, crossheads and within the bearings, and little trouble from these sources is anticipated.

9. Apart from the gate pivot assemblies, the only parts of the machinery below water level are the lower part of the links and the shaft and bearing assemblies attached to the gate arms. Corrosion problems are therefore reduced to a minimum and protection is given by carefully selected painting and metal spraying procedures and the use of suitable corrosion-resisting materials. Spherical bearings at the main hinge points allow a large tolerance in erection and for any subsequent movements. The cost was the lowest of the schemes considered and all the major components except the bearings were capable of being manufactured in the United Kingdom. Maintenance work is expected to be minimal. All the major components are designed to last throughout the life of the Barrier and such maintenance work that will be necessary can be largely carried out on the piers or at the site.

10. Factors affecting the power required of the machinery (Fig. 1)

(a) Shearing of the silt deposits in the gap formed between the curved surfaces of the gate and the sill.

(b) The affect of silt deposits accumulating on top of the gate segment.

(c) Buoyancy.

(d) Out of balance of machinery weights.

(e) Out of balance of gates and counterweights.

(f) Friction.

(g) Dynamic hydraulic forces.

(h) Rate of gate closure.

(j) The effect of water entrained in the gate segment.

11. The magnitude of the moments acting on the gates arising from these factors varies with the gate angle. Fig. 1 shows the estimated values for most of the conditions listed.

12. Of major importance in the design of machinery in a project such as the Thames Barrier, is establishing accurate information on the forces to be overcome. Some are readily determined by calculation, such as (c), (d) and (e) above, while good approximations can be calculated in others, such as (f) and (j). For those conditions which could not be calculated, laboratory tests and model studies were necessary, and factors (a) and (g) were analysed in this manner.

13. Even when all these forces have been estimated, there is no guarantee that the working

conditions will be exactly the same as those assumed during testing and calculation, and in determining the power of the machinery a large factor of safety was allowed to cater for unknown conditions.

14. When movement of a Rising Sector Gate is first initiated, a considerable force may be required to shear through the silt if the sill recess is completely filled, particularly if the silt has had time to consolidate. As referred to in Paper 4, samples of silt were taken, each allowed to settle, and the force required to shear it measured. It was established that, even with silt which had been left to consolidate over long periods, this force was unlikely to exceed 600 N/m^2 (12.5 lbf/ft^2). The available thrust from one set of main machinery alone allows for a shear force of 2400 N/m^2 (50 lbf/ft^2) over the whole curved surface of the Rising Sector Gates, thus giving a load factor of at least four, (or eight with both sets of machinery operating).

15. Silt deposits on top of the Rising Sector Gate have little effect on the initial gate movement since the weight acts vertically below the bearing centre and so exerts little or no moment on the gate. As the gate rotates, the silt will tend to slide or flow towards the trailing edge of the gate and exert a moment assisting gate closure. Testing and maintenance will require the regular raising of the gate, and any silt not dispersed by the river flow will be dumped in the sill scallop. As there will be little time for the silt to consolidate again, it will offer little resistance to the return of the gate into the sill scallop. When the gates are closed against a surge, the flow of water in the gap between the gate and the scallop will effectively scour away any silt deposit.

16. The counterweights in the gate arms have been arranged not only to balance the weight of the gate segment, but also to counteract the out-of-balance machinery weights.

17. When the Barrier is closed against a surge, the sudden arresting of the tidal flow causes a further rise in the water level, referred to as the reflected wave effect. To reduce this effect, consideration was given to the stopping of movement of the 61 m Rising Sector Gates before they reached the fully closed position so that water would spill over the top edge of the gate; or by raising the gate beyond the closed position to open up a gap between the lower edge of the gate and the top of the sill to allow water to undershoot the gate.

18. Model studies showed that when water was allowed to spill over the top of the gate, the hydrodynamic forces generated on the gate could be unacceptably high depending on: the upriver and downriver levels; the height of the water above the gate edge; the gate angle etc; and also indicated that forced oscillation of the gates could occur (see Paper 11). Although the magnitude of the forces could be controlled by adjusting the height of the gate edge and hence the amount of overspill, it is felt that if

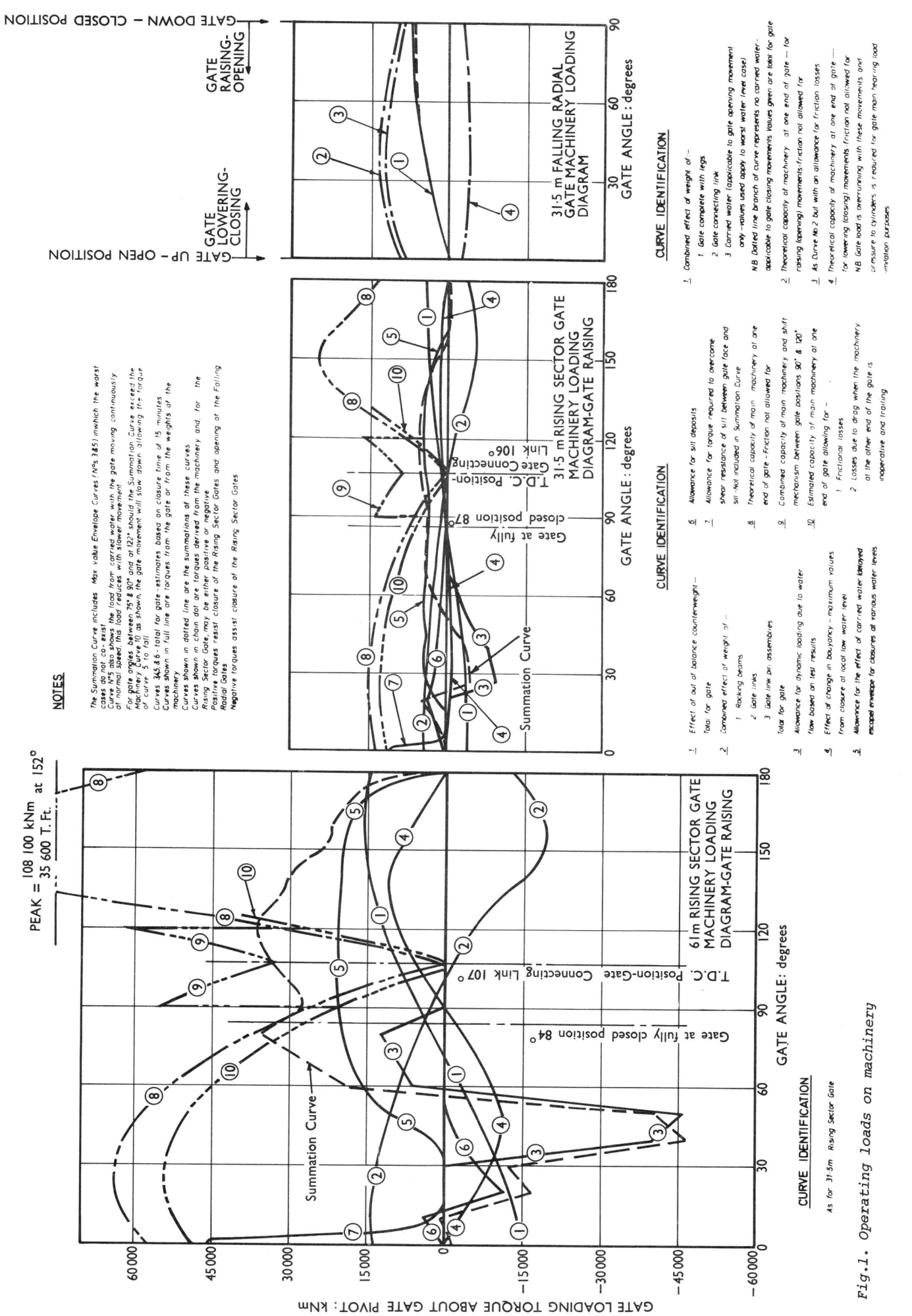

Fig.1. Operating loads on machinery

NOTES

The Summation Curve includes Max value Envelope Curves (N°s 3 & 5) in which the worst cases do not co-exist
Curve N°5 also shows the load from carried water with the gate moving continuously at normal speed, this load reduces with slower movement

For gate angles between 75° & 90° and at 122° should the Summation Curve exceed the Machinery Curve 10 as shown, the gate movement will slow down allowing the torque of curve 5 to fall

Curves 3,4,5 & 6 - total for gate - estimates based on closure time of 15 minutes
Curves shown in full line are torques from the gate or from the weights of the machinery

Curves shown in dotted line are the summations of these curves
Curves shown in chain dot are torques derived from the machinery and for the Rising Sector Gate, may be either positive or negative
Positive torques resist closure of the Rising Sector Gates and opening of the Falling Radial Gates
Negative torques assist closure of the Rising Sector Gates

CURVE IDENTIFICATION

1   Effect of out of balance counterweight —
    Total for gate
2   Combined effect of weight of —
    1 Rocking beams
    2 Gate links
    3 Gate link pin assemblies
    Total for gate
3   Allowance for dynamic loading due to water flow based on test results
4   Effect of change in bouyancy - maximum values from closure at local low water level
5   Allowance for the effect of carried water (delayed escape) envelope for closures at various water levels

6   Allowance for silt deposits
7   Allowance for torque required to overcome shear resistance of silt between gate face and sill - Not included in Summation Curve
8   Theoretical capacity of main machinery at one end of gate - Friction not allowed for
9   Combined capacity of main machinery and shift mechanism between gate positions 90° & 120°
10 Estimated capacity of main machinery at one end of gate allowing for —
    1 Frictional losses
    2 Losses due to drag when the machinery at the other end of the gate is inoperative and trailing

CURVE IDENTIFICATION

1   Combined effect of weight of —
    1 Gate complete with legs
    2 Gate connecting link
    3 Carried water (applicable to gate opening movement only - values used apply to worst water (wet case)
    N.B. Dotted line branch of curve represents no carried water applicable to gate closing movements Values given are total for gate

2   Theoretical capacity of machinery at one end of gate — for raising (opening) movements-friction not allowed for

3   As Curve No 2 but with an allowance for friction losses

4   Theoretical capacity of machinery at one end of gate — for lowering (closing) movements friction not allowed for
    NB Gate load is overrunning with these movements and pressure to cylinders is required for gate main bearing load during simulation purposes

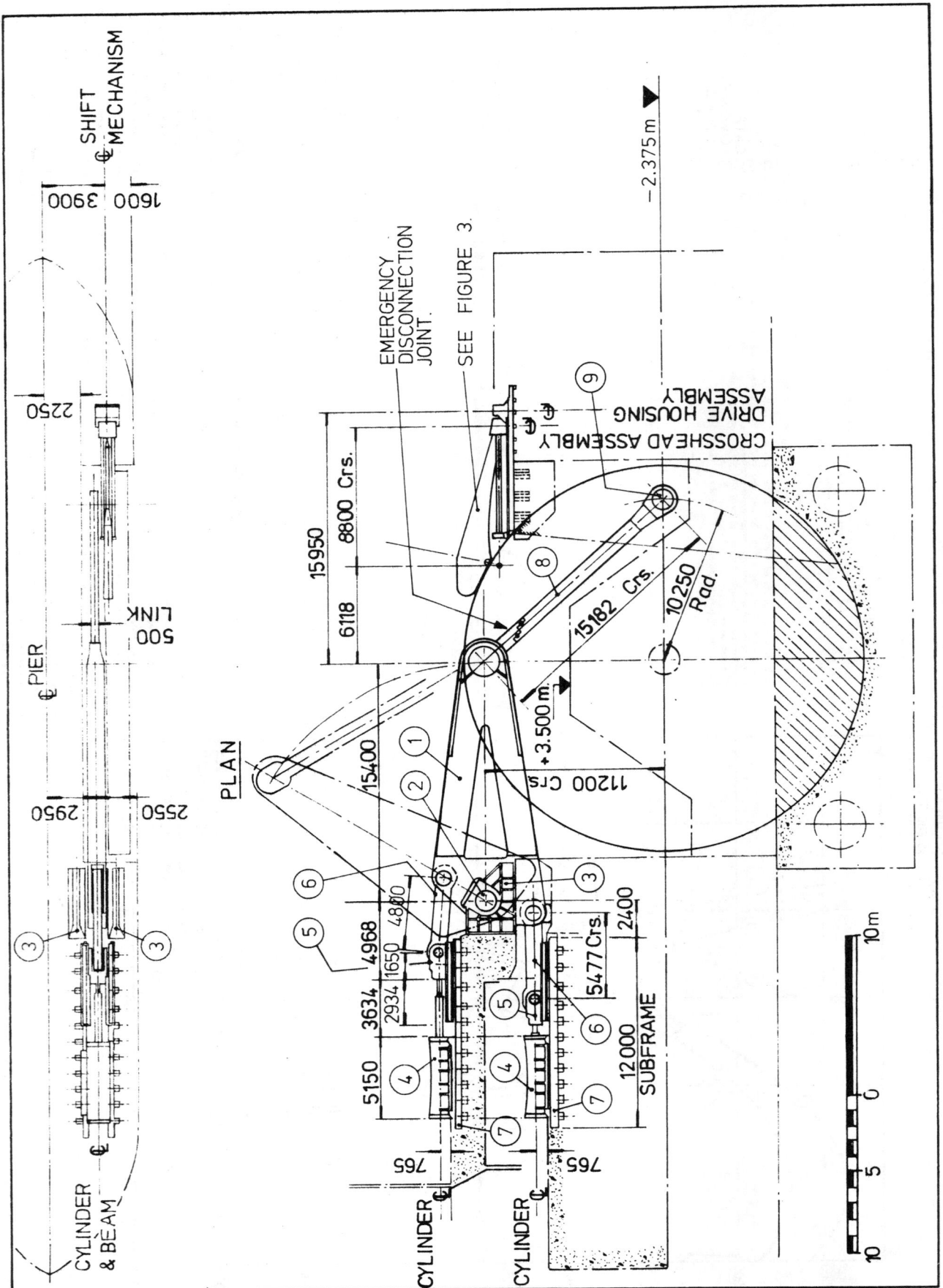

Fig.2. Main machinery for 61.0 m rising sector gate

overspilling were permitted, overloading of the machinery might occur inadvertently and therefore the stopping of the 61 m Rising Sector Gates before the fully closed position is reached to allow overspilling, is not proposed.

With undershooting, however, the hydrodynamic forces are much less and will not affect the machinery. Also the gates will be moved and held by the shift and latch mechanisms, at this gate position, which prevent gate oscillation.

19. Although there are large ports in the gate to allow water to flow out of the gate as it rises out of the water, and into the gate as it is lowered into the water again, a difference between the water levels inside and outside the gates must occur to induce the flow of water. The rate of water flow and the differential in water levels is dependent on the gate speed, angle and tidal height. The weight of water could be large and could create large moments resisting the gate movement. This moment could become significant towards the end of the gate closure cycle if one set of machinery only is operational, (see Fig. 1), but as the hydraulic pump capacity varies with the load, the effect would only be to slow down the gate speed with a corresponding reduction in the differential water level, thus reducing the moment on the gate.

20. As the time from the warning of a surge to the completion of the Barrier closure could be as short as one hour, the machinery is designed to close each gate in 15 minutes and the complete Barrier within 30 minutes. The performance of the variable flow hydraulic power packs allows full closure of a gate within the 15 minute period with the hydraulic cylinders developing maximum thrust throughout their stroke, and in half that time if the thrust requirement remains below half the maximum.

DESCRIPTION OF MACHINERY

Rising Sector Gate Main Machinery

21. Fig. 2 shows the machinery for the 61 m Rising Sector Gates. The 31.5 m Rising Sector Gate is the same in principle but to a smaller scale.

22. The fabricated steel rocking beam (1) is supported on a pivot shaft assembly (2) having bearings fitted at each side of the beam which are housed in brackets (3) rigidly bolted to the pier. Hydraulic cylinders (4) connect to the rocking beam at either side of the pivot shaft through crossheads and connecting rod assemblies (5 and 6). The hydraulic cylinders are 1100 mm bore x 3130 mm stroke, and develop a thrust of 15.21 MN (1526 tons) at 17.24 N/mm^2 (2500 lb/in^2) oil pressure. The reaction from the hydraulic cylinders is transmitted into the concrete through fabricated subframes (7) which are bolted down by pretensioned high tensile bolts and embedded in the concrete floor. At the opposite end of the rocking beam, a link (8) connects to a pin and bearing assembly (9) fitted to the gate arm.

23. To raise the gate, the upper cylinder pulls, and the lower cylinder thrusts, the rocking beam causing it to rotate about its pivot, raising the gate link and so rotating the gate. Each main machinery unit is capable of exerting a torque of 48000 kN m (15805 ton ft) on the 61 m gate at the zero gate angle position (fully open), the torque increasing to a maximum at approximately 30° gate angle but reducing again at larger angles due to the geometry of the linkage. The torque becomes zero when the gate link lies in line with the gate main bearing at approximately 107° gate angle. A similar torque curve takes place between the 107° and the 180° (maintenance) positions (see Fig. 1).

Rising Sector Gate Shift and Latch Mechanisms (Fig. 3)

24. To rotate the gate through the link dead centre position the Shift and Latch mechanism is used. Other duties of this mechanism are to lock the gate in the Open, Closed, or Maintenance positions, and to move the gate beyond the fully closed position to open up a gap between the lower edge of the gate and the sill to allow undershooting. This latter requirement applies only to the 61 m Rising Sector Gates. The mechanism can also be used, in conjunction with an identical unit at the other end of the gate, to assist the main machinery to initiate gate movement should the main machinery have insufficient power to do so on its own. This possibility is, however, very unlikely. In an emergency, the shift and latch mechanisms can also raise the 61 m gate to the closed position should the main machinery be out of action for any reason. This again is a most unlikely situation.

25. Since one of the functions of the shift and latch mechanism is to lock the gates when the hydraulic power has been shut down, its design is such that after it has been locked onto the gate, the locking action is independent of the hydraulic power supply. The mechanism is more rigid than the main machinery and when locked on to a gate, effectively prevents gate oscillation.

26. Referring to Fig. 3, the mechanism comprises two screwed shafts (1) each driven by a hydraulic motor (2) with a gear train connecting the shafts to synchronise their rotation. The shafts pass through nuts contained within the crosshead (3) so that rotation of the shafts imparts axial movement to the crosshead. Connected to the crosshead through a self-aligning bearing is the shift arm (4), the opposite end of which is supported on wheels (5) running on rails provided round the periphery of the gate arms. At the gate end of the shift arm are the latches (6) which can be engaged or disengaged to pins (8) in the gate arm through a system of links and levers actuated by hydraulic cylinders (7).

27. To engage a latch to a gate pin the hydraulic motors are started, the shift arm moves out and the latch automatically finds and engages to the nearest pin, or, when required, to the

next pin. When a latch is disengaged the shift arm automatically retracts. The shift mechanism can move the gate through an angle of 30° and pins are provided in each 61 m Rising Sector Gate arm at each 30° of gate angle up to 90° and at the undershooting, fully open, and maintenance positions. The hydraulic power to operate the latches is independent of the main hydraulic system and is provided by a power pack (9) mounted on each crosshead. The 61 m gate shift mechanism is capable of generating a pull or thrust of approximately 3.40 MN (340 tons) with a hydraulic supply pressure of 17.24 N/mm^2 (2500 lb/in^2), the reaction being transferred from the screwed shafts through spherical roller thrust bearings in the thrust housing (10), to the fabricated subframe bolted to and concreted into the pier floor.

Falling Radial Gate Machinery (Fig. 4)

28. As the falling radial gates rotate through only 90° a simpler machinery arrangement is used. Referring to Fig. 4), double-acting hydraulic cylinders (1) connect to each end of a gate through crossheads (2) and connecting rods (3). Automatic locks (4) are provided to lock the gate in the open position. To lower a gate from its parked position, the automatic locks are disengaged; the rams push the gate away from the parked position until the weight of the gate is sufficient to maintain the lowering movement which is then controlled by regulating the flow of oil discharged from the cylinders. In the event of a failure of both sets of main machinery it is possible to lower a gate manually using a hand-operated pump to move the gate to the point where gravity will continue the movement.

Gate Trunnion Shaft and Bearing Assemblies

29. Fig. 6 shows a trunnion shaft and bearing assembly for the 61 m Rising Sector Gates. The trunnion shaft (1) is bolted to the trunnion shaft support structure embedded in the adjacent pier by a double row of bolts screwed into the shaft flange and by a group of high tensile bolts passing through the centre of the shaft.

30. The shaft is manufactured from a 1% Mn, 1% Cr, Ni, Mo, steel forging. Because of the complex shape of the shaft the stresses were analysed by model testing and a photo-elastic study. A scale model of the shaft was made in steel and the stresses under load measured by the use of strain gauges. The results from this test were used to finalise the dimensions of the shaft before the photo-elastic study was carried out. This study not only confirmed the acceptability of the stresses but showed the stress pattern throughout the shaft and provided useful information on the shaft deflection. The model test and photo-elastic study were carried out by Dr. B.H. Baines of the University of Manchester.

31. A cover (2), which is aluminium sprayed for corrosion protection, is shrunk on the shaft flange and projects down to the trunnion bearing. A rubber seal (3), between the cover (2)

and the bearing (4), completes the sealing arrangement which prevents the river water from coming into contact with the trunnion shaft. Two annular seals (5) are bolted to cover (2), and a cover plate (6) is bolted to the river side of the gate arm to prevent water from entering into the bearing housing. Provision is made to drain out any leakage but, since it is not possible to guarantee that the seals (5) will remain effective throughout the life of the Barrier, the bearings are designed to operate without deterioration even should the bearing chamber become flooded. The bearing is pressed on to the tapered shaft by studs (7) which are tightened up to a predetermined load hydraulically. The interference fit between the bearing and the shaft is controlled by adjusting the thickness of the spacer plate (8) during the trial assembly of the bearing on the shaft. A groove is provided round the centre of the bearing bore, to allow easy removal of the bearing after the trial assembly by oil injection. The outer race of the bearing fits into a housing in the gate arm so that the arm, and hence the gate, is free to rotate on the stationary shaft.

32. To protect the bearing from damage should a ship collide with the gate arm, thrust plates (9) are fitted, the inner thrust plate also serving as the clamp plate for the bearing. The impact load from a collision passes through the thrust assembly on to the end of the shaft.

33. The 31.5 m Rising Sector Gate trunnion shaft and bearing assembly is similar to the 61 m assembly except that the group of bolts passing through the centre of the shaft is omitted.

34. The gate arms of the falling radial gates are not circular and permit a simpler bearing assembly to be used. The bearing is fitted into the gate arm, and the supporting shaft extends outwards from each side of the arm into a bracket bolted to the outside of the piers.

Bearings

35. The journal bearings throughout the Barrier, including the gate main bearings, required special attention. They had to be capable of remaining in one position for long periods without damage; a number were sited below water level and, while seals were fitted, absolute water tightness throughout the Barrier life was improbable and therefore they were required to work satisfactorily even when flooded. They had to be insensitive to damage from shocks or low frequency vibration and to have a life equal to that of the Barrier. Replacement of even the smaller bearings would mean sections of the machinery being out of action for a considerable time due to the large size and weight of the components concerned, thus reducing the reliability of the Barrier. Replacement of the Rising Sector Gate main bearings would be a major engineering job involving gate removal, and consequently the Barrier would be ineffective, possibly for some months, during this operation.

PLAN

700

② ②

③

① ①

⑤

HYDRAULIC MOTORS

+ 8.00 m

1489

⓵ DRIVE

1032

⓵ CROSSHEAD

⑩

⑨

6280 TRAVEL

⑦

④

⑥

⑧

⑥

⑦

+ 7.925 m

8800

ELEVATION

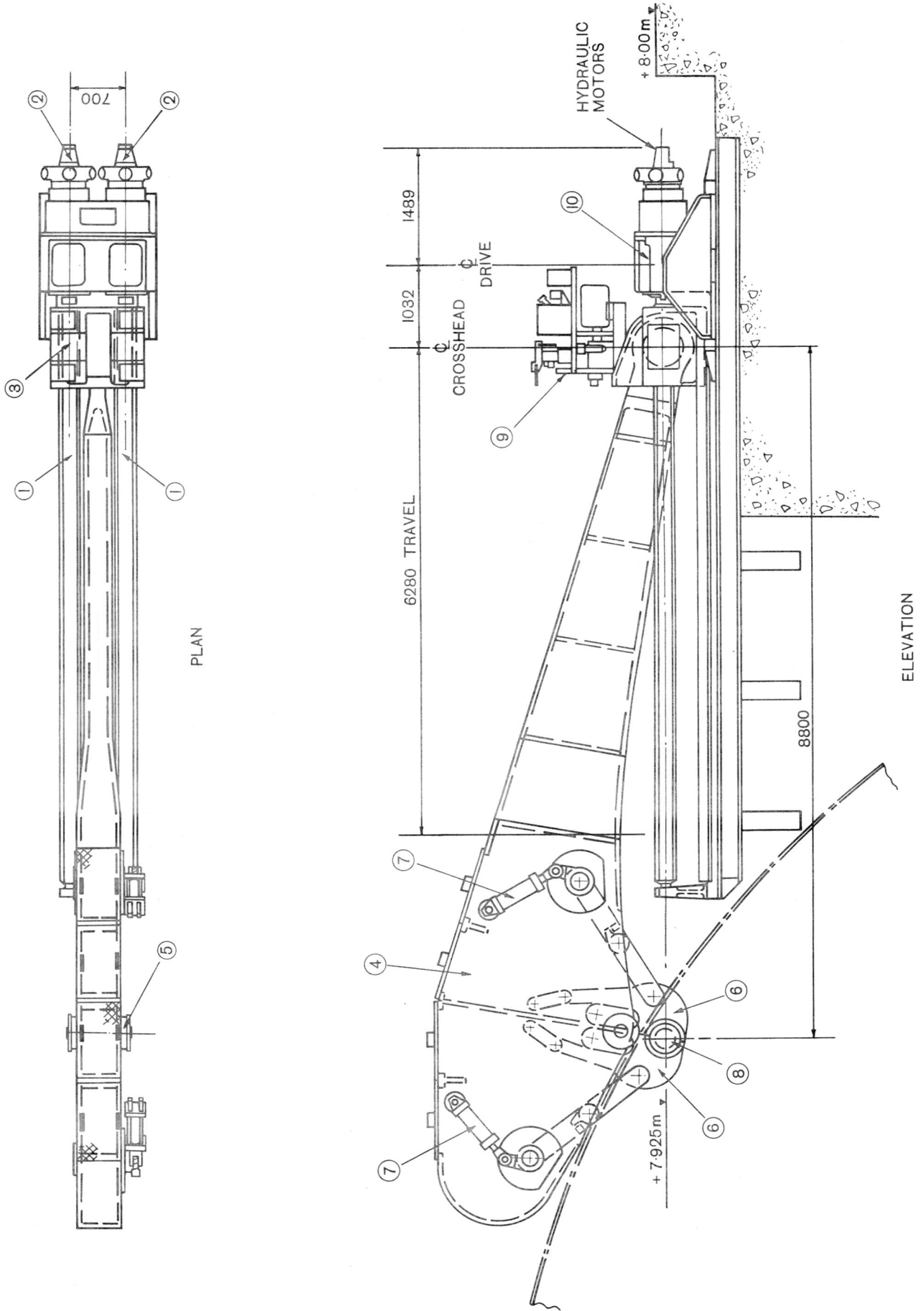

Fig.3. 61.0 m rising sector gate shift and latch mechanism

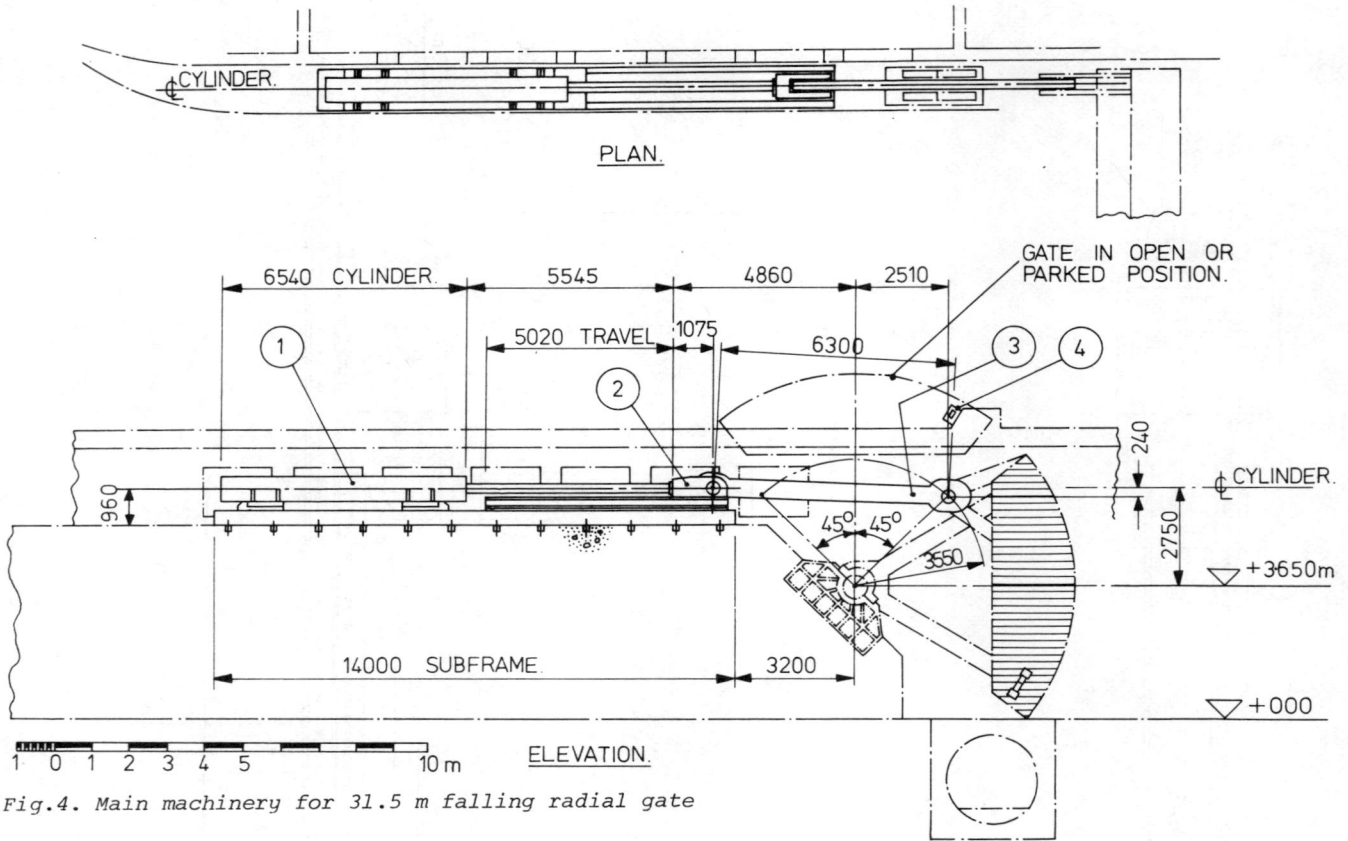

Fig.4. Main machinery for 31.5 m falling radial gate

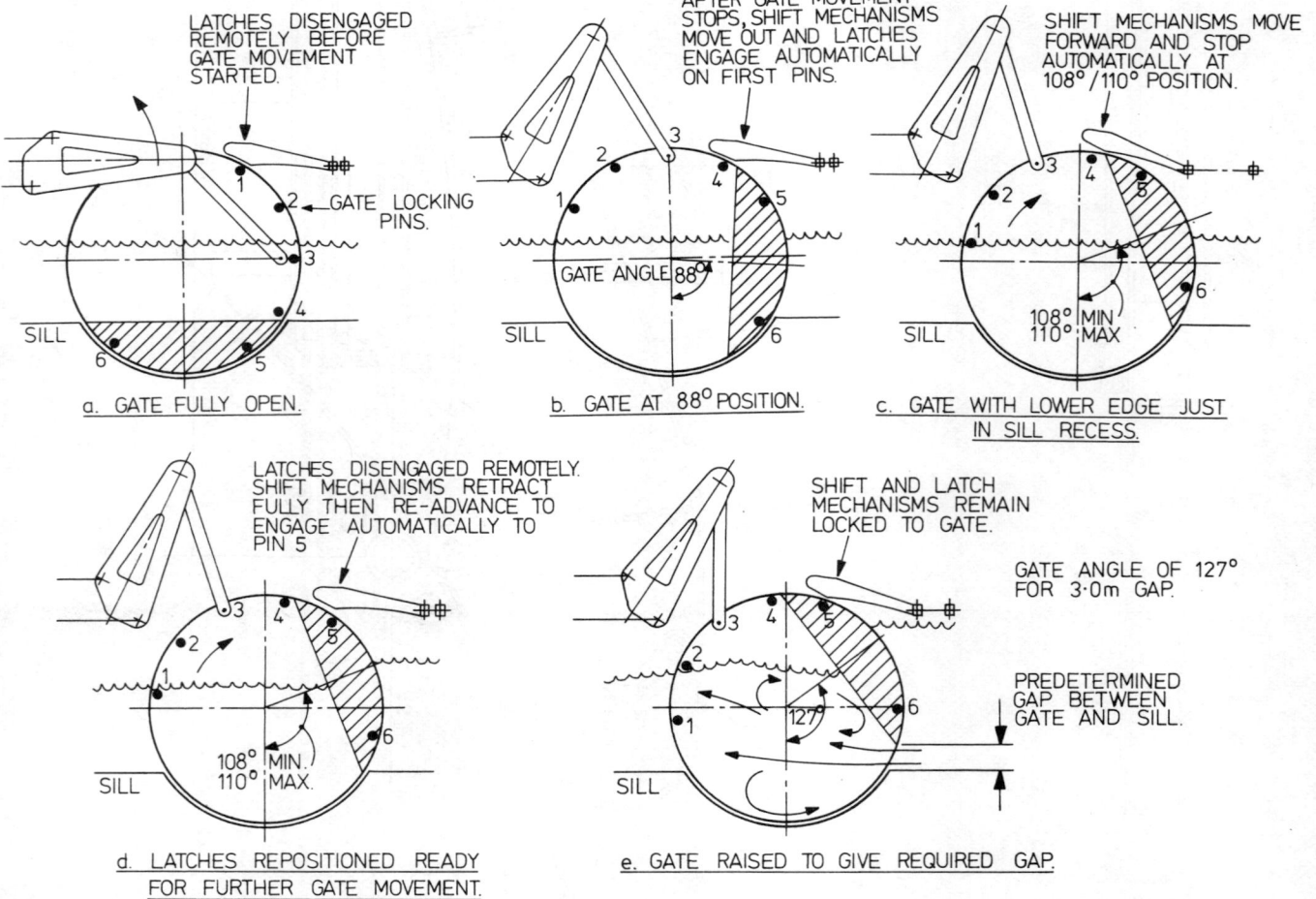

Fig.5. Gate operation sequence to undershooting position

36. The bearings which have been designed specially for the Barrier, are all of the plain bearing spherical type, supplied by Merriman Inc. and are of their Lubrite design, (see Fig. 6). The inner ball is of a Ni, Al, corrosion resisting bronze, having trepanned holes all over its spherical surface into which a solid dry lubricant is injected. In addition the whole surface is coated with a thick layer of the lubricant. The outer race is of corrosion resistant stainless steel. As a secondary means of lubrication, and to protect the shaft and housing from corrosion, the bearing will also be grease packed. The largest bearing supplied is the 61 m gate trunnion bearing shown in Fig. 6 and it is capable of supporting a radial load of 50 MN (5000 tons).

37. The lubrication of the crosshead guides presented another problem. If grease lubrication was used large amounts of dirt and grit would adhere to the grease during the long periods of idleness. To avoid the use of grease, the sliding surfaces had to be self-lubricating and protected from corrosion. In this case linear bearings, consisting of a stainless steel backing plate to which woven Teflon is bonded as a dry lubricant, are fitted to the crossheads, and the guide surfaces are coated with aluminium bronze to prevent corrosion. Wipers are fitted at each end of the crosshead to remove accumulations of dust and grit as the crosshead moves along the guides.

DUPLICATION OF MACHINERY

38. Although the number of working hours per year for the machinery will be small and reliability has been a prime factor in the choice of the machinery, maintenance operations will inevitably be necessary from time to time with the result that a machinery set may be out of action at a critical time. Duplication of the main machinery on all gates was therefore essential. The machinery set at each end of a gate is of adequate power to move the gate on its own.

39. For normal operation both sets of machinery will be used together for gate closure, so providing a large reserve of power and reducing the time of gate closure to a minimum. However should one of the machinery sets be out of action the supply and return connections to the hydraulic cylinder(s) can be interconnected allowing the pistons to move freely when the gate is driven by the other machinery.

40. On the Rising Sector Gates, if a machinery set is out of action and cannot be moved for any reason, then the emergency disconnection joint incorporated in the gate link can be freed in a relatively short time to disconnect the gate from the locked machinery. This disconnection joint is of bolted construction with shear pins fitted, and the use of hydraulic nuts on the bolts enables their removal without the use of large spanners or great force. When the joint is disconnected the lower part of the link, which remains attached to the gate, is lowered onto a concrete ledge protruding from the side

of the pier, and is free to move with the gate when it is rotated by the other machinery set.

HYDRAULIC SYSTEM

41. The hydraulic power for the machinery is supplied by electric motor-driven power packs comprising a variable capacity main pressure pump, fitted with a small pump which provides servo pressure to the flow control unit attached to the main pump. A boost pump driven by a separate motor supplies hydraulic fluid at low pressure to the main pump. The excess flow from the boost pump combines with the excess servo pump flow and passes through the main pump case to an air blast oil cooler fitted with a thermostatically controlled fan. The solenoid valves controlling the machinery are supplied with pilot pressure from a separate pilot pump mounted on the power pack. The hydraulic fluid reservoir mounted over the pump units is equipped with a thermostatically controlled heater and fluid level and temperature switches. The acceleration and deceleration of the gates and machinery is controlled by the rate of increase and decrease of flow from the main pump.

42. Due to the large volume of oil in the operating machinery cylinders, a decompression unit is included in the piping to the cylinders to allow the compressed volume of oil to be bled off at a controlled rate. This eliminates hydraulic shocks in the system due to the sudden expansion of the oil when the solenoid control valves open the pressurised side of the cylinders to the oil tank. Also mounted on the decompression unit are holding valves which prevent the gate from running away, i.e. moving under external forcing loads at a rate greater than that dictated by the oil flow from the hydraulic power packs.

43. One duty power pack is provided for each machinery set. On all piers, (other than 4 and 8) where the two duty power packs on each pier are the same size then only one standby power pack is provided to serve both systems. On piers 4 and 8, where the duty power packs are of different sizes, a standby power pack is provided for each system. On the north and south abutments where there is only one duty power pack, a standby is provided.

44. A substantial part of the hydraulic system is duplicated with the standby units so that a valve failure can be by-passed by changing to the standby power pack.

45. As stated previously the hydraulic system operating the gate latches is independent of the main system. The power packs required are not duplicated but incorporate two variable pumps working in parallel so that failure of one pump will result only in an increase in the time required to raise or lower a latch. A manually-operated oil pump is provided to allow disengagement of the latches should the power pack be completely out of action, since a gate could not be closed if a latch remains engaged to the gate. The hydraulic cylinders operating the

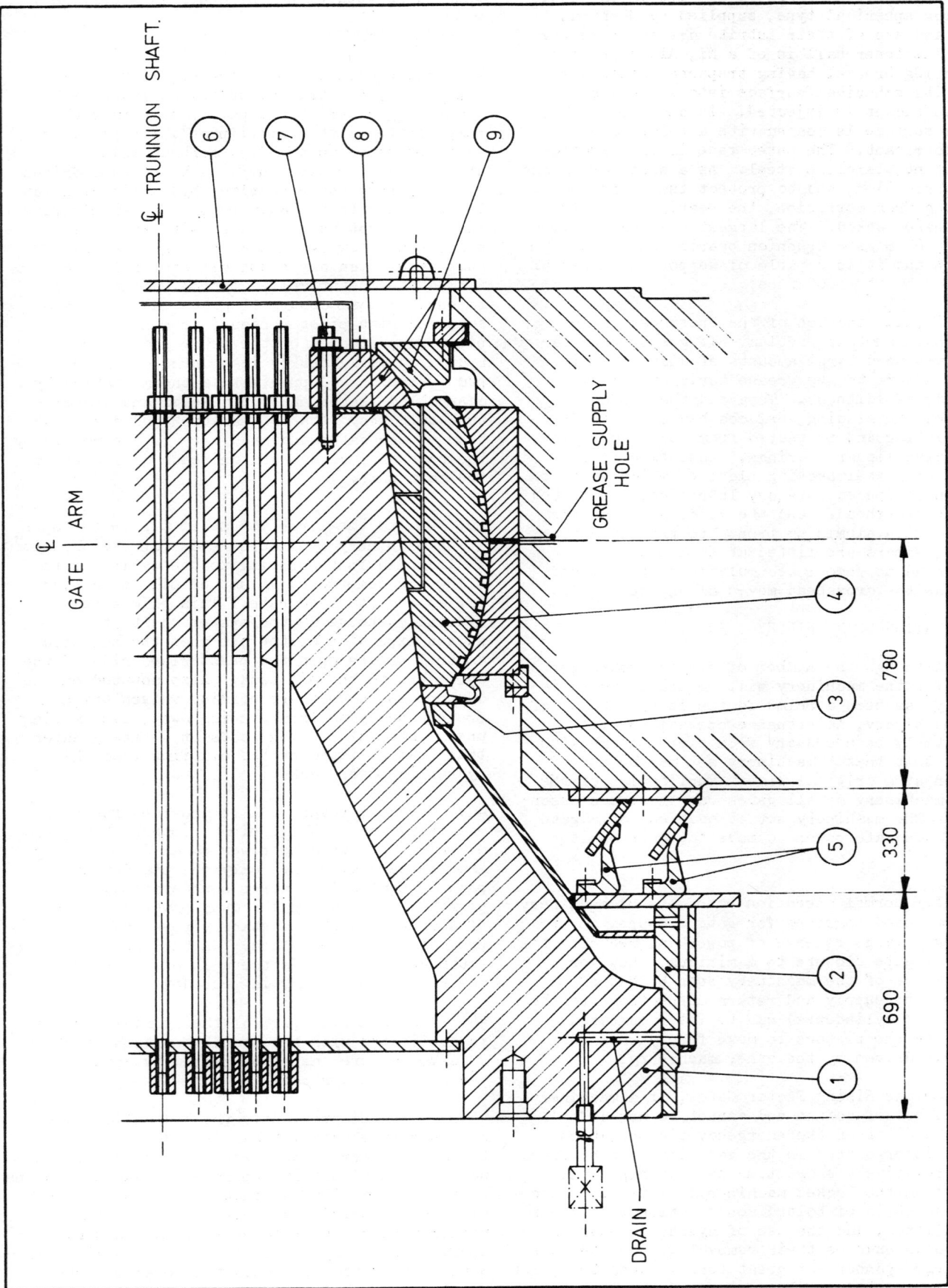

Fig.6. Trunnion shaft assembly for 61.0 m rising sector gate

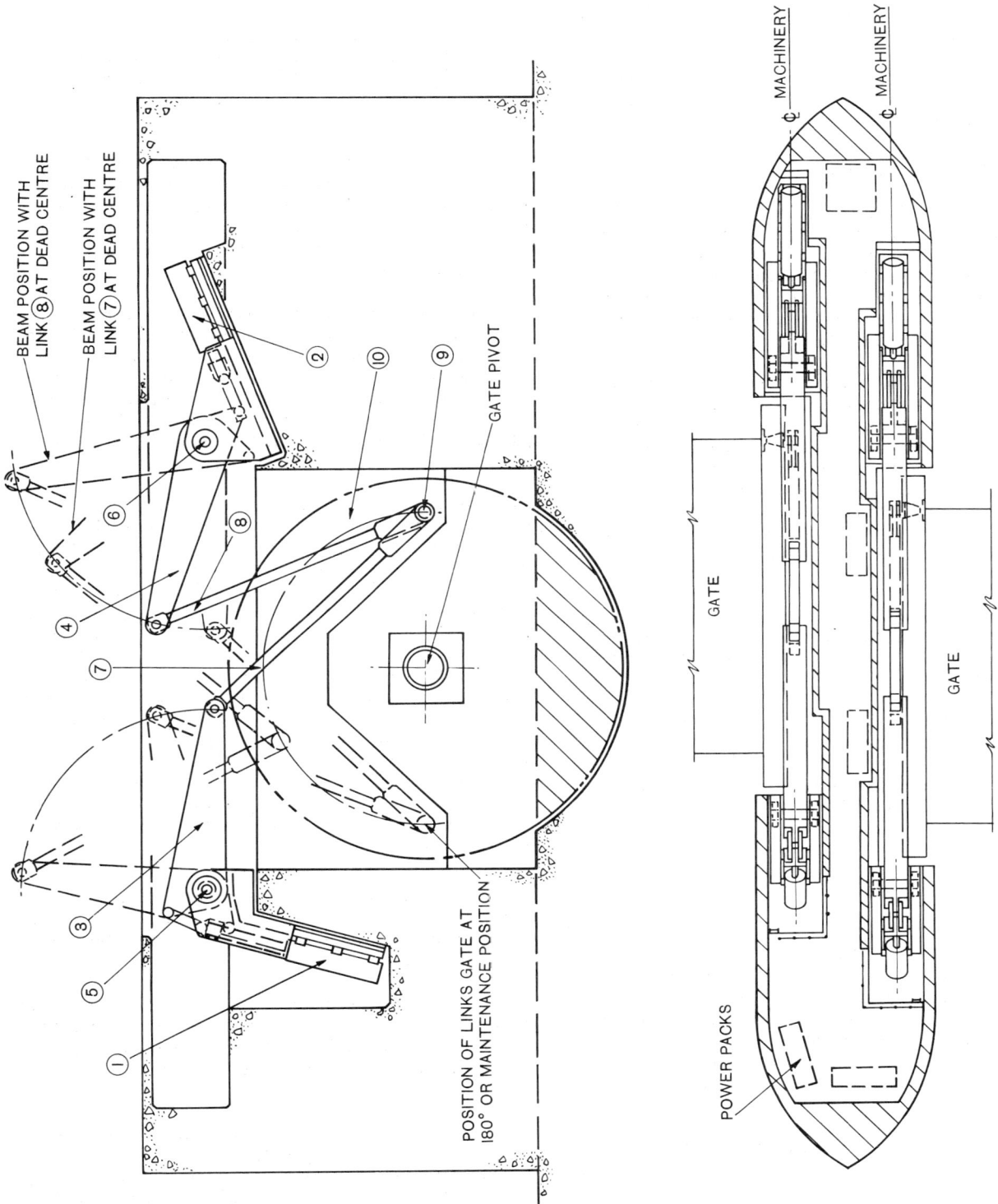

BEAM POSITION WITH LINK ⑧ AT DEAD CENTRE

BEAM POSITION WITH LINK ⑦ AT DEAD CENTRE

② ⑥ ④ ⑦

⑩ ⑧ ⑨

GATE PIVOT

POSITION OF LINKS GATE AT 180° OR MAINTENANCE POSITION

⑤ ③ ①

MACHINERY

MACHINERY

GATE

GATE

POWER PACKS

Fig.7. Dual rocking beams and link drive

latches are duplicated and the service cylinder can be selected by means of valves mounted on the power pack.

46. The piers are equipped with an oil distribution system incorporating a pump and distribution valve panel, with storage tanks for clean oil and for used or contaminated oil. The capacity of the oil storage tanks allows the whole of the oil to be drained out of an oil system including the power pack reservoirs and hydraulic cylinders if necessary.

PROCEDURE FOR GATE CLOSURE

47. Normally the Barrier gates will remain in the open position to allow free passage through the Barrier to shipping. While the Barrier is open, the Barrier staff will, as a matter of routine, carry out regular checks on all equipment and systems to ensure that they are always in a state of readiness for closing the Barrier.

48. When conditions indicate that a tidal surge is building up to a dangerous level, an early warning is given which will normally be several hours in advance of the order to close the Barrier although, in certain circumstances, the warning period could be as short as one hour. This advance warning allows time for essential pre-closure procedures and checks to be carried out before the actual closure of the Barrier is initiated.

49. A brief summary of the procedures necessary from receipt of a surge warning to the closing of the Barrier and for reopening it again, are given below:

(a) Check that all electrical and control systems are operational.

(b) Start up the main machinery power packs and check for pressure availability. Select the duty power packs.

(c) Start up the latch power packs and check for pressure availability.

(d) Check each Rising Sector Gate in turn. Disengage the latches from each end of the gate, move the gate a few degrees, using the main machinery only, to overcome the resistance from any compacted silt below the gate and to ensure that the gate is free to move when the close Barrier signal is given. Return the gate to the fully open position again but leave the latches disengaged.

(e) Disengage the locks from the falling radial gates and lower the gates to the closed position

(f) The Barrier is now ready for a full closure and remains in this stage until the signal to close the Barrier is given or until the surge warning is cancelled.

(g) If the surge warning is cancelled the falling radial gates are raised, the latches on the Rising Sector Gates are locked on to the gate pins and all systems are shut down.

(h) If the order to carry out a full closure is given, then closure of the Rising Sector Gates is initiated at predetermined intervals. (Referring to Fig. 5(a), the status is shown of a Rising Sector Gate before the signal to close is given).

(j) Using the main machinery only, the gate is rotated to the 88° position, where it automatically stops, (see Fig. 5(b)). When all gates are at 88° the Barrier is closed and London is protected against the surge.

50. To minimise the reflected wave effect referred to earlier, the 61 m Rising Sector Gates may be further raised to open up a gap between the lower edge of the gate and the top of the sill. This is carried out using the shift mechanisms only.

51. For this operation, valves are operated to interconnect the supply and return pipes of the main machinery hydraulic cylinders together and to connect them to a make up oil tank so that the piston is not hydraulically locked but is free to move if the gate is rotated. Thus the cylinders free-wheel when the gate is moved by the shift and latch mechanisms.

52. The shift mechanisms are advanced and engage automatically with the first gate pins encountered: pins (4). Further movement of the shift mechanisms rotate the gate to the 108° position, see Fig. 5(c), where it automatically stops with the lower edge of the gate just within the sill recess. The latches are then engaged to pins (5), see Fig. 5(d), by alternately disengaging the latch at one end of the gate when the shift mechanism will automatically retract fully. The shift mechanism is then advanced again and engages with pin (5). This procedure is repeated with the shift and latch mechanism at the other end of the gate. Moving the latches alternately is to ensure that the gate is always locked and is not moved by any external forces acting on it.

53. The procedure is repeated on all four Rising Sector Gates and when complete the machinery is ready to move the gate to the undershooting position.

54. The 61 m Rising Sector Gates remain in this position while the surge level rises downriver of the Barrier, until conditions are optimum for undershooting to minimise the reflected wave effect.

55. The height of gap required is preselected and the shift mechanisms are then operated to rotate the gate, movement stopping automatically when the required gap is obtained, see Fig. 5(e).

56. When the required undershooting period has been completed the gates are moved back to the 88° position under the action of the shift and latch mechanisms by reversing the previous procedure. Provision is made for the latches to bypass pins (5) as the shift mechanisms are advancing to enable the latches to engage with pins (4).

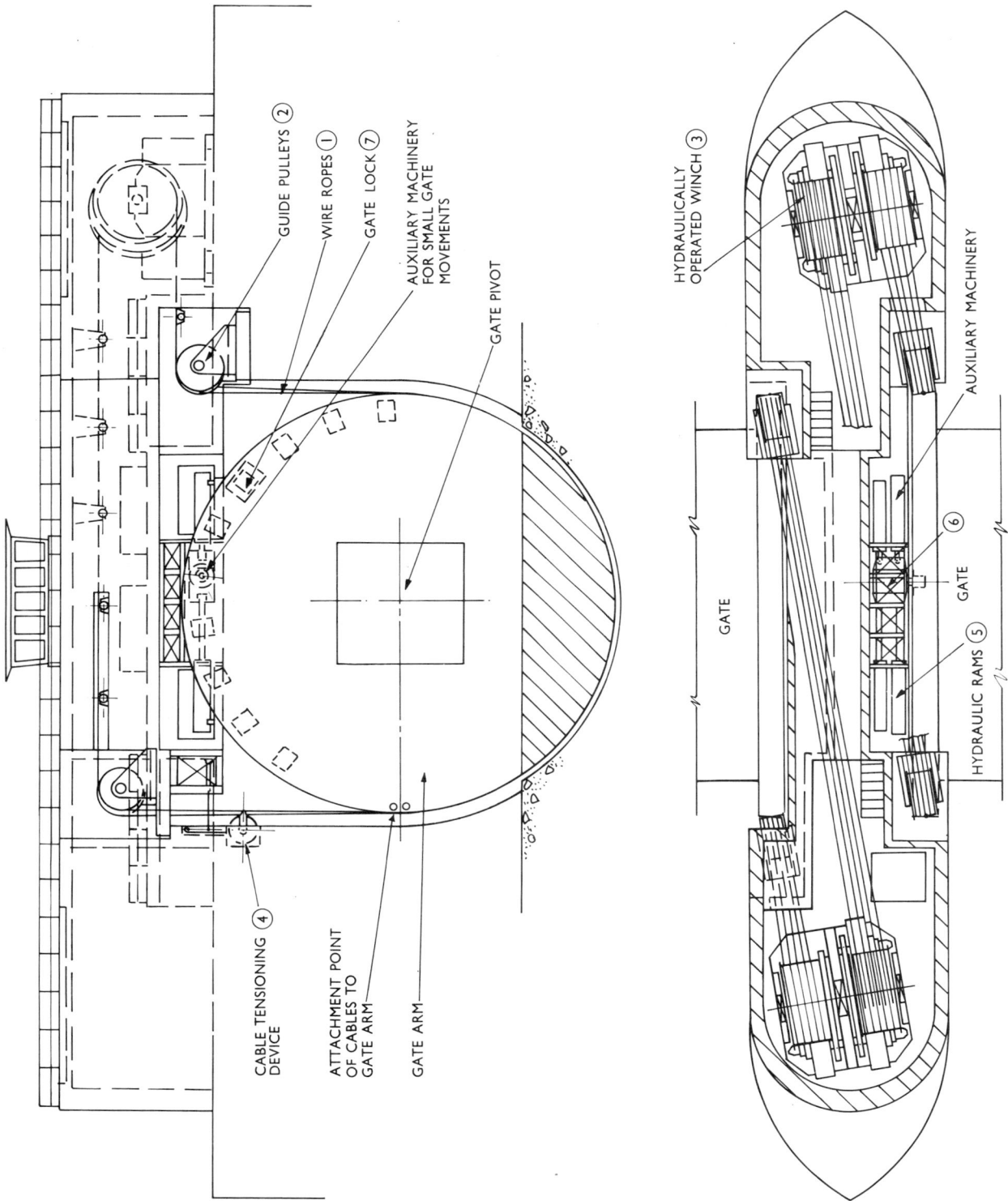

GUIDE PULLEYS ②
WIRE ROPES ①
GATE LOCK ⑦
AUXILIARY MACHINERY FOR SMALL GATE MOVEMENTS
GATE PIVOT
CABLE TENSIONING DEVICE ④
ATTACHMENT POINT OF CABLES TO GATE ARM
GATE ARM

HYDRAULICALLY OPERATED WINCH ③
AUXILIARY MACHINERY
GATE
GATE ⑥
HYDRAULIC RAMS ⑤

Fig.8. Wire rope drive

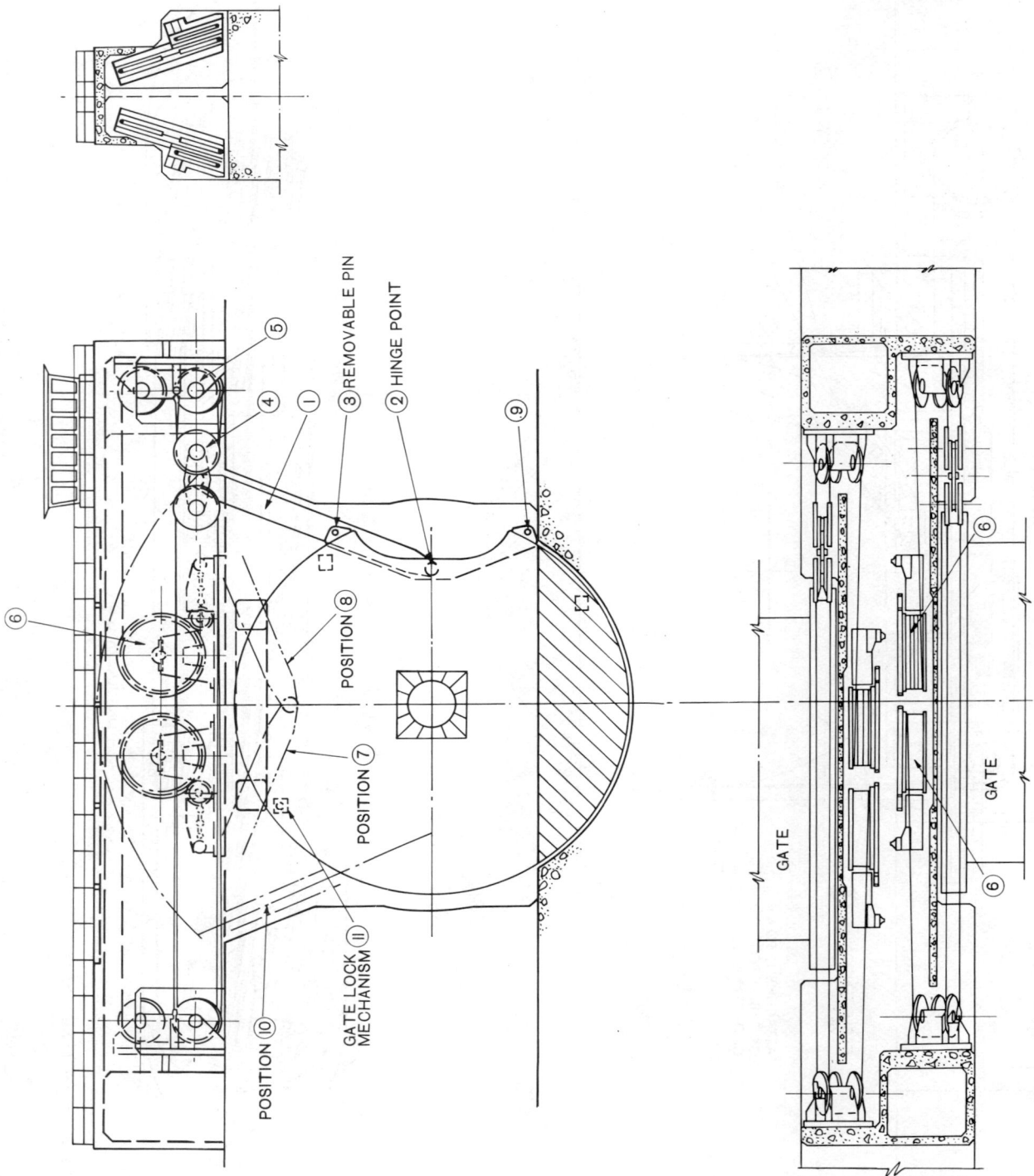

Fig. 9. Wire rope and lever drive

57. When the surge level has fallen to the level of the water upriver of the Barrier the latches are disengaged and the Rising Sector Gates are lowered to the fully open position using the main machinery and the latches engaged to the gate pins (1). The falling radial gates are fully raised and are automatically locked in the raised position. All systems are then shut down.

RAISING THE RISING SECTOR GATES TO THE MAINTENANCE POSITION

58. The gate is raised to the 88° or 90° position by the main machinery, the latches are engaged to gate pins (4), Fig. 5, and the main machinery hydraulic cylinders set to free-wheel.

59. The gate is rotated by the shift mechanisms to the 118° position, this being the maximum angle possible by the shift mechanism at full stroke, taking the gate link beyond the 107° dead centre position. The latches are then disengaged and the main machinery is operated to lower the rocking beam. As the force on the gate from the link now acts on the opposite side of the gate pivot centre, the gate raising movement continues until the gate reaches the 180° or maintenance position, and the latches are engaged to gate pins (6), Fig. 5, to lock the gate in position.

60. The gate is lowered back to the fully open position again by reversing the above procedure. Here again the latches must bypass pins (5) to engage with pins (4) for the shift mechanisms to move the gate back over the dead centre position.

FREQUENCY GATE CLOSURE

61. Closure of the Barrier against a tidal surge is unlikely to be necessary more than once or twice per year initially, although tidal surges will become more frequent in later years. This frequency of operation is inadequate to ensure that all the equipment and systems remain in an operational state and ready for operation. It is therefore intended that each gate is rotated to its closed position once per month, not only to prove that all the equipment and systems are satisfactory, but to ensure that the operating personnel are fully trained in the operational procedures. Raising the Rising Sector Gates to the maintenance position once per year is also intended for the same reasons.

APPENDIX

TYPICAL MACHINERY SCHEMES CONSIDERED

1. Many machinery schemes were investigated before adopting the machinery described in this paper. A selection of these schemes is shown on Figs. 7 to 11 and a brief description of each is given.

Dual Rocking Beam and Link Drive (Fig. 7)

2. This is a variant of the machinery adopted and differs from it in operation in that it could raise a Rising Sector Gate to the maintenance position without the use of auxiliary machinery.

3. Double-acting hydraulic rams (1) and (2) connect to rocking beams (3) and (4) through crosshead and guide assemblies to cause the beams to rotate about their pivots (5) and (6). Gate links (7) and (8) connect the opposite ends of the rocking beams to a shaft and bearing assembly (9) attached to the gate arm (10). The position of each machinery set relative to the gate pivot is such that when one link lies in line with the gate pivot and cannot exert a torque on the gate, the other link lies almost at right angles to the pivot and exerts near maximum torque, (see dotted outlines).

4. At the start of movement to raise the gate, both links drive the gate. As gate link (8) approaches its dead centre position, its hydraulic cylinder is made to free-wheel as with the present machinery, and when past the dead centre position the movement of the hydraulic cylinder is reversed to lower the rocking beam again and continue the raising of the gate. This procedure is repeated for the machinery driving link (7) when it passes over the dead centre position, and gate motion continues to the 180° or maintenance position. The procedure is reversed to lower the gate.

Wire Rope Drive (Fig. 8)

5. This machinery arrangement has five 90 mm diameter wire ropes (1) passing underneath and attached to each gate arm. Pulleys (2) above the gate arms guide the wire ropes to a large hydraulically operated winch (3) which incorporates twin 4.3 m (14 ft) diameter drums. A winch for each gate is sited at opposite ends of the pier.

6. The gate is rotated by winching the wire ropes on to one drum while paying them out from the other, and a tensioning device (4) equalizes the loads on the cables. Rotation of the gate through 180° was possible.

7. Auxiliary machinery for moving the gate through small angles to facilitate installation and replacement of the cables, and for locking the gate when the main machinery was shut down and proposed, but these were not developed to the practical design stage. Hydraulic rams (5) acting parallel to the gate were proposed for moving the gate, and a ram (6), acting at right angles to the other rams engaged or disengaged the unit to ports in the gate arms. Another ram (7) was proposed for locking the gate when ram (6) was disengaged and rams (5) were retracting ready for a further operational stroke.

8. As the wire ropes were partly in the water, corrosion at the splash zone was expected to limit their life to ten years maximum. Thus, during the life of the Barrier, twelve sets of wire ropes would require to be replaced at least six times, which would incur large costs and would reduce the reliability of the Barrier during those periods.

MULE IN STOWED
POSITION

MACHINERY
ROOM

MHWS

MLWS

MULE
IN USE

GATE
PIVOT

SURGE
TIDE

ELEVATION

STANDBY
UNIT

MULE
IN USE

POWER
PACKS

PLAN

3760

SECTION C-C

SECTION A-A

SECTION B-B

65 DIA.
RACK PINS

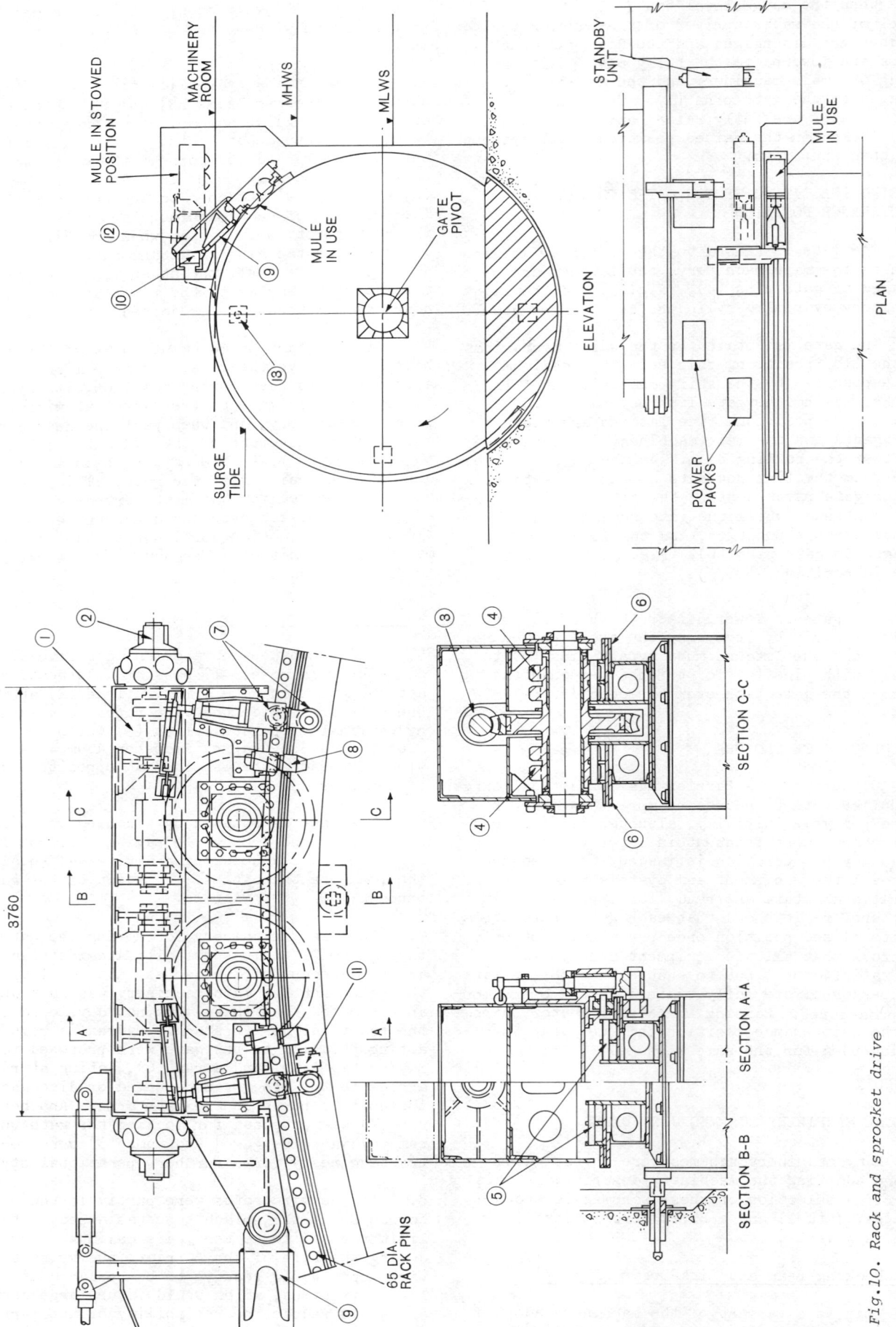

Fig.10. Rack and sprocket drive

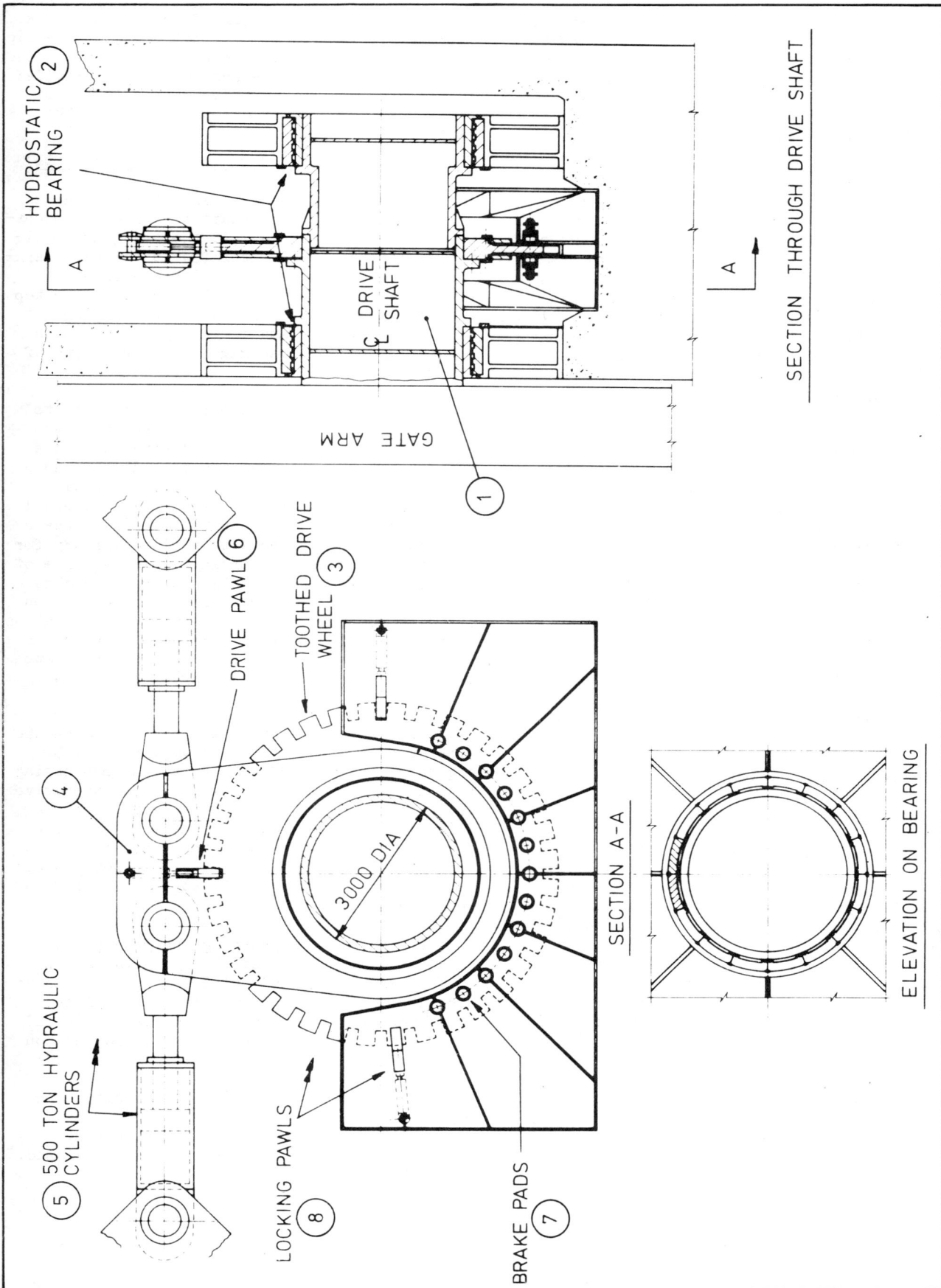

Fig.11. Hydraulic cylinder and ratchet drive

Wire Rope and Lever Drive (Fig. 9)

9.   This machinery scheme is a development of the wire rope scheme (Fig. 8). The wire ropes are sited clear of the water and by incorporating a pulley block arrangement, the number of wires wound on to the winches was reduced from five to one, with a corresponding reduction in the torque requirement.

10.   Lever (1) is hinged to the gate arm at (2) and locked to it by a removable pin at (3). Hinged at the top of lever (1) is a pulley block arrangement (4) having two pulleys at each end. Bolted to each pier end is a matching system of pulleys (5). A wire rope is attached to one of the pulley brackets at (5) and passes over the pulleys at (4) and (5) to form a four wire pulley block, and from there over another guide pulley to the winch drum of winch (6). A similar arrangement connects the opposite side of lever (1) to the other winch unit.

11.   To rotate a gate, the wire rope attached to one side of lever (1) is wound in, while that attached to the other side is paid out, pulling the lever from the position shown in solid line on Fig. 9 to position (7) shown in dotted outline. This represents a gate rotational movement of 90°. To move the gate a further 90° to the maintenance position, it is necessary to first lock the gate by mechanism (11), remove the pin at (3) and reverse the winches to move lever (1) back to position (8) where it is locked to the gate at position (9). The winches are then operated as previously to complete the gate rotation to 180° when link (1) lies in position (10). To lower the gate again the whole procedure is reversed.

Rack and Sprocket Drive (Fig. 10)

12.   The driving unit (1) (called a mule), for this proposal, is a double unit with hydraulic motors (2) mounted at each end. Each motor drives a worm and wheel (3) which in turn drives sprockets (4) at each side of the worm wheel. The sprockets engage with pin type racks formed round the gate arm, the racks being of sufficient length to allow 180° of gate movement. Flanges (6) protrude from the sides of the racks, and wheels (7) at each corner of the mule run on top and underneath the flanges locating the mule relative to the rack and taking the vertical reaction from the sprocket teeth. Rollers (8) running on the edges of the flange, locate the mule sideways. The tangential tooth load from the sprockets is transmitted through link (9) to beam (10) cantilevered out from the side of the pier.

13.   When not in use or for servicing, the mule can be lifted away from the gate arm by rotating the lower wheels (8) clear of the flanges (see 11) and raising the mule by retracting hydraulic ram (12). The mule can then be withdrawn into the pier. A locking mechanism (13) is provided to lock the gate in the 0°, 90° and 180° positions when the mule is disconnected from the gate.

Hydraulic Cylinder and Ratchet Drive (Fig. 11)

14.   The drive to the gate in this proposal is through the gate pivot shaft (1) which is mounted on two bearings (2). At the shaft mid position a toothed drive wheel (3) is bolted, flanked by two side plates (4) which are free to rotate on the wheel hubs. Connected to the top of these side plates are two double-acting hydraulic rams (5) and a drive pawl (6) which can be engaged or disengaged into the slots of the wheel.

15.   To rotate a gate, the pawl is engaged to the wheel, one ram retracts and the other extends causing the assembly and hence the gate to rotate through a small angle. At the end of the stroke the rams are unloaded, the pawl disengaged and the motion of the rams reversed to return them to their initial position where the pawl is re-engaged with the wheel ready for a further operational stroke. This sequence of operations is repeated until the gate reaches the required angle. Brake pads (7) bearing on the sides of the driving wheel, prevent rotation when the driving pawl is disengaged during the ratcheting operation. Pawls (8) lock the gate positively when the ratcheting operation is complete.

16.   The gates deflect under load and cause the gate arms to move by a small amount at right angles to the axis of gate rotation, preventing the use of a rigid connection between the drive shaft and the gate arm. Investigation into a suitable universal coupling which would allow for this movement, support the gate, locate it axially and transmit the drive, did not produce an acceptable design likely to last the life of the Barrier without maintenance.

17.   The bearings supporting the gate pivot shaft in Fig. 11, are of the hydrostatic type and a feasibility study of this type of bearing was carried out by the University of Salford. The diameter of the bearing was 4.27 m (14 ft) and was considered too experimental for use on the Barrier.

18.   The very low coefficient of friction obtainable with hydrostatic bearings was not essential for the Barrier, while a bearing failure might prevent a gate from being closed.

9. Services

S. W. PRATT, FIEE*

INTRODUCTION

1. The previous paper has described the machinery and hydraulic systems for driving the gates and the provisions to be made for alternative sources of hydraulic power through duty and stand-by power packs. The paper has also described the procedures to be adopted for driving a gate from one end only in the event of failure of motive power at the other end.

2. All of the power packs and auxiliaries have electric motor prime movers. Without electric power the Barrier would be completely immobile. Therefore the sources of electrical power and systems of distribution must have duty and standby capabilities at least equal to those provided for the hydraulic systems, and, together with back-up services, must have a basic reliability compatible with the viability of the Barrier.

3. The design and specification of the power systems and services is biased towards the conventional. The specifications stress that only well tried and proven equipment and components may be used and that prime importance must be given to reliability and easy maintenance.

POWER LOADS

4. The power demand loads for operating the Barrier have been assessed as follows:-

Base Loads

Shoreworks	- miscellaneous power, heating and lighting	600 kW
Piers, Abutments, Tunnels	- miscellaneous power, heating and lighting	500 kW
	Total Base Load	1100 kW

Peak Loads

Duty and Stand-by power packs starting	900 kW	
Duty and Stand-by power packs running		500 kW
Falling Radial Gates closing simultaneously	200 kW	
Rising Sector Gates closing at 1 minute intervals		900 kW
Fire Pumps		200 kW
Total simultaneous peak loads		1600 kW
Total Barrier peak demand load		2700 kW

*Specialist Associate, Rendel, Palmer and Tritton

POWER SOURCES

5. Three separate sources of power will be provided each independent of the others and capable of meeting the peak demand of the Barrier and so arranged that each source can act as stand-by to the one selected for duty operation.

6. The London Electricity Board will be providing two of the sources from separate connections from their networks on either side of the river. One 11000 volt connection will be made into a substation on the South Bank and will be an exclusive cable from a bulk supply station. All terminal connections will be arranged so that they will not be affected by flooding up to a level of 7.2 metres O.D. The other connection will be provided into a sub-station on the North Bank from the local 11000 volt ring main distribution system.

7. Although the separate connections of the London Electricity Board are considered to be firm under most conditions of operation, complete shut-down due to industrial dispute cannot be overlooked. Therefore the third source of power will be provided from diesel engine driven generator sets accommodated in a generator station on the South Bank. Three sets will be provided, each of 1500 kW rating; two sets will each feed half of the Barrier load with the third set on stand-by. Under extreme emergency conditions, and with suitable load shedding, one generator will be able to provide sufficient power to close the Barrier.

POWER DISTRIBUTION

8. Power will be distributed at 11000 volts from the shore sub-stations, through the duplicate subways and access tunnels, to two transformer stations on each pier and abutment. The transformer stations will be connected to medium voltage switchrooms located at the upriver and downriver ends of the piers and abutments. The medium voltage switchrooms will accommodate the switchgear, and motor starters for the gate machinery and for the services.

9. Alternate connections from the North and from the South shore sub-stations, which will be 'handed' on adjacent piers, and interconnections between the medium voltage switchrooms, will make power available to at least one switchroom on each pier in the event of isolation of one tunnel or failure of two of the three separate sources of power.

Thames Barrier Design. Institution of Civil Engineers, London, 1978, 111-116

111

10. A line diagram of the 11000 volt distribution system is shown in Fig. 1, and a cross-section of the tunnel unit in Fig. 2.

CONTROL OF THE GATES

11. Control of the machinery and of the operation of each gate will be initiated from local control panels, accommodated in a control room incorporated in the superstructure of each pier and abutment, and remotely from the Central Control Room in the Control Building on the South Bank.

12. Two local control panels will be provided for each gate. Interconnecting cables between the local control rooms via the access shafts and tunnels, and the connections to the machinery and hydraulic systems, will enable operation of the driving machinery at both ends, or alternatively at the north or south end, from the control room on either side of the gate.

13. Operation of the gates and machinery and also of the Barrier electrical power distribution system, will normally be carried out from the Central Control Room and control will only be transferred to the local control rooms in the event of system failure or for maintenance purposes.

14. All control systems will operate on 50 volts d.c. and will include electro-magnetic relays of the Post Office and British Rail approved types.

15. The control and indication equipment fitted to the local control panels will be repeated and integrated into one console in the Central Control Room. Supplementary panels will give the operator instantaneous information of the state of operation, gate position and fault conditions. The latter information will enable the operator to evaluate the fault and to act as follows:-

Carry on under surveillance or,

Switch from duty to stand-by equipment as appropriate or,

Abandon drive from the faulty gate end and switch to drive from the other end only.

A maintenance patrol will be on call for investigation of faults.

The state of the electrical systems will be represented on a mosaic mimic diagram built into a console fitted with equipment for the remote operation of control circuit breakers and the generating sets.

COMMUNICATIONS

16. The main Communication Centre will be adjacent to the Central Control Room on the 7th floor of the Control Building. The communication facilities will include:-

i. PABX telephone system for the South Bank buildings.

ii PAX telephone system connected to 200 extensions throughout the Barrier complex.

iii. Direct lines from Post Office networks to strategic positions in the Control Building, and Gate Houses on the North and South Banks.

iv. Direct Post Office lines from tide level sensors situated at remote sites downriver and upriver. Information will be displayed on chart recorders. Critical locations will have duplicate connections either by a secondary Post Office line or radio link.

v. Telex system.

vi. VHF radio telephone connections to patrol staff equipped with portable units. 'Leaky' feeder cables will be provided in subways, tunnels and piers.

vii. Public Address systems throughout the Barrier complex.

viii. Magnetic tape recording of all communication circuits during emergency conditions. The Public Address and VHF systems will also be used to augment the fire alarm system and to direct staff to predetermined stations. Local communication will be provided on the piers between various levels for control of the cranes.

17. The Post Office will bring lines as appropriate from both North and South sides of the Thames with routeing through the access tunnels as necessary. Duplication and separation will be provided wherever possible.

18. A separate emergency communication centre will be provided on the South Bank fitted with VHF equipment, Public Address equipment and a direct line to the Post Office Exchange. The gatehouses will be similarly equipped.

EARLY WARNING AND NAVIGATION CONTROL

19. Priority telephone lines through the Post Office networks will give direct connections to the Storm Tide Warning Service at Bracknell and other Tide and Weather Stations situated downriver and on the East coast. In addition there will be a duplicate circuit to Bracknell via UHF radio link.

20. An evaluation of tide and weather data indicating the possibility of a Barrier closure will initiate priority procedures for operation and control. These will include starting the hydraulic power packs on the piers, priming and proving hydraulic and control systems, starting generating plant and the selection of the duty source of electric power. At the same time Public Authorities will be notified.

21. The piers will be fitted with navigation lights at their up-stream and down-stream ends each comprising red crosses and green arrows to indicate navigable or closed channels. The navigation lights will have a range of approximately two miles and will be controlled from the

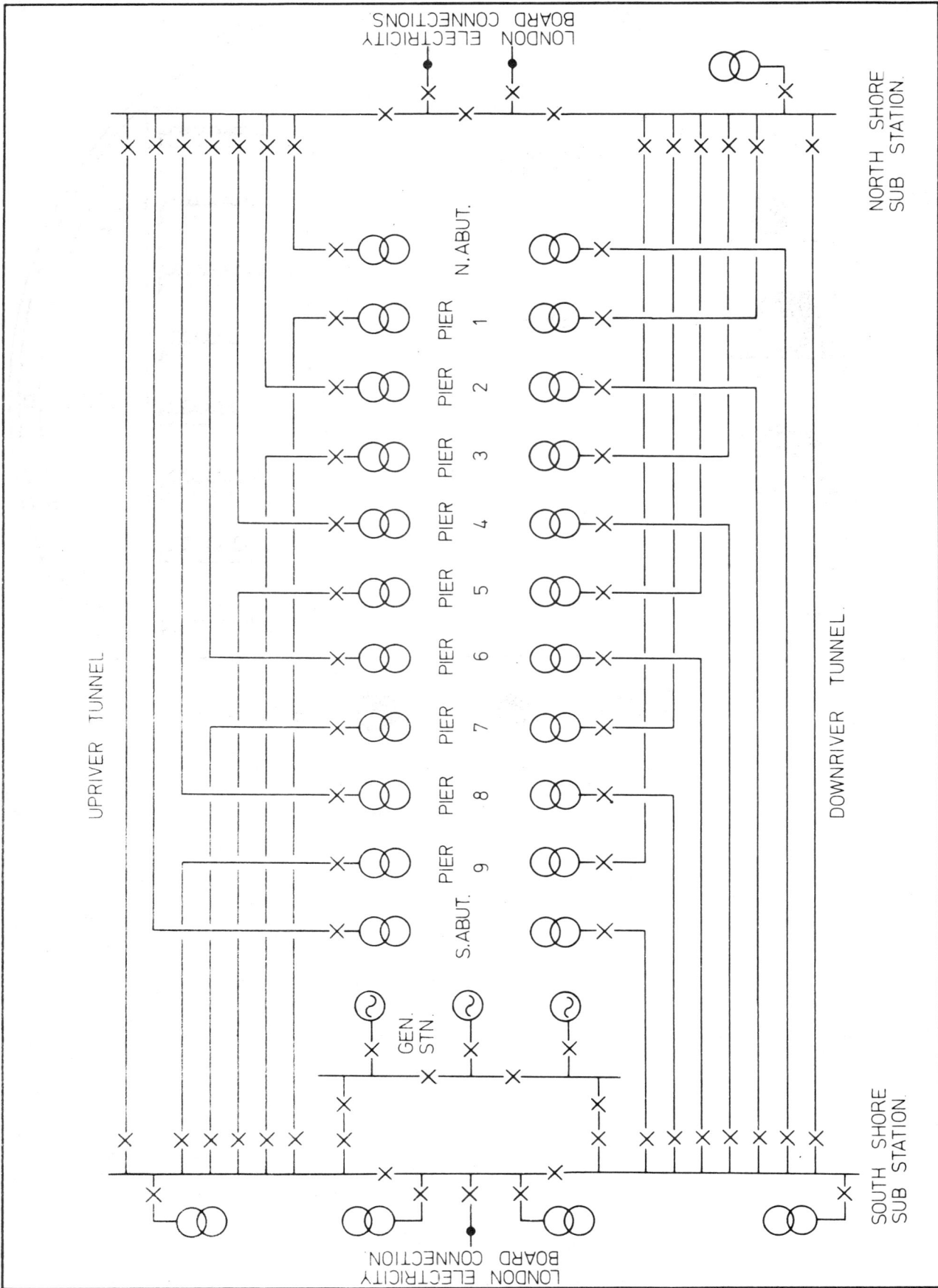

Fig.1. 11 000 volt distribution system

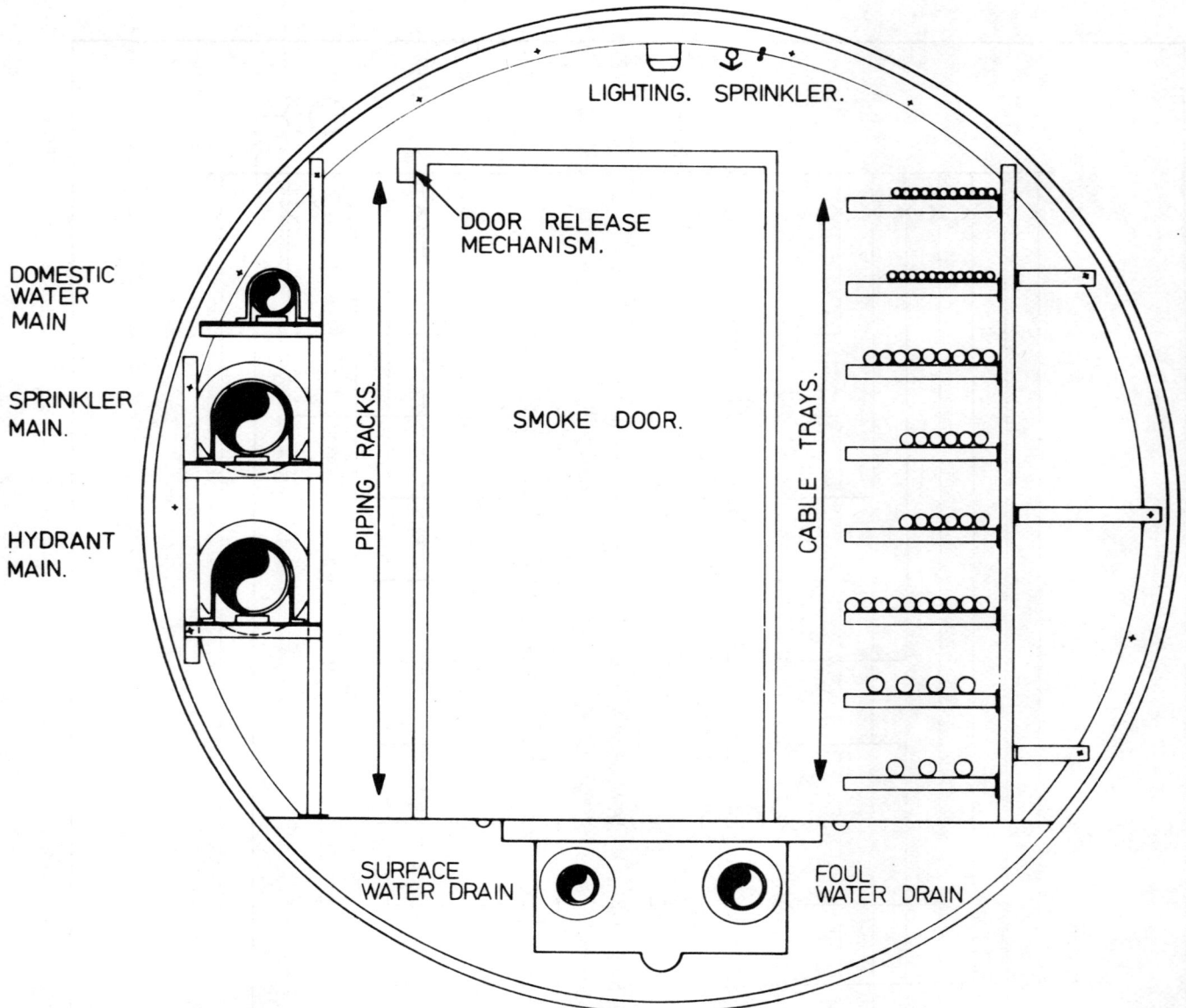

Fig.2. Tunnel section

PLA Control Centre situated on the South Bank near to the Barrier. The Barrier Control Room will have an over-ride control to switch the navigation lights to red under emergency conditions.

22. The PLA have extended their radio and radar systems to give optimum surveillance of the up-river and downriver approaches to the Barrier and in addition, they will establish illuminated signals at strategic points on the river banks to give advance warnings to river traffic.

23. PLA launches will be available during times of partial or total closure to give assistance or directions to river traffic as may be necessary.

FIRE PROTECTION

24. A period of not less than 15 minutes has been estimated by the London Fire Brigade from the time an alarm is given until they arrive at the scene of a fire on the piers. The G.L.C.'s security and maintenance personnel will maintain a 24-hour patrol of the Barrier, but the patrol time-cycle is unlikely to provide full surveillance for fire protection purposes. Therefore, to avoid major incidents, fires will be detected and suppressed automatically at source.

25. The design of the services and structures will incorporate materials to minimise flame propagation. Flame-retardant insulation and sheathing materials will be used for all cables. Bund walls will be provided around oil storage tanks and transformers.

26. The tunnels and subways will be fitted with fire-retardant smoke doors at the junctions with the piers and abutments and also at midway positions, to give isolation of tunnel lengths not greater than 30 metres. The doors will be normally open and will close automatically following the initiation of the fire detection equipment.

27. Discussions with the London Fire Brigade and other specialists have indicated that the most effective method of fire suppression in the

tunnels and piers would be water saturation, although gas suppression would be acceptable in enclosed areas where electrical equipment is located.

28. Three separate systems of water suppression will be provided:-

(i) Sprinklers

Sprinkler heads will be fitted in the tunnels, subways, access shafts, buildings on the North and South Banks, basements and underground car parks.

(ii) Water Spray Projectors

High velocity water spray projectors will be located at oil storage tanks, the hydraulic machinery, hydraulic pipe routes and at selected positions in the Generator Houses adjacent to fuel oil auxiliaries, tanks and pipes.

(iii) Hydrant and Hose Reels

Hydrant points for use by the Fire Brigade will be located at tunnel and subway intersections, on the piers and throughout the shore installations at entrances to buildings. Dry risers and hose reels will be provided in the Control and Workshop Buildings.

29. Water will be provided from the Thames Water Authority's network into two separate storage tanks, each of 170 cubic metres capacity, located under the Workshop on the South Bank. Duplicate pump sets will be provided in separate chambers at the east end of the Generator Houses at low level to ensure flooded suction connections.

30. Separate duplicated pipework systems will be provided for the sprinklers and water spray projectors, and for the hydrants. Pipe and valve interconnections in the piers and abutments will ensure hydrant supply in the event of isolation of tunnel sections. The hydrant pipework will be extended to the North Bank to provide direct connection to the Fire Brigade's mobile appliances and pumps, in the event of complete failure of the pumps on the South Bank. The systems will be maintained under minimal pressure through compressed air cylinders and jockey pumps; operation of a hydrant will cause a drop in the maintained pressure and initiate the automatic starting of the fire pumps.

31. Gas suppression by CO_2 or Halon as appropriate, will be provided in areas where electrical equipment is located or where water systems are impractical due to restrictions of space.

32. Operation of the sprinkler and water spray systems will be initiated by direct heat action on glass pressure bulbs, and the gas suppression systems will be initiated by smoke and/or heat detectors.

33. Main cable routes will include a recently developed small diameter cable which has an inherent characteristic of sensitivity to small changes in temperature and can initiate an alarm system.

34. Heat and smoke detectors will be fitted in all areas for purposes of alarm and surveillance and a zonal diagram panel will be situated in the Control Building to indicate the position of initiation of the water and gas systems, smoke and/or heat detectors, and manual call buttons.

PIER AND TUNNEL DRAINAGE

35. Rainwater on exposed decks on piers and abutments and the discharges from sprinkler heads and water spray projectors will become oil-contaminated and cannot be disposed directly into the river. Gullies and falls at all levels will connect in to main down-pipes which will discharge into underfloor sumps located at the tunnel junctions. The floor levels of tunnel sections will also fall to the sumps.

36. Each sump will be fitted with duplicate submersible pumps and sensing devices, and will be connected from separate electrical sources entering from opposite tunnel sections; high level alarms will be included. Water will be pumped through interconnecting pipes from one sump on one pier to the sump on the next pier towards the South shore where it will be collected in an oil separator.

37. Foul drainage from the piers will be handled and transferred to the South shore in a similar manner but in a separate system.

TUNNEL VENTILATION

38. Each section of the access tunnels between piers and abutments will be individually power ventilated via the vertical access shafts.

39. Two-stage axial flow fan sets will be located in Fan Rooms at the head of the access shafts. Single stage operation will provide a minimum of two air changes for transient occupancy of the tunnels, but if required, the ventilation rate can be doubled by operating the fans as a two-stage unit. When running as a single stage unit one fan will serve as stand-by to the other.

40. Provision will be made in the ventilation control system for the fans to be selected locally, or adjacent to smoke doors, to run as single or double stage units and to be reversible for supply or extraction as required.

41. The access subways between the abutments and the sub-stations on the North and South Banks, and the underground car park on the South Bank will be separately ventilated.

42. Operation of any heat and smoke detector will automatically close the smoke doors and shut down the ventilation fans associated with that section of tunnel. Similar arrangements will be provided for the access subways, car parks and other areas.

ELECTRICAL AND MECHANICAL SERVICES

43. Electrical services in the shore buildings, tunnels and piers will be conventional.

Illumination will be provided from fluorescent luminaires in most areas, but high bay metal halide units will be provided in the Generator Houses and Workshops. Security and emergency lighting systems will be incorporated in the general lighting systems but will be separately connected to priority circuits powered from stand-by battery inverter units.

44. The shore buildings will be heated by low pressure hot water systems feeding radiators and convectors, and will be fed from electrode boilers in the Control Building and Workshop.

45. Air conditioning will be provided for the Control Floor of the Control Building and in the local control rooms in the piers and abutments.

46. Space heating will be provided in the piers to maintain an internal temperature of not less than 5°C in an ambient of -10°C.

47. Manual 'break glass' contacts will be provided at salient positions to augment the Fire Alarm System and Security and Intruder Alarms will be incorporated.

48. The shore sub-stations, Control Building and the piers will be provided with steel piles for earthing purposes and the electrical systems will be connected to them.

49. The stainless steel roofs on the piers and other exposed steelwork, and similarly the shore buildings, will be bonded to earth via the structural reinforcement for purposes of lightning protection.

CATHODIC PROTECTION

50. Paper No. 16 on Protective Treatment, mentions the decision to augment the paint coating of the steelwork with cathodic protection. The design of the system for protecting the internal surfaces of the Rising Sector Gates had to take account of the complexity of the internal steelwork, which for the 61 metre gates and arms, comprises nearly 300 compartments formed by bulkheads and stiffener plates with a surface area of approximately 13,500 square metres.

51. These complexities led to the selection of a system of sacrificial anodes as the most practical method of providing protection, and since the water in the Thames can vary in resistivity according to season and tide, from about 55 to 5000 ohms/cm, the choice of anode material was limited to zinc. This choice was further governed by the possibility of silt accumulation inside some of the compartments burying the anodes.

52. The design aims at an anode life in excess of 15 years, and a screw clamp fixing cannot be relied upon to maintain an efficient electrical connection over this period. Therefore, a thermit welded bond cable will be fitted to each anode and welded to a non-stressed adjacent steel member.

53. A total of about 1100 anodes will be distributed in the compartments of each 61 metre gate and the total weight will just exceed 11 tonnes. Each anode will be limited to 10 kg to facilitate handling and installation.

54. Since the internal protection potentials can only be measured when the gate is submerged, provision will be made for monitoring the potential of the first installed gate by reference electrodes.

55. The external surfaces of the gates will be protected as far as practicable by an impressed current system from anodes mounted on the piers. Cathode connections will be brought from either end of a gate onto the adjacent pier by a flexible cable.

56. The anodes will be platinised niobium elements mounted in recesses in the piers where they will be protected from debris carried by the tidal current.

57. Automatic control of output will be provided to cater for the wide range of water resistivity.

58. The residual gap between the curved underside of a gate and the sill scallop, is insufficient to permit the practical installation and renewal of any form of anode which would be immune to mechanical damage when a gate is rotated. For this reason the underside of a gate is unprotected except for a short distance from the edge where current may spread into the gap.

ACKNOWLEDGEMENTS

59. The Author gratefully acknowledges the advices of Dr. L.L. Schrier on the Cathodic Protection of the Barrier gates and of Dr. R.H. Golde on the Lightning Protection of the Barrier structures and buildings; also the advices of the C.E.G.B. and London Fire Brigade on the Fire Protection in services tunnels and other high risk areas. Acknowledgement is also made of the kind cooperation of the London Electricity Board, the Post Office London Telecommunications Region and the Thames Water Authority on the provision and routeing of incoming services.

10. Structural model testing of a rising sector flood gate

P. J. DOWLING, BE, DIC, PhD, MICE, MIStructE*

G. W. OWENS, MSc, DIC, MICE†

INTRODUCTION

1. When the first Author of this Paper was approached by Rendel Palmer & Tritton in April 1972 to advise the Consulting Engineers on the feasibility of carrying out a programme of model testing for the Rising Sector Gates of the Thames Barrier, he recommended (ref. 1) that two models should be constructed. The first model to be constructed and tested was an Araldite model to an overall scale of 1:25, and the second was a steel model to a scale of 1:6.

2. The purpose of the smaller model was to provide an experimental check on the linear elastic finite element modelling used in the analysis of the actual gate. Previous experience in the Civil Engineering Structures Laboratories of Imperial College with this type of model, of such complicated structures as ships, had proved the usefulness of such models as design tools to complement, and indeed sometimes replace, expensive finite element analyses.

3. However, it was considered even more important to carry out an ultimate load test on a model of welded structural steel construction to obtain information on the load factors against collapse of the gates. The consequences of failure of this major structure would be so catastrophic, both in human and financial terms, it was deemed essential to determine the ultimate load-carrying capacity with a greater certainty than could be established using available theoretical procedures. As the design of the gates commenced at about the same time that the box girder bridge failures occurred, analytical and design approaches for predicting failure loads for plated structures were very much in a state of infancy and flux and so the need for experimental analysis was quite clear. The Consulting Engineers and their Clients accepted the recommendations.

CONSTRUCTION AND TESTING OF THE ARALDITE MODEL

4. Two scales were used in the construction of the small model. The scale used for overall dimensions was 1:25 while that used for thickness was 1:10. It was necessary to use a smaller scaling factor for thickness to ensure that the

sections used would be sufficiently large to enable sufficient control to be maintained over variations in thickness, and also to ensure that the behaviour would not be influenced by buckling of the components so that the response of the structure would be linearly elastic. It is permissible to use two different scale factors in this way as long as the structure is thin-walled. The difference in factors can then easily be accounted for by dimensional-analysis, as there are single scale factor constants for stresses, strains and deflections between model and prototype.

5. The material chosen for the model was epoxy resin casting Araldite 219 manufactured by CIBA. Previous experience with this material (ref. 2), had shown that its time-dependent properties were more satisfactory for testing than that of conventional modelling materials such as Perspex. It also has the advantage that it can be cast flat to any required thickness, and moulded to more complicated shapes. Alternatively, the plate can be moulded in its semicured state – a technique developed for modelling ships' hulls (ref. 2). This latter procedure was used for the curved face of the Araldite model.

6. In constructing the model all components present in the actual structure were reproduced, although certain idealisations in shape were made for particular components, e.g., where bulb flats were used as stiffeners in the prototype they were reproduced as ordinary flats in the model. The complex nature of the model can be seen in Fig. 1.

7. A total of 700 electrical resistance strain-gauges was attached to the model. This instrumentation had a considerable influence on the fabrication sequence. In all some 250 electrical resistance strain-gauge elements were attached to the inside of the gate to which no access was possible once construction was completed. It was therefore necessary at each detailed stage of the construction to consider the implications of placing a particular component, on the subsequent accessibility needed for placing gauges. In many instances construction was interrupted to place gauges and wiring.

8. Fig. 2 shows the completed model under test. As the aim of the test was to measure the linear elastic response of the model, a simpli-

* Reader in Structural Steel, Imperial College
† Lecturer in Structural Steel, Imperial College

Thames Barrier Design. Institution of Civil Engineers, London, 1978, 117-124

117

fied loading system of point loads applied at the intersection of the webs and cross-frames was used. Any pattern of load could thereafter be approximated by applying the principle of superposition. As the model behaved linearly with no comparative loss of stiffness caused by loading stiffened flanges in compression rather than tension, it was only necessary to load the model in one direction. It was decided to apply the load to the curved flanges, the direction in which the main head loading would be applied in practice. A few supplementary tests were carried out using patch loads to load the stiffened flange directly between the cross-frames and webs and measurements of torsional rigidity were made. Loading in each case was provided by the dead weight of frames surrounding the model. The frames were supported in the off-load position by a movable platform underneath the model. Further details of this test series are contained in ref. 3.

CONSTRUCTION AND TESTING OF THE STEEL MODEL

Model Fabrication and Instrumentation

9. The second model was built to a scale of 1:6, the minimum scale at which previous experience on box girder tests (ref. 4), had suggested that representative material could be obtained. At this scale realistic welding procedures could be adopted in the manufacture of the model, and imperfections representative of those occurring in prototype structures could be expected.

10. Because of pressure of work on the Imperial College workshops it was decided to subcontract the manufacture of the model to Research Models and Equipment Ltd. The Authors were responsible for the model idealisation and overall supervision of the fabrication.

11. The cellular nature of the model presented very serious access problems. It was therefore decided to construct the model in the form of sub-assemblies on which as much work as possible would be carried out prior to the final assembly.

12. Fig. 3 shows the model during assembly. On the extreme right can be seen a completed web sub-assembly, waiting to be incorporated into the main model. In the centre middleground a technician can be seen attaching strain-gauges inside one of the edge chambers. This had to be carried out after all welding in the vicinity had been completed, because of the risk of damage to the strain-gauges. It demonstrates the difficulties of access that occurred throughout the fabrication of the model. On the left an intermediate web sub-assembly is being aligned before welding it into position.

13. A total of approximately 1000 strain-gauge elements was attached to the model. The results of the earlier studies on the Araldite model were used to ensure that these were placed in the most suitable positions.

14. Throughout the fabrication process demec readings were taken on portions of the structure to record the growth in residual strains. From these readings an estimate could be made of the residual stresses induced in the structure by the cutting and welding operations. On completion of the model, a representative selection of geometric imperfections was measured.

15. The fabrication and instrumentation of the model Sector Gate and one gate arm, was completed in six months.

Proof Test of Sector Gate under Reverse Head Loading

16. The loading condition for the first test on the structure is shown in Fig. 4. This loading condition is not in any way concerned with using the Barrier for the flood prevention, but looks at the possibility of its use for amenity purposes, where it retains a certain amount of water in the upper tidal basin. The upstream level is determined by environmental considerations, being the minimum level to cover satisfactorily the foreshore; the downstream level is based on a low water spring tide level.

17. The Consultants requested that a proof loading of not less than the dead load plus 1:5 times live load be applied to the structure in this reverse direction, i.e., putting the flat side in compression. It was a requirement of this test that no permanent damage be done to the structure that might affect its behaviour in the main load test to collapse.

18. Under the reversed loading, due to the holes in the flat side, the pressure is resisted by the curved skin, which is in tension from the overall bending. Because the absolute magnitude of the local bending stresses is small, their effect on the tensile strength may be discounted, and it becomes feasible to idealise the loading as a number of discrete point loads.

19. It would strictly correct to apply these loads to the curved skin, but the undesirability of attaching tension lugs to the curved side (in compression in the main test), together with the fact that the stresses induced by the point loads are small and quickly dissipated away from their point of application, led to the solution shown in Fig. 5.

20. The model was tested with the flat side downwards and horizontal. For this test the gate arm was omitted. Three of the jack's lines of action are convergent on the centre of curvature of the curved skin, and to minimise congestion in this area, the model was positioned so that the centre of curvature was below floor level.

21. The reaction system was designed to measure separately the horizontal and vertical components of reaction. The vertical system of tie rods was arranged so that the reactions from individual webs could be monitored. The horizontal reaction was provided by the sloping beams shown in the foreground of Fig. 5. These were adjustable so that after each increment of load, the

Fig.1. View of Araldite model during placing of curved flange longitudinal stiffeners

Fig.2. General view of Araldite model under test

Fig.3. General view of steel model under construction

SECTION AT CENTRE OF GATE SPAN

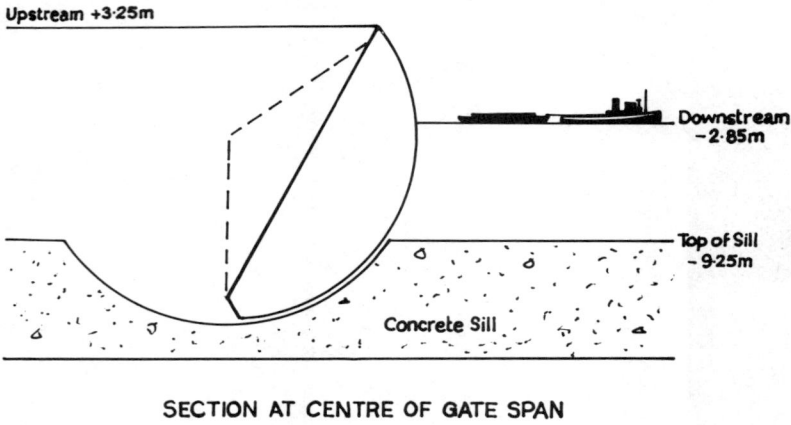

Fig.4. Reverse head loading for steel model

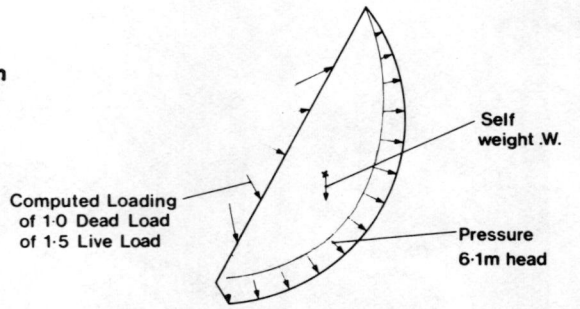

LOADING DIAGRAM REVERSE HEAD TEST

Fig.5. General view of steel model during reverse head test

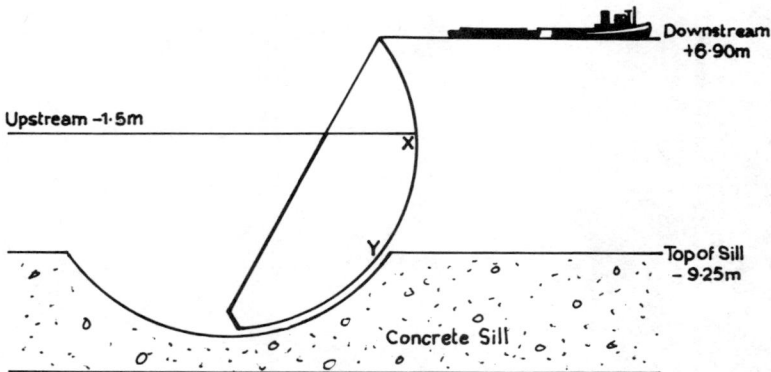

SECTION AT CENTRE OF GATE SPAN

Fig.6. Main head loading for steel model

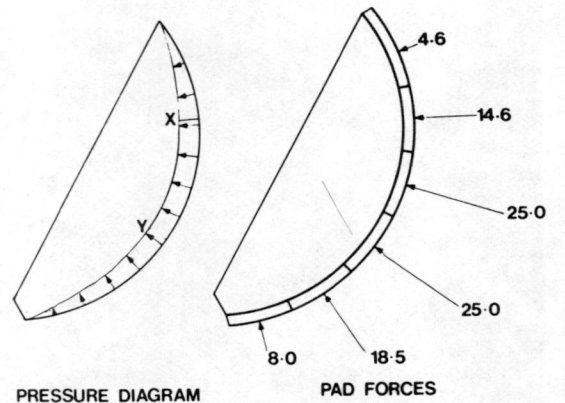

LOADING DIAGRAM MAIN TEST

120

model could be returned to its no sway position. Without this adjustment it is likely that the vertical reactions would have been distorted.

Main Head Test on Sector Gate and Attached Gate Arm

22. The loading condition for the main head test, shown in Fig. 6, arises when the Barrier is being used for its primary function of flood prevention. The pressure diagram presumes seepage between the gate and the sill.

23. Under this main flood prevention loading, the pressure is resisted by the curved surface which is in compression from the overall bending. Because of the possible interactive nature of failure of the curved surface under in-plane and lateral loading, it was considered essential to reproduce the true distributed pressure over all the curved surface. Not only was it required to apply the pressure correctly to the undeformed geometry, but the medium used had to be capable of maintaining that pressure as plate and stiffener buckling developed.

24. The two solutions for applying a distributed load were a bag and a pad; an example of each is shown in Fig. 7. A bag (on the left) basically consists of a steel plate to the edges of which a reinforced latex rubber membrane is attached. The void thus formed is filled to an average depth of 50 mm with water. It is loaded at its centroid. It is capable of maintaining a uniformly distributed pressure even when the loaded surface is substantially deformed. The principle disadvantages of a bag are that it cannot apply a pressure gradient to the surface and it is both costly and difficult to make.

25. A pad consists of a steel plate, suitably stiffened to ensure that it does not significantly deform under load. To the steel plate, timber profiles are glued to reproduce the undeformed shape of the pressurised surface. The pressurising medium is an open-celled synthetic foam. Pads are capable of applying a pressure gradient and are cheaper and easier to manufacture than a bag. Their main disadvantage is their inability to maintain a consistent distribution of lateral pressure as the loaded surface deforms.

26. It was decided therefore that pads would be used in the support regions which were considered to be less critical, while the bags would be used on the centre regions where buckling was expected.

27. Fig. 8 shows the model test. In the foreground can be seen the provisions made to ensure overall lateral stability (the model is 4 m above the test floor because the Sector Gate is now attached to a gate arm at the far end). At the left-hand end of the model can be seen an array of pads. These are loaded by tiewires that pass through small holes in the model. In the central portion of the model can be seen the complex rig associated with the loading bags. Each bag was loaded by an individual jack, all the jacks reacting against an arch structure; numerous straps and an outside frame were

necessary to ensure the stability of the entire system.

28. The structure failed at the quarter span position, close to the change in skin plate thickness and stiffening. Fig. 9 shows an overall view of the curved side after failure and Fig. 10 shows the final deformed state of the centre web, taken after the model had been cut into sections. Further details of the tests on the steel model are contained in ref. 5.

SUMMARY OF RESULTS AND CONCLUSIONS

Araldite Model (ref. 3)

29. The model was shown to be a representative model of the prototype gate and was further shown to produce a linear elastic response for the range of loads used in the test. In particular, the scaled second moment of area of the model was within 5 per cent of the prototype value.

30. The model test results demonstrated the high transverse bending rigidity resulting from the cross-sectional configuration of the model. Much of this stiffness was provided by the Vierendeel action of the transverse diaphragms, as the stresses measured in the diagonal bracing members were very low indicating that little truss action was taking place.

31. Comparison between the model behaviour and that predicted by a finite element analysis based on the same idealisations as used for the early analysis of the prototype, showed that adjustments were needed to the original finite element idealisation to give good agreement with the observed behaviour. In particular, adjustments were needed to the treatment of transverse rigidity of the curved skin plate to account for the arching action observed to occur in the model; this had been neglected in the early analysis.

32. This arching action of the skin reflected the contribution to shear stiffness made by the sloping walls of the curved skin. Model results suggest that the skin contributed up to 23 per cent of the total shear stiffness, the remainder being contributed by the webs. A further consequence of this arching action was to produce high corner reactions on the outer supports at each end.

33. After the necessary modifications were made by the Consultants to the finite element rigidities to cover this arching action, the agreement between model results and analytical results was generally good. Fig. 11 shows a typical example.

34. One consequence of the high transverse rigidity displayed by the model was the extreme sensitivity of the distribution of reactions to lack of fit-up in the bearings. It was recommended that this aspect needed careful consideration in design.

Fig.7. Alternative methods of applying distributed load: fluid filled loading bag (left); foam rubber loading pad (right)

Fig.8. General view of steel model under main head test

Fig.9. View of half model after failure with centre of span in bottom left hand corner: contours drawn on local co-ordinate system at 2, 4, 8, 16 and 32 mm; solid lines represent inward deflexions, dotted lines outward deflexions

Fig.10. Portion of central web of model taken after failure and after sectioning: contours drawn in similar way to Fig.9

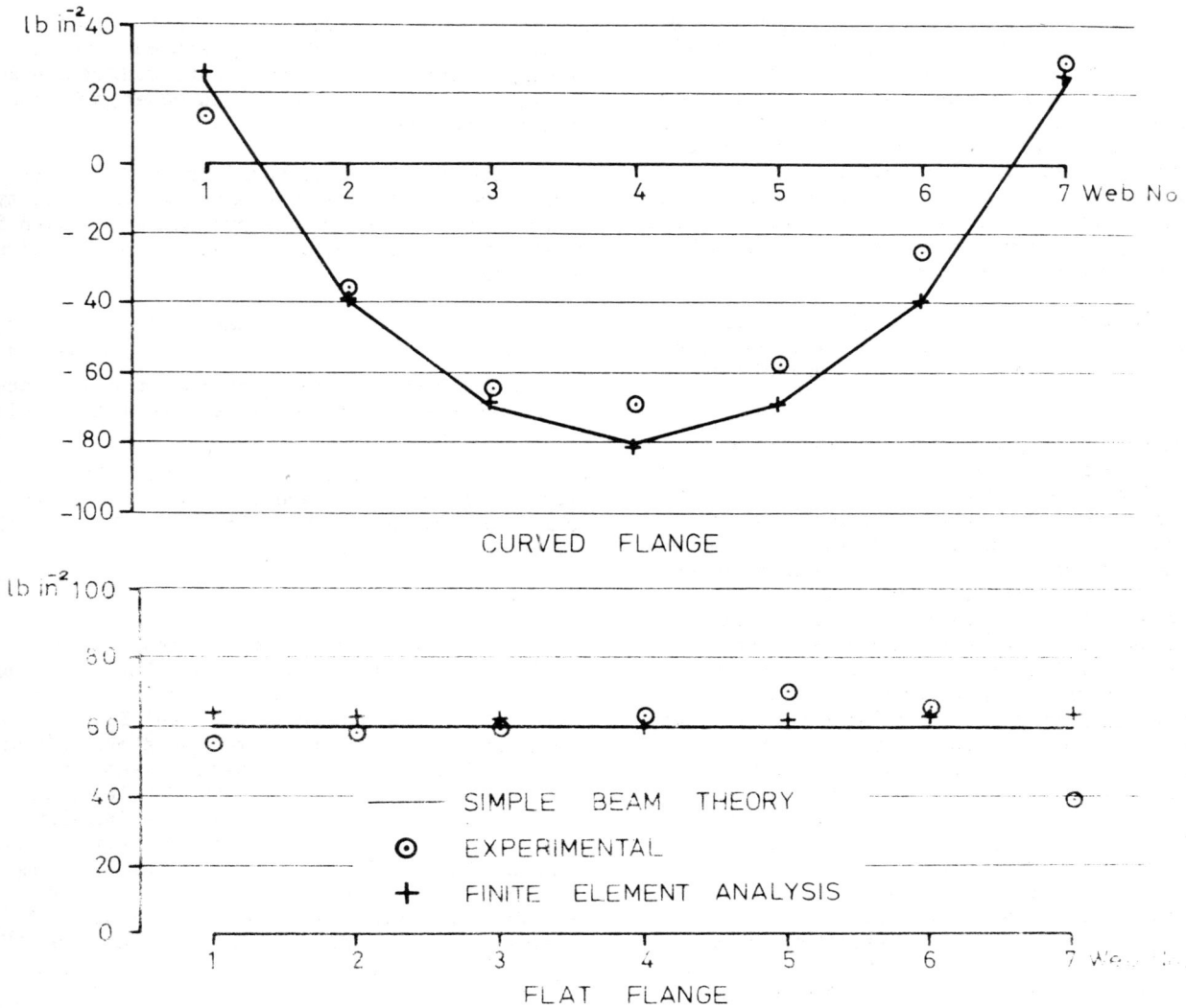

Fig.11. Araldite model: comparison of extreme fibre stresses for mid-span uniform movement test

35. The distribution of shear stresses over the depth of the webs seemed to be affected greatly by the presence of the openings in the webs. Maximum web shear stresses were measured in panels which were at the same level as the openings next to the supports.

36. Results of the torsion test showed that the gate possesses a considerable torsional rigidity which can be conservatively estimated by thin walled torsion beam theory, neglecting the contribution of the inner webs and accounting only for the outer plates.

37. It was concluded that sufficient experimental evidence had been made available for comparison to establish the validity of the finite element analysis used in the final global analysis of the prototype gate.

Steel Model (ref. 5)

38. The maximum initial lack of flatness of the flange stiffeners in the direction of stiffener span was 1/623 of the span towards the stiffener outstand and 1/358 of the span towards

the plate. The compressive residual stresses measured in the curved skin plates were an average of 114 N/mm². In each case the imperfections were somewhat larger than would be expected in the actual gate.

39. The steel model tests confirmed the findings of the Araldite model tests with respect to the distribution of loads in the bearings at the hinged end of the gate.

40. The model withstood a reverse head loading of 1.8 times live load, and 1.2 times dead load without any significant reduction in overall rigidity.

41. In the main load test, collapse of the model was caused by buckling of the curved skin at the quarter-point nearest the gate arm (at the change in section) under a main head loading corresponding to 2.17 times dead load and live load. Subsequent calculations by the Consulting Engineers confirmed that the critical section under the applied loading was the quarter-point rather than the centre.

42. The curved flange collapsed inwards, i.e., with the plate panels in compression, despite the fact that the stiffener outstand yield stress was very much less than the plate yield stress at the failed section. When this factor is considered, together with the influence of the water pressure, and the effects of the imposed overall curvature, there is strong evidence to suggest that this would be the likely mode of failure of the curved flange of the prototype under factored main head loading.

43. A final test on the model, (see ref. 5), to determine its shear strength, showed that the model had a load factor of 2.34 against shear failure.

44. Early yielding occurred in the end web panels adjacent to the web openings as a result of the high stress concentrations which occur in these zones, even under a load factor on dead and live load of 1.0. It was recommended that the possibility of fatigue failure under repeated loading should be checked for the prototype. If unserviceability limit load is defined as that load which produces permanent deformation of the structure which interferes with the satisfactory performance of the structure and necessitates major repair, it may be concluded that the model gate was still serviceable at loads exceeding the factored unserviceability loading. This was so despite the presence of local yielding in the webs, as this localised yielding had no noticeable effect on the rigidity of the structure. It was recommended that the deflections expected in the prototype under service loading should be checked to ensure that a satisfactory clearance between gate and sill is maintained, and that the satisfactory working of the operating mechanism is not affected.

45. The mode of collapse of the webs was greatly influenced by the regular positioning of the openings along their length. In the final test the progressive spread of horizontal shear failure in the webs at the level at which the openings occurred, eventually precipitated collapse of the flanges when shear failure reached the change in cross-section at the quarter-point nearest the hinged end. It was recommended that consideration be given to eliminating the openings near the supports in the final design.

46. The measured stresses in the gate arm bearing casting and the gate arm itself, were all well within acceptable limits.

47. Although the model differed in some details from the exactly scaled prototype, the behaviour of the model is representative of that of the prototype. It could be used to compare the validity of available methods for calculating the ultimate strength of such a structure. This can be done by applying the various methods to estimate the strength of the model using model dimensions, measured material properties, measured imperfections, and the experimental

loading in the calculations. Those methods, shown to give a satisfactory estimate of the strength could then be used with confidence to design the prototype gate, and other gates of similar construction.

48. Although the results of the model tests cannot be directly extrapolated to the prototype without going through the procedure outlined in the previous paragraph, the general indications were that:

(a) the load factors against collapse of the prototype are likely to be more than adequate;

(b) high stresses would occur in the web panels under service load conditions unless the design was altered;

(c) the detailed design of the bearings at the hinged end required careful consideration to guard against the possibility of failure of the structure in that region.

ACKNOWLEDGEMENTS

49. The Authors would like to thank their following colleagues: Mr. Peter Carr who played a leading role in the tests on the Araldite model; Drs. John Harding and Paul Frieze who participated in the tests on the steel model; Mr. Jack Neal and the laboratory staff for their enthusiasm and skill; Mr. Ron Rappley and his workshop staff who constructed the Araldite model. Thanks are also due to Mr. Peter Guile and Mr. Nowell Scott of RME for constructing the steel model in such a short time and to such a high standard. Finally a special word of thanks is due to the staff of RPT and the GLC for their excellent co-operation throughout the project.

REFERENCES

1. DOWLING, P.J., Preliminary Proposals for Model Testing of Rising Sector Gates - Thames Barrier Project. Report for Messrs. Rendel, Palmer & Tritton, Imperial College, London, 1972.

2. DOWLING, P.J., Tests on a Model of the OCL Far East Container Ships. Report for OCL, Civil Engineering Department, Imperial College, London, 1970.

3. DOWLING, P.J. and CARR, P., Tests on an Araldite Model of a Rising Sector Gate. CESLIC Report TB1, Imperial College, London, June, 1974.

4. DOWLING, P.J. et al., The Experimental and Predicted Collapse Behaviour of Rectangular Steel Box Girders. Proceedings of the International Conference on Steel Box Girders. Institution of Civil Engineers, London, 1973.

5. DOWLING, P.J. et al., Tests on a Steel Model of a Rising Sector Gate. CESLIC Report TB2, Imperial College, London, April, 1974.

11. Hydraulic model studies of the rising sector gate: hydrodynamic loads and vibration studies

D. A. CROW, BSc(Eng)*

R. KING, PhD, BSc(Eng)†

M. J. PROSSER, MA, MIMechE‡

INTRODUCTION

1.   This Paper describes tests carried out at
the British Hydromechanics Research Association
Laboratories on a 1:20 scale elastically similar
(hydroelastic) model of the 61 m Thames Barrier
Rising Sector Gate (see Figs. 1 and 2 and ref. 1).
A brief summary is also given of work carried
out at BHRA on earlier Rising Sector Gate models.

2.   The primary objectives of the work on all
models was to evaluate the hydrodynamic forces
acting on the gate structures and to establish
the limits of the vibrational stability of the
gates to any flow induced vibration which could
occur.

3.   The first model was tested during the early
development of the Rising Sector Gate design and
was constructed in wood to a 1:20 scale. As
this could not model the elastic response of the
prototype, this preliminary test programme was
limited in scope, but nevertheless, was able to
indicate the conditions for which more detailed
tests were required. As a result of this work,
it was decided to construct a 1:40 scale model of
a section of the gate span together with a 1:20
scale hydroelastic model of a complete 61 m
Rising Sector Gate.

4.   The supports of the 1:40 scale model were
designed so that the model would vibrate in the
same manner as a mid-span section of the gate,
and at natural frequencies related to the primary
modes of vibration of the prototype (see Paper 7).
The hydroelastic model was designed to reproduce,
as closely as possible, the scaled dynamic behaviour of the prototype for structural vibrations, and included a spring system designed to
represent the stiffness of the operating
machinery.

PRELIMINARY TESTING USING THE 1:20 SCALE
HYDRAULIC MODEL

5.   From the preliminary work done with this
inelastic model of the gate, the following
observations were made:

(a)   For conditions of flow in the upriver
direction over the crest of the gate, two

distinct flow patterns were observed. These
depended on the gate position and the upriver
and downriver water levels, and are referred to
as the 'surface jet', which occurred for small
head differences across the gate, and the
'diving jet', which occurred for larger head
differences (see Figs. 4 and 5). In the case of
the diving jet, a further distinction could be
made between a ventilated and an unventilated
(clinging) nappe.

(b)   The change from surface jet to diving jet
was accompanied by a sudden increase in the torque tending to lift the gate. This effect is described as the hydrodynamic lift or hydrodynamic
lifting torque. During surface jet conditions
the flow separated from the gate crest; the
forces on the gate were basically hydrostatic
and the hydrodynamic lift was zero. For the
diving jet condition, the nappe curved downwards
more sharply and re-attached to the flat back of
the gate. The hydrodynamic lift was due to the
curvature of the streamlines accompanied by a
pressure gradient, resulting in an area of low
pressure in the separated region beneath the
nappe.

(c)   Sustained oscillations of the gate rotating
against the machinery ('mass oscillations') were
demonstrated. The tests were limited to one
gate angle (25°) but covered a range of stiffnesses of the restraining spring system simulating the machinery. They were carried out with
approximately 5 ft head difference across the
gate within the diving jet region. By raising
the natural frequency of the system, it was
possible to prevent mass oscillation.

6.   A detailed report on this work is given in
ref. 2.

FLOW PATTERNS INVESTIGATED USING THE 1:20 SCALE
HYDROELASTIC MODEL

7.   The various types of flow first demonstrated
in the earlier 1/20 hydraulic model, were confirmed by both the 1/40 sectional model and the
1/20 hydroelastic model. The variation in flow
patterns over the gate was shown by fixing the
gate crest height and downriver level, and progressively lowering the upriver level. With a
small head difference across the gate, the surface jet flow pattern occurred. In this case the
flow separated from the crest. The curvature of

---

*Senior Research Engineer, The British Hydrome-
chanics Research Association
†Senior Research Engineer, BHRA
‡Principal Research Engineer, BHRA

the free streamlines was small and so pressures on the gate were approximately hydrostatic.

8.   As the upriver level was lowered, a point was reached where the flow dived through the surface and re-attached at a lower point on the flat back of the gate (diving jet). The more pronounced curvature of the free streamlines gave rise to a low pressure region acting on the back of the gate, with a resultant torque tending to raise the gate. Further reduction of the upriver water level exposed the diving jet so that it became a clinging nappe. Finally, at some ill-defined upriver level, the jet became a ventilated free nappe.

9.   The point at which a clinging nappe becomes ventilated is not well reproduced on a reduced scale model, and in the experiments reported here the nappe was artifically ventilated during each test.

10.   If the upriver level was now raised, a point was reached where the flow pattern suddenly returned to the surface jet type, but in this case the change-over took place at a higher upriver level than did the change from surface to diving jet. The extent of the hysteresis effect was found to depend on the gate crest height and on the downriver level. By carrying out experiments at various downriver levels, a hysteresis region could be plotted for each gate crest height, (see Fig. 6).

11.   The upper and lower extremes of the two boundaries appear to converge, but this could not be fully examined during the tests. At the lower downriver levels the flow over the gate was so small that the changes in flow pattern became imperceptible, while the upper extreme was limited by the pumping capacity of the rig.

12.   During the tests to determine the change-over point between one flow pattern and the other, the water levels were changed carefully and gradually. Occasionally the change-over would occur earlier than anticipated due to some random disturbance in the approach flow. The boundaries shown should therefore be regarded as extremes achieved under near-ideal conditions. Practical flow conditions with random disturbances would tend to bring the two boundaries closer together (i.e. reduce the size of the hysteresis area).

FORCES ON THE GATE

13.   Because of the symmetry of the gate, the investigations could be treated as dealing with a two-dimensional system where only the forces acting in a cross-sectional plane of the gate need be determined. For any given crest height and flow condition, the gate will be in equilibrium between the forces acting on the gate due to hydraulic effects, gravity, and the restraining forces at the pivots and the operating mechanism (see Fig. 7).

14.   The forces acting on the gate (i.e. forces other than the restraining forces) may be classified as either static or dynamic:

(i) the static forces consist of the gravitational forces on the gate structure, Mg and $M'g$; and the hydrostatic or buoyancy forces acting on the submerged parts of the structure, B & B'.

(ii) the dynamic forces consist of the quasi-steady force, L, being a steady force due to the flow over the gate, and the unsteady forces due to turbulence in the flow.

These tests were confined to measurements of the cable tension, T, which simulated operating machinery restraint.

15.   Although no detailed investigation was made, the quasi-steady dynamic force L was attributed principally to the low pressure region under the nappe, and to a lesser extent to the reaction on the lower part of the gate due to the deflection of the flow.

16.   The variation of model cable tension with upriver water level was recorded for constant values of gate crest height and downriver water levels. A typical result is shown in the graph A of Fig. 8. The measured cable tension consisted of the force due to the quasi-steady hydrodynamic effect, which was correctly scaled, and the hydrostatic buoyancy force, which because of unscaled structural thickness, etc. was not to correct scale and therefore had to be the subject of separate adjustment.

17.   To determine the model buoyancy forces, tests were made to determine the cable tension as a function of water level under static conditions, i.e. with upriver, downriver, and water level inside the gate all at the same level. A typical curve is shown as B in Fig. 8. The curve shows that the still water cable tension increases as the water level rises, until a maximum is reached when the gate crest is submerged; i.e. the buoyancy force B reaches its maximum value. Further increase in water level causes a reduction in the cable tension since now only the buoyancy of the counterweights, B', is increasing. Referring to Fig. 7, it may be seen that the value of cable tension measured in this way will be almost equal to the buoyancy force in a flowing water test for the corresponding upriver water level. The fact that the downriver water level is higher does not affect the result, since the hydrostatic forces on the curved face do not contribute to the torque and hence not to the cable tension. The water level inside the gate was found to remain practically the same as the upriver water level.

18.   The procedure to obtain the required prototype values of hydrodynamic torque was to subtract the still water cable tension from the total measured cable tension (in model values), so producing the lower graph C of Fig. 8, and then to multiply this result by the force scale factor (equal to the cube of the linear scale factor). This value was then multiplied by the radius of the prototype gate arm to give the value of the hydrodynamic torque.

SCUPPER HOLE

WATER INLET PORT

NA

SECTION BB

1220

1012

100

106

6

6 6 6

12 6

a

a

a

a

6

6

6

6

610 RAD

NOTES

1 ALL DIMENSIONS ARE IN MILLIMETRES.

2 CIRCULAR APERTURES
MARKED a 30 mm DIA
MARKED b 30 mm DIA

3 MATERIAL : DARVIC (ICI)
ASSUMED YOUNGS MODULUS   3300 MN/m^2
ASSUMED DENSITY   1·36 × 10^{-6} kg/mm^3

4 MASSES
MASS OF DARVIC IN GATE   132·71 kg
MASS OF ADDED BRASS & LEAD   92·29 kg
TOTAL MASS OF GATE   225·00 kg
MASS OF EACH ARM   16·65 kg
MASS OF EACH COUNTERWEIGHT   112·50 kg
TOTAL MASS OF MODEL   489·30 kg

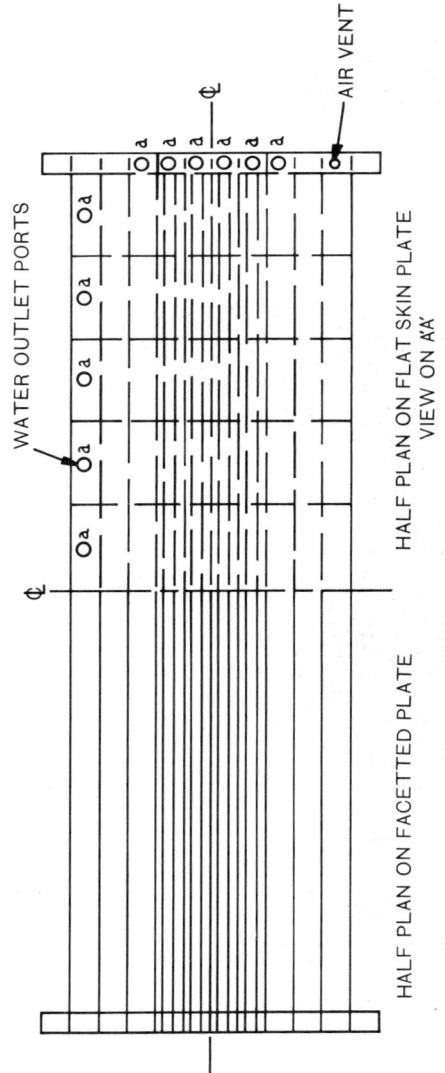

A'

A'

1220

270

340

114

75

310

310

310

ALL TRANSVERSE DIAPHRAGMS - 4 mm

RIGID CONNECTION

PIVOT ₵

4 mm

6 mm

4 mm

A

A

A'

A'

B

B

3250

SCUPPER HOLE

HINGED CONNECTION

₵ PIVOT

CENTRE OF GRAVITY COUNTERWEIGHT

B

B

75

310

310

310

75

114

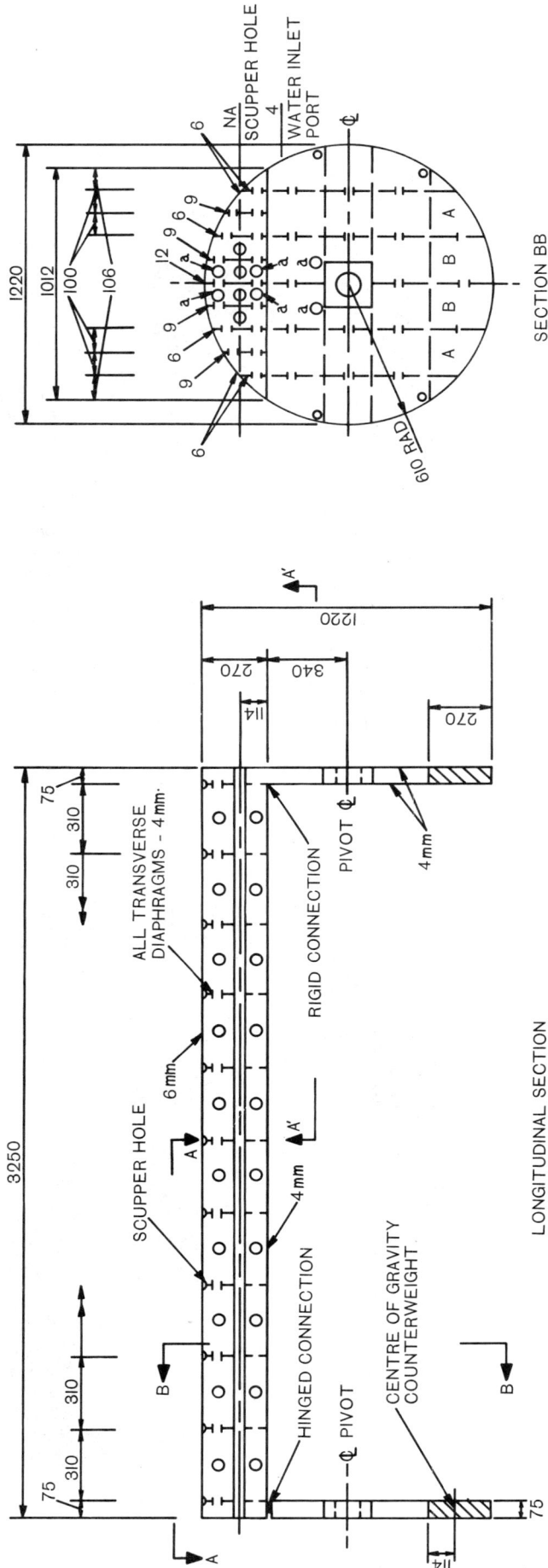

LONGITUDINAL SECTION

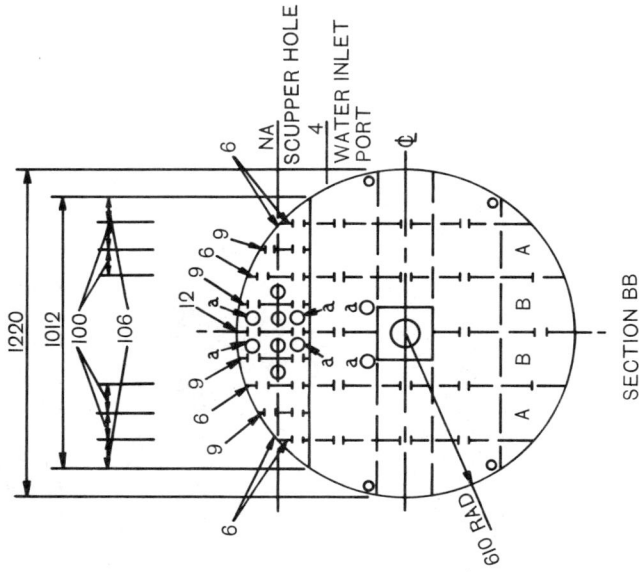

₵

WATER OUTLET PORTS

O a

O a

O a

O a

O a

₵

a

a

a

a

a

a

AIR VENT

HALF PLAN ON FACETTED PLATE

HALF PLAN ON FLAT SKIN PLATE
VIEW ON AA'

Fig.1. Hydroeleastic model of rising sector gate

(b) Interior of gate

(d) Installation in flume

(a) Gate during construction

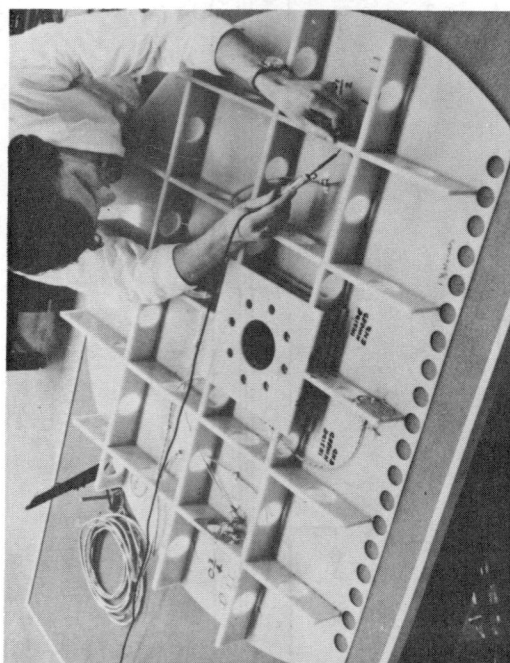

(c) Installing strain guages

Fig.2. General views of hydroelastic model

19. It was found that by plotting the maximum values of hydrodynamic torque against the head over the gate crest for a given height of gate crest, a smooth curve could be drawn through the points. Three of these curves are shown in Fig. 9. All the curves were of the same general pattern, but with a tendency to steepen as the gate crest height increased.

20. An empirical expression was obtained for the value of maximum hydrodynamic torque for the range of conditions covered by these tests:

$$T_{max} = (0.28H_c + 2.25)H^{2.28} \times 10^3 \text{ kN m}$$

(For gate crest between -1.2 m O.D.N. and +3.0 m O.D.N., the accuracy will be approximately ± 5,000 kN m)

T_{max} = max hydrodynamic torque (kN m)

H_c = gate crest height (m, O.D.N.)

H = head over gate crest (m)

VIBRATION TESTS ON THE 1:40 SCALE SECTIONAL MODEL

21. A dynamically similar sectional model was constructed to continue the studies of the vibration problems of the Rising Sector Gate. The model was designed so that the natural frequencies of the structural modes of vibration and the vibration against the machinery could be separately controlled, and a wide range of test conditions was covered in order to obtain a thorough understanding of the vibrational behaviour of the gate (ref. 3).

22. The cross-section of the model was to a scale of 1/40 but the width was limited to 1 ft, i.e. 1/5 of the gate span. The mass of the section was scaled and the elasticity for the two motions was provided by an external system of springs and linkages (see Fig. 3). Hydrostatic bearings were used to minimise damping due to friction. Tests were carried out on the tangential and radial modes separately and on the tangential and radial modes together. The tangential mode refers to the vibration of the gate against the machinery (mass oscillation) and the radial mode, the principal structural mode of vibration. The principal conclusions from these tests were as follows:

(i) Certain combinations of gate frequency (tangential mode), gate angle and upriver and downriver water levels, gave rise to unstable mass oscillations for flow in the upriver direction. The interaction between the gate movement and the flow patterns was similar to that observed on the 1/20 hydraulic model. The dependence on frequency was studied and a stability boundary was found in terms of frequency, for constant gate angles and water levels.

(ii) Damping due to the water (hydraulic damping) was low for the tangential mode and high for the radial mode. In the case of radial vibrations, the damping depended on the gate angle, or how much of the gate was in the sill. At gate

angles below about 30°, measuring from the gate open position, the damping was greater than critical, and no free periodic motion was possible.

(iii) Some random excitation of the radial mode occurred, but this was of very low amplitude.

THEORY OF HYDROELASTIC MODELS

23. The scaling criteria for hydroelastic vibrations of structures in free surface flow is obtained by means of dimensional analysis. It is argued that for free surface flow, various sources of excitation such as waves, will be scaled according to the Froude law, i.e., for a 1/20 scale model, velocities will be reduced by $\sqrt{20}$. Frequencies associated with the flow will be increased by $\sqrt{20}$.

24. The theoretical reasoning for the design of the hydroelastic model is given in Paper 7. In addition, an overriding requirement is that the material should be available in sheet sizes, thicknesses, etc, and amenable to normal fabrication techniques, so that the construction of the model is a feasible proposition.

25. The experience of the Delft Hydraulic Laboratory with a model of similar size and scale indicates that rigid P.V.C. is a suitable material for the construction of the model (ref. 2). "Darvic" is the I.C.I. equivalent of "Trovidur", the material used at Delft, and has the advantages of local availability, wide range of thicknesses, and more published data of its properties.

26. The relevant properties for the steel prototype (suffix p) and the P.V.C. model (suffix m) are:

Young's Modulus

$$E_p = 207,000 \text{ MN/m}^2 \ (30 \times 10^6 \text{ lb/in}^2)$$
$$E_m = 3,300 \text{ MN/m}^2 \ (0.48 \times 10^6 \text{ lb/in}^2)$$

Density

$$\rho_p = 7.85 \times 10^3 \text{ kg/m}^3 \ (0.283 \text{ lbs/in}^3)$$
$$\rho_m = 1.36 \times 10^3 \text{ kg/m}^3 \ (0.049 \text{ lbs/in}^3)$$

27. By increasing the model thickness to a little over three times that of the scaled prototype, the required model structural stiffness was achieved, and by adding non-structural mass distributed along the gate the mass was correctly scaled.

CONSTRUCTION OF THE HYDROELASTIC MODEL

28. The hydroelastic model, which was designed by Rendel, Palmer & Tritton, is shown in Fig. 1. The model was made of Darvic, with brass sheet and lead shot used to make the necessary density corrections. The lead was loaded into four tubes located on the neutral plane of bending, thereby not affecting the bending stiffness. Fig. 2 shows the model during construction with the brass sheets being attached to the diaphragms, details of the model during the in-

BEARINGS 1 2 3 ARE HYDROSTATIC

ROTATION (A) ABOUT THE BEARING 1 PRODUCES GATE MOTION IN A RADIAL DIRECTION. ROTATION (B) ABOUT BEARING 2 PRODUCES GATE MOTION IN A TANGENTIAL DIRECTION. FREQUENCIES IN THESE TWO MODES MAY BE ALTERED BY ADJUSTING THE SPRINGS k AND K.

THE GATE ANGLE IS CHANGED BY ROTATING THE WHOLE ASSEMBLY ABOUT AN EXTERNAL PIVOT COAXIAL WITH BEARING 2.

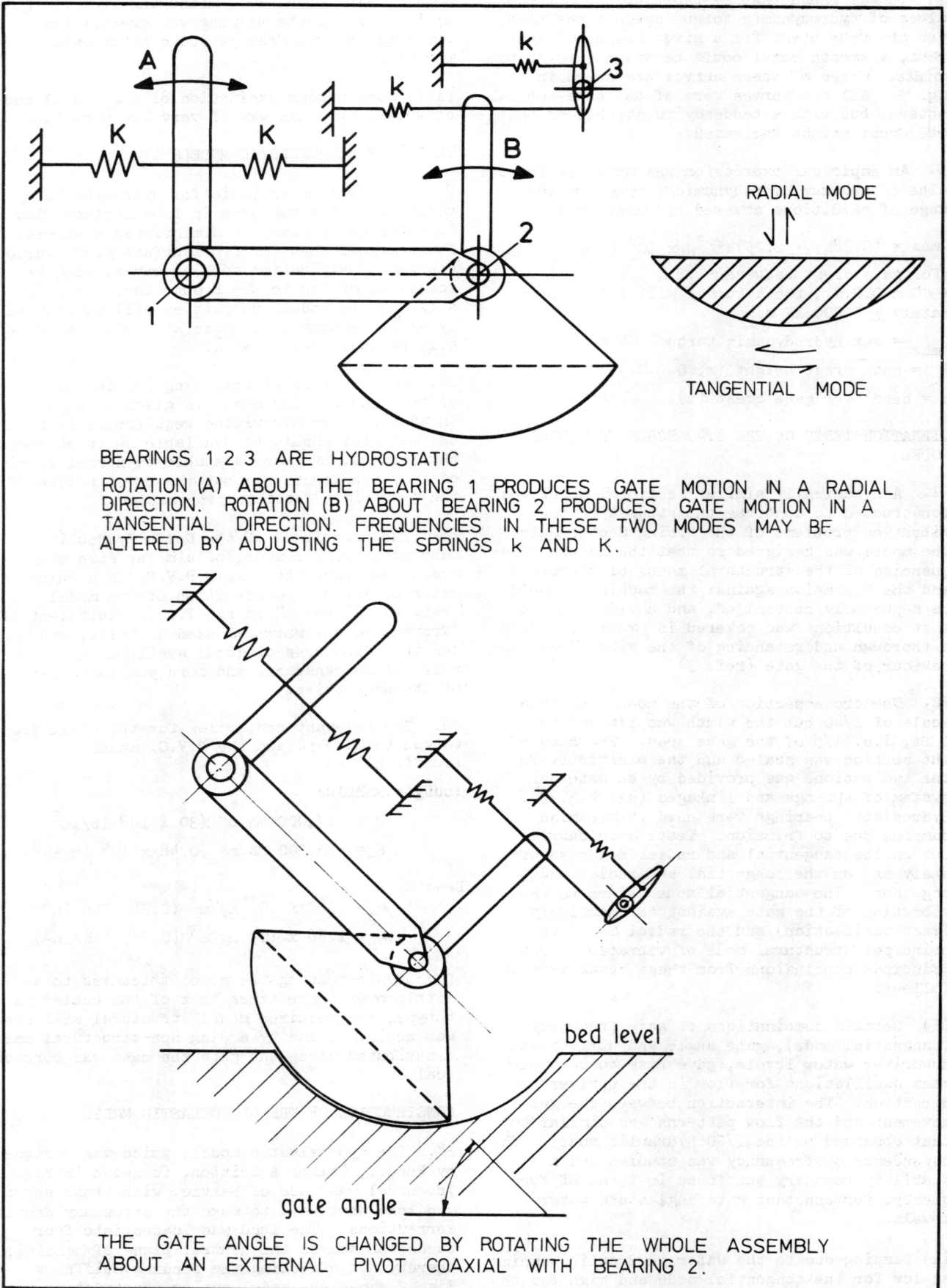

Fig.3. Sectional model

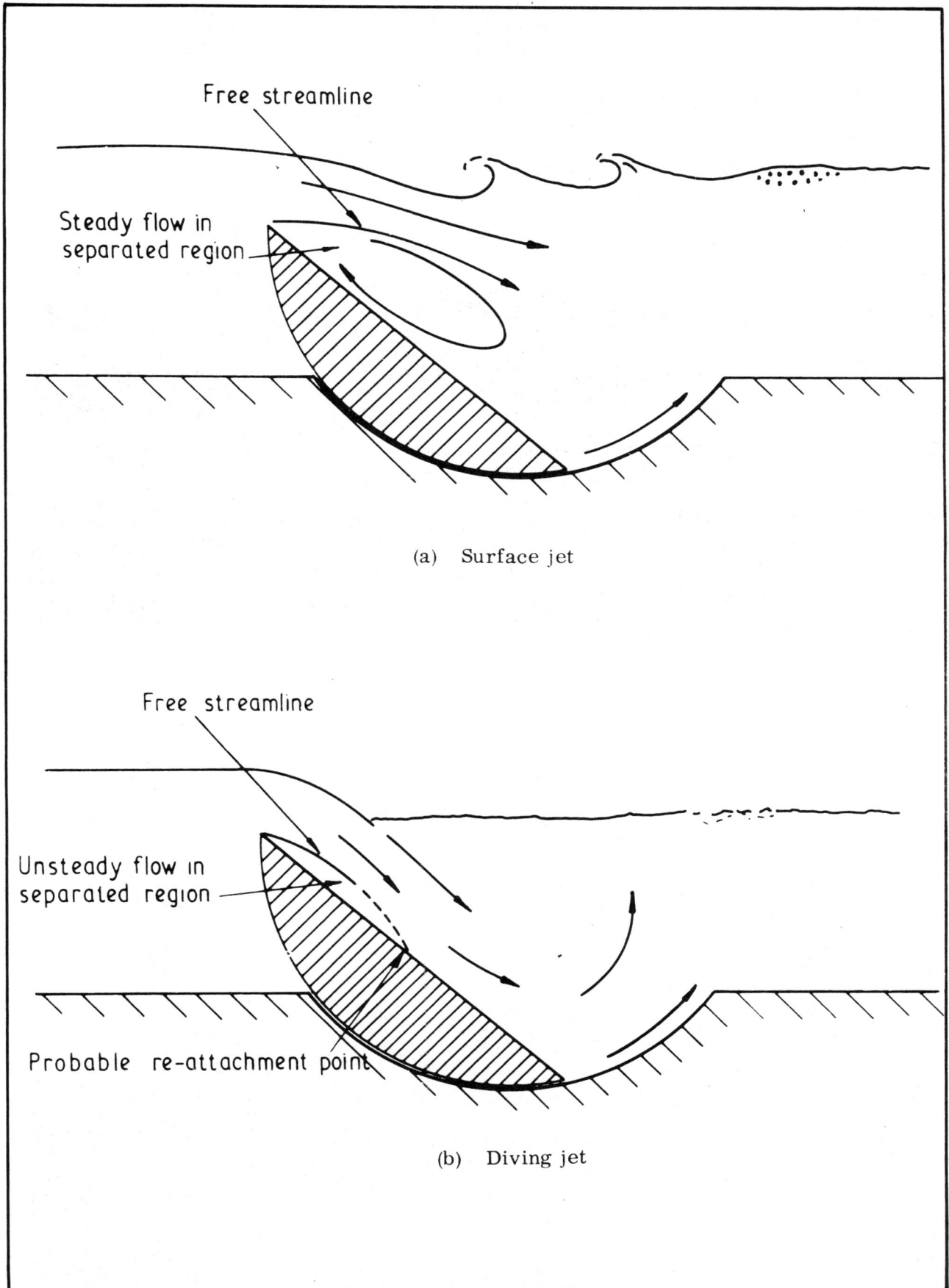

Free streamline

Steady flow in
separated region

(a)   Surface jet

Free streamline

Unsteady flow in
separated region

Probable re-attachment point

(b)   Diving jet

Fig.4. Flow patterns for surface jet and diving jet

Fig.5. Surface jet (top); diving jet (bottom)

strumentation stage, and a general view of the model being installed in the flume.

29. The model was supported in the flume on self-aligning ball bearings housed in the arms. The bearing shafts were held in massive steel frameworks bolted to the concrete floor of the laboratory. The gate could be retained in any required angular position by steel cables wound round the rims of the arms and attached to the overhead steelwork. Tension springs were fitted between the cables and the steelwork to represent the stiffness of the gate operating mechanism.

30. The test channel consisted of a 10 ft wide flume with an available flow rate of about 30 ft^3/s. Water levels upstream and downstream of the gate could be adjusted by means of two adjustable weirs: one varying the proportion of the total flow to be diverted to a bypass channel, the other controlling the downstream level.

31. The instrumentation of the model consisted of strain gauges bonded to various parts of the structure, accelerometers mounted at the centre of the gate, and displacement transducers on the rims of the arms.

PROVING TESTS ON THE HYDROELASTIC MODEL

32. Proving tests were carried out in order to be satisfied that the model would adequately represent prototype behaviour. Prototype mode shapes and frequencies obtained by mathematical analysis were available for comparison with model response (see Fig.8 of Paper 7). This data is for the gate in air.

33. As certain structural simplifications had to be made in the model, a similar analysis was performed for the model gate as designed, resulting in slightly different frequencies. The Table below compares the frequencies of the first three structural modes. The first column shows frequencies scaled from the values computed for the prototype, the second column shows frequencies calculated for the model as designed, and the last column shows the frequencies found by tests on the model. The differences shown in the Table and the methods used for calculating these frequencies, are discussed in Paper 7.

Table 1. Comparison between calculated and measured frequencies of first three transverse modes

	Frequencies Hz		
	$\sqrt{20}$ prototype	Model calc.	Model test
1st mode	6.48	6.09	5.5
2nd mode	12.25	11.68	9.2
3rd mode	18.4	16.92	15.4

34. The method of testing was to excite the various normal modes of the gate and so detect the mode shapes and to measure their frequencies. A normal mode of vibration of a structure occurs when all moving parts of the structure reach their maximum displacement simultaneously, or all moving parts of the system are moving in phase with one frequency. Ideally such tests would be carried out using an elaborate test rig comprising banks of servo-controlled hydraulic jacks coupled to a computer. The present tests did not justify the use of such sophisticated equipment and it was thought that sufficient information could be gathered using simpler and more direct tests. Two methods of excitation were used:

(i) Transient - in which the gate was given a sharp blow or suddenly released from a deflected position.

(ii) Driven - in which continuous periodic excitation was applied to one point of the gate. The frequency of excitation could be varied.

35. The response of the gate was detected by simultaneously recording the outputs of three strain gauges on a U-V recorder. The gauges used were chosen so that they indicated respectively the bending of the two arms and of the mid span of the gate. The simultaneous traces of the outputs indicated the phase relationships between the bending moments at the three gauge positions.

36. The first mode was most easily excited by pushing the counterweight of the hinged arm so that all three parts of the gate were made to vibrate at the same frequency. The phase relationship between the arms and the gate span indicated that at this frequency (5.5 Hz) the mode shape was the same as the first mode obtained by analysis, (see Fig. 8 of Paper 7).

37. Excitation of the second mode was more difficult and evidence of a frequency of 9.2 Hz could only be obtained by pushing on the counterweight of the fixed arm. For this excitation the fixed arm appeared to vibrate at 9.2 Hz while the gate span and hinged arm vibrated at 5.5 Hz. The mode shape could not be seen.

38. The third mode appeared to be excited by striking the centre of the gate span. In this case a clear signal was obtained from the gauge at the gate centre (15.6 Hz) while the signals from the arms showed components at 15.6 Hz and 5.5 Hz. Again, the mode shape could not be inferred from the trace.

39. Only the first torsional mode, i.e. the mass oscillations of the gate on the stiffness of the operating mechanism, could be excited by transient excitation. The outputs of the Linear Variable Displacement Transducers (L.V.D.T.'s) at the rims of the two arms, showed clear exponentially decaying oscillations. Both ends vibrated in phase at a frequency of 3.23 Hz. This frequency, corresponding to a prototype in-air frequency of about 0.7 Hz, was chosen at

Fig.7. Diagram of forces acting on gate

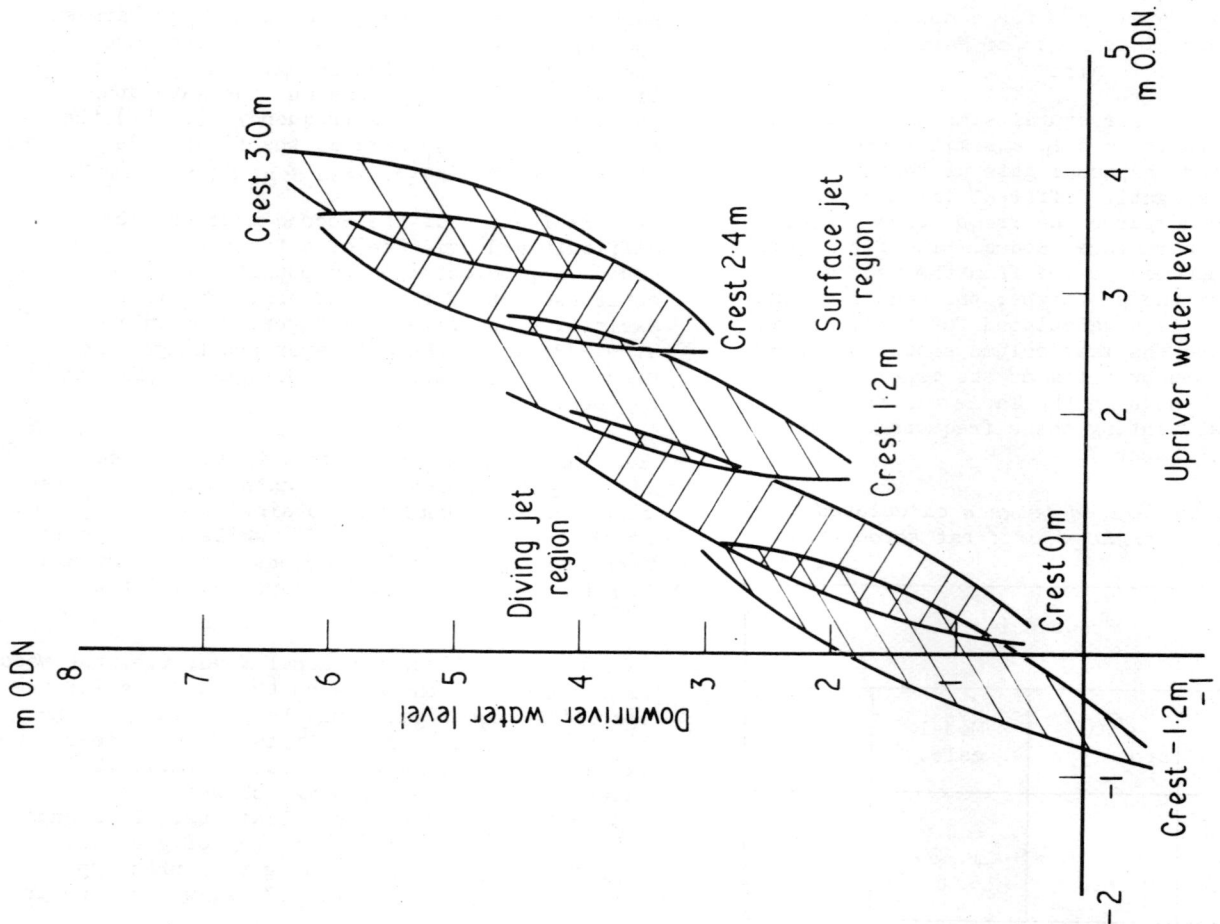

Fig.6. Hysteresis regions for various crest heights

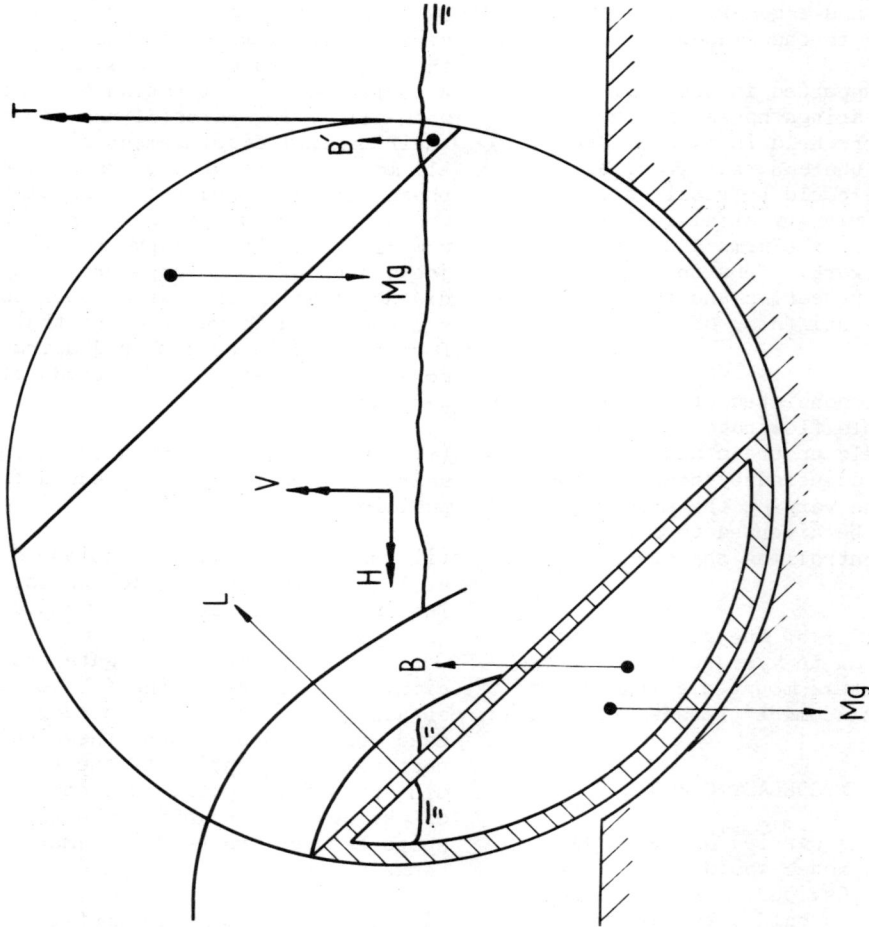

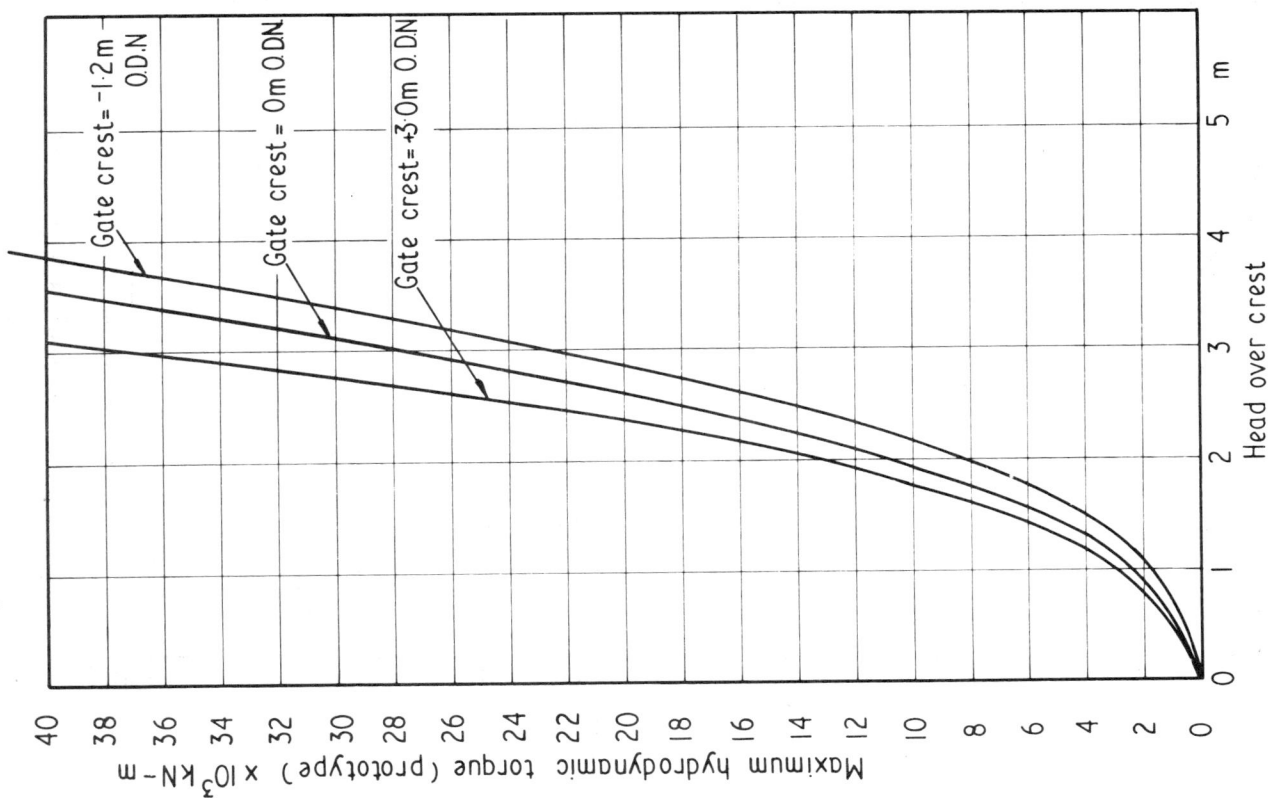

Fig.9. Maximum hydrodynamic torque plotted against head over crest

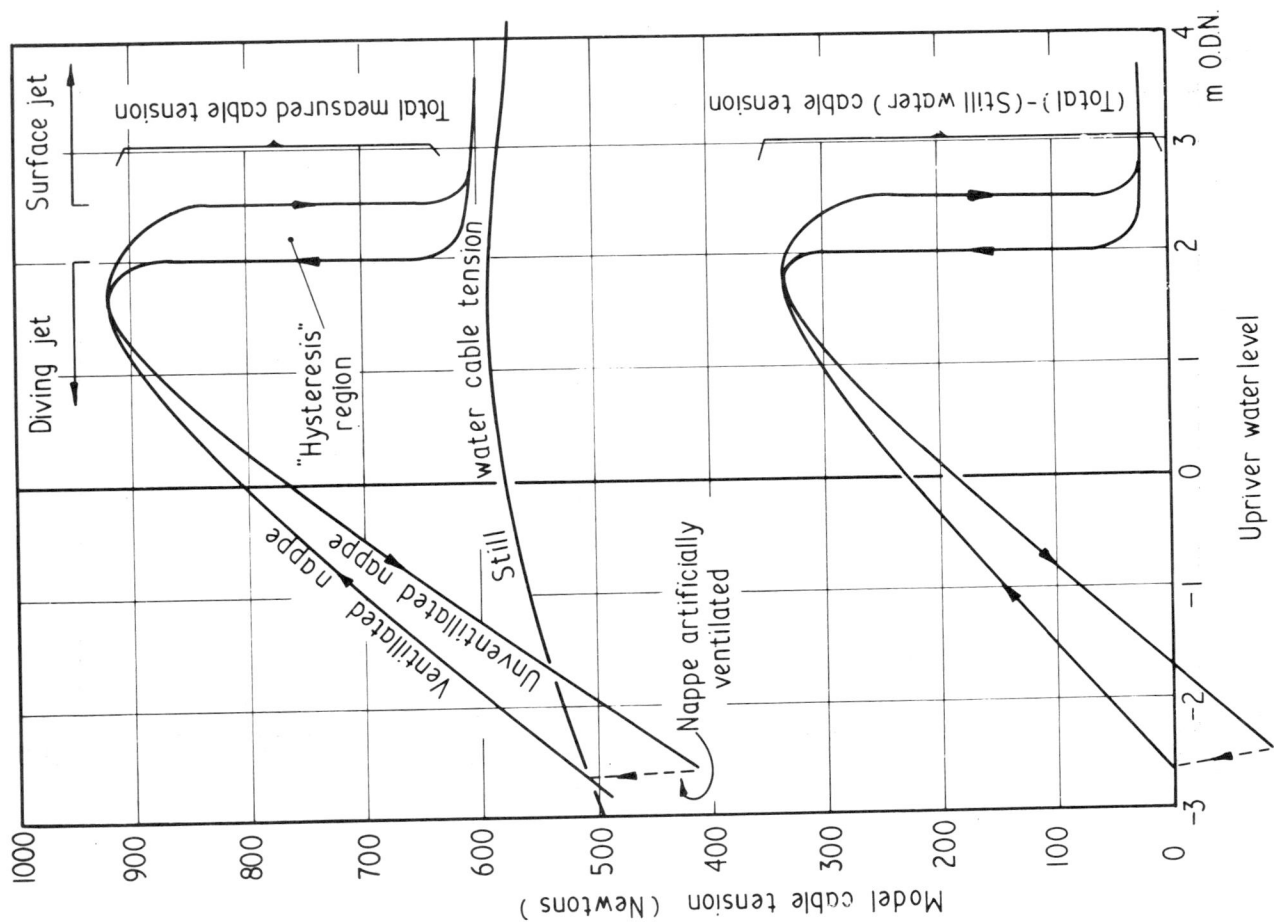

Fig.8. Typical curves of cable tensions measured on model

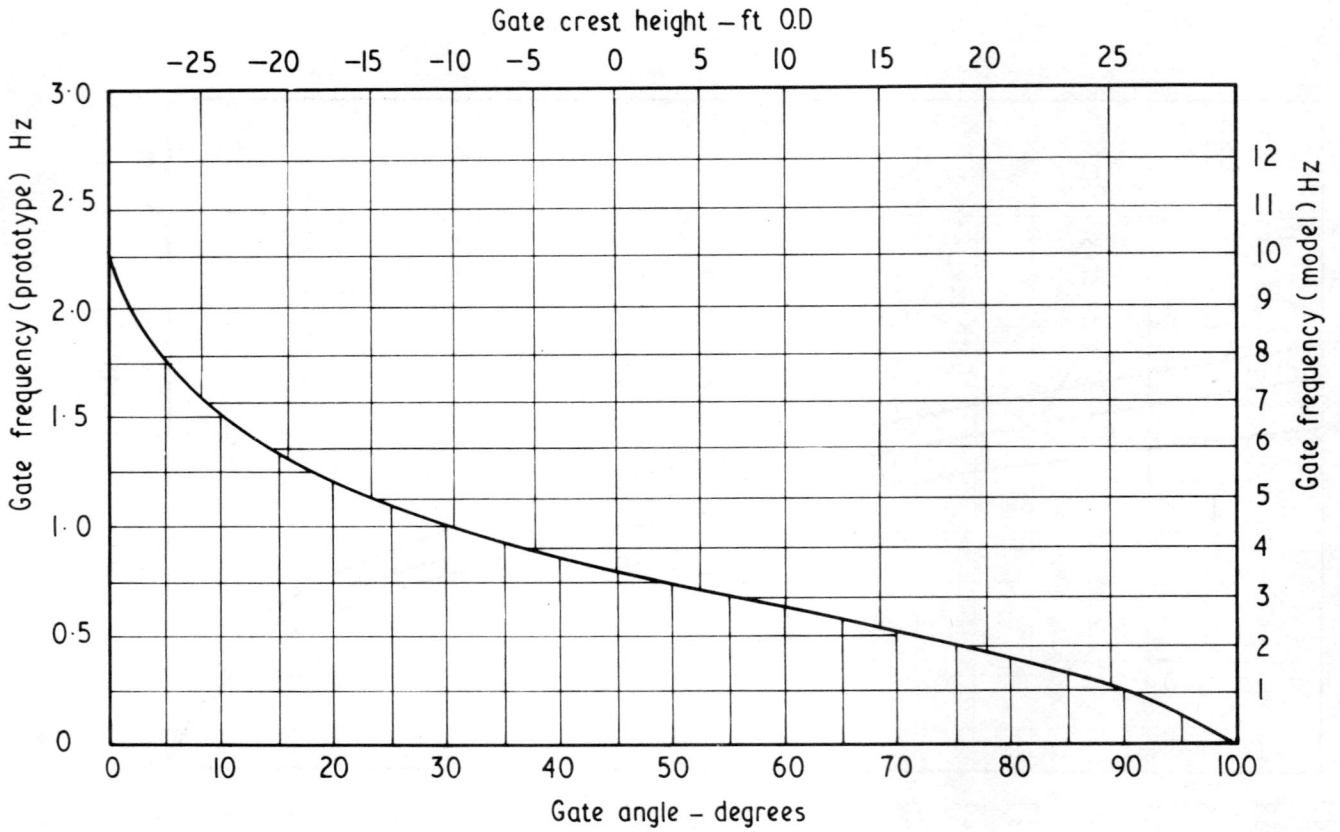

Fig.10. Variation of natural frequency of mass oscillations in air with gate angle

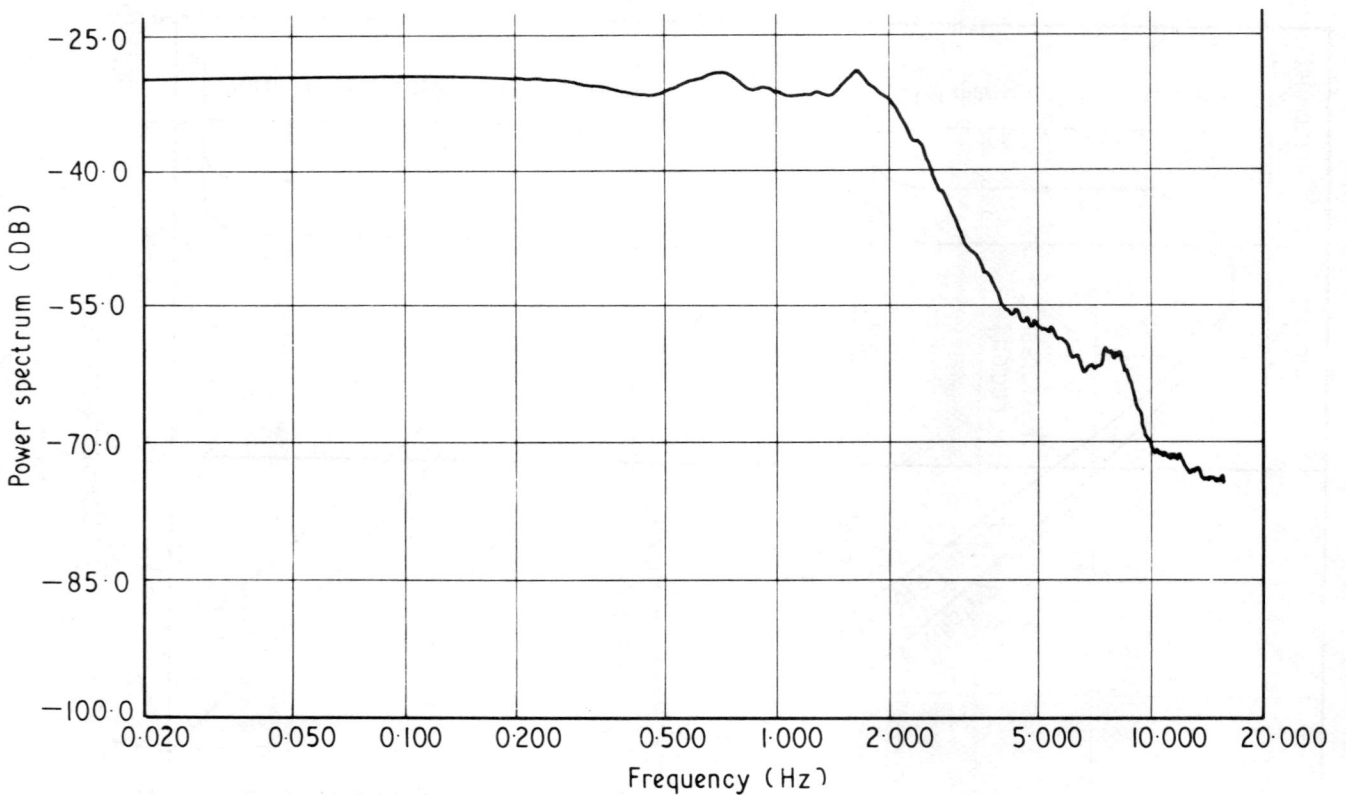

Fig.11. Typical power spectral density of transverse vibrations

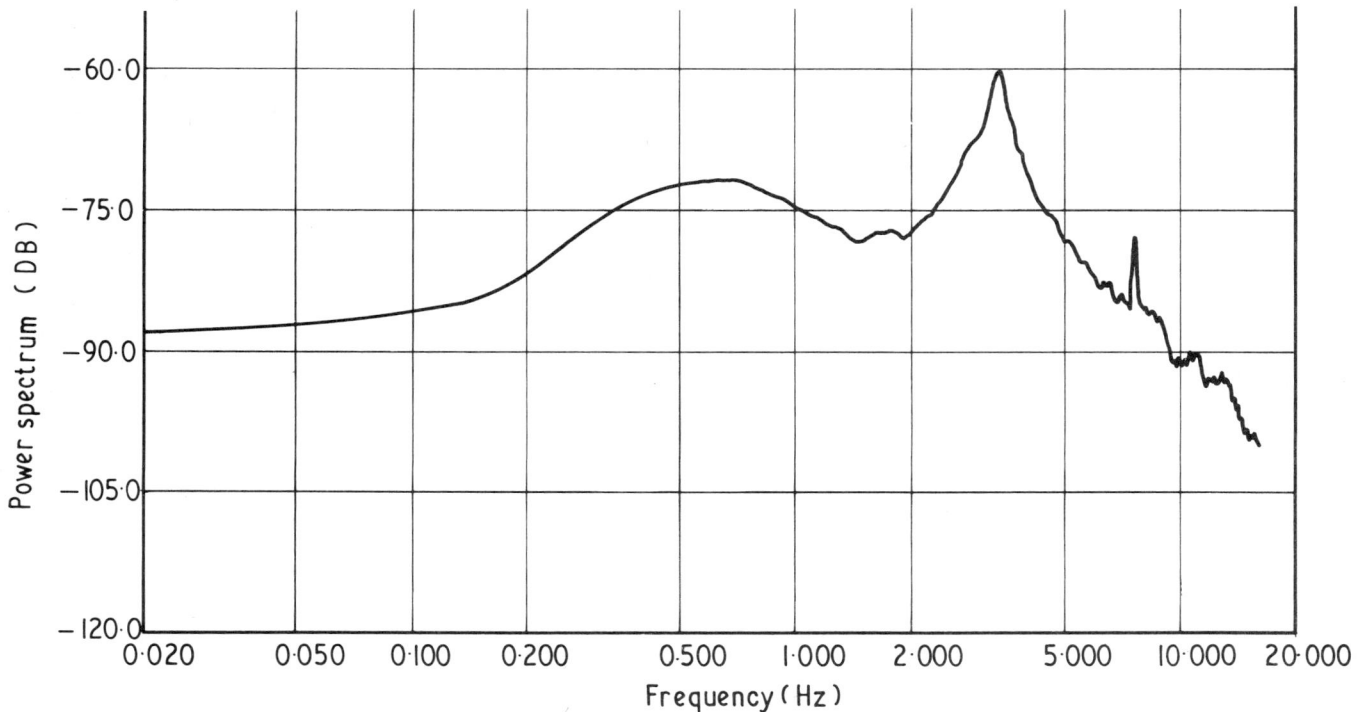

Fig.12. Typical power spectral density of torsional vibrations

the time as being a typical prototype condition, but was subsequently varied for the mass oscillation tests to cover the calculated range of prototype operating natural frequencies, see Fig. 10. With continuous periodic excitation, two resonances were found: the first at 3.1 Hz and the second at 13 Hz. The 3.1 Hz resonance was the natural frequency of mass oscillation of the gate on the suspension springs, with the two arms moving in phase. The second peak, at 13 Hz, had the two arms 180° out of phase.

40. Transient and driven tests were made with the gate in water, in an attempt to establish the mode shapes and frequencies for the submerged case and to relate these to the modes in air.

41. Tests on the transverse vibrations were generally unsatisfactory due mainly to the high damping effect of the water. In the case of the transient tests this meant that the traces decayed very rapidly so that frequencies could not be accurately measured, while for the driven tests, signal levels were low and not easily distinguished from background noise.

42. From the transient tests, frequencies at about 2 Hz, 4 Hz and 5.5 Hz were detected, with the 2 Hz frequency apparently corresponding to the first mode shape. The mode shape at 4 Hz was also fairly clear but was unlike any of the in-air modes recorded.

43. The torsional vibration tests were repeated with the gate totally submerged. The corresponding resonances were found at 2.2 Hz and 10 Hz.

TESTS ON THE HYDROELASTIC MODEL IN FLOWING WATER

44. Thirteen test conditions were chosen to give a representative coverage of the possible operating conditions. The response of the gate was detected by eleven transducers and the signals recorded on magnetic tape for subsequent analysis. A record was also made of the output of a water level recorder mounted downriver (upstream) of the gate. For each test condition, a 10 minute recording was made and subsequently analysed to provide graphs of Power Spectral Density in the frequency domain, and overall rms values. Particular gauges were calibrated to enable the rms values (in mV) to be related to an rms value of displacement.

45. Power Spectral Density analysis of the transducer signals enabled graphs of the distribution of energy in the frequency domain to be plotted. A peak, or concentration of energy in a narrow frequency band would indicate excitation of a normal mode in the structure. Typical examples of the P.S.D. curves are shown in Figs 11 and 12. With few exceptions the curves of all the other tests, which are not reproduced here, resembled that of Fig. 11 where the energy was seen to remain at a steady level up to about 1.5 Hz and then to fall away rapidly at higher frequencies. The general shape of the response curves resembled the curves obtained from the water level recorder, indicating that the response was mainly due to the changes in water level on the downriver side of the gate.

46. A peak in the torsional response, varying between 2 and 3 Hz, was observed and identified

137

as excitation of the first (mass oscillation) mode of torsional vibrations, (see Fig. 12). Although only a low level of vibration was recorded, the peak appeared consistently, and further investigation was considered advisable.

Further Tests on the Mass Oscillation (Torsional) Mode

47. The main conclusion drawn from the tests described so far was that some excitation of the mass oscillation (torsional) mode occurred for most test conditions. In view of this and the results from the sectional model (see Fig. 3 and ref. 3), a series of tests was performed to investigate the stability of the gate in the mass oscillation mode.

48. A blanket coverage of all possible conditions was considered to be unnecessary and the test programme was limited to the following conditions:

(i) Three gate angles; 45°, 60°, 75°.

(ii) Two natural frequencies; 0.36 Hz and 0.30 Hz, prototype in air. (c.f. 0.7 Hz of earlier tests).

49. A test procedure was therefore devised which would enable stability boundaries to be plotted for the two frequencies and three gate angles. For each condition to be tested, the gate was rocked backwards and forwards with an amplitude of 2 degrees. If the amplitude increased, the condition was classed as unstable, and vice-versa. For all tests the nappe was ventilated.

50. For the natural frequency setting of 0.36 Hz, a typical trace of the circumferential motion at the gate arm for a 'stable' condition is shown in Fig. 13. The amplitude of the motion varied in a random manner, but the predominant frequency was about 1 Hz, corresponding to the gate natural frequency in water. As may be seen, occasional peak excursions from the mean position were about four times the rms motion. In the particular test illustrated, peak to peak amplitudes of nearly 4 inches (prototype) were observed.

51. At the lower pre-set frequency 0.3 Hz, a wide range of unstable conditions was found and it was possible to draw a 'stability boundary' on the D/R against U/R graph, separating the unstable operating conditions, (see Fig. 14). From these tests, four main points were observed regarding torsional (mass oscillation) vibration:

(a) The most unstable condition occurred well into the diving jet region.

(b) For surface jet conditions, the gate was heavily damped.

(c) The vibrations were self-excited or self-controlled, i.e. the driving force occurred as a result of gate motion, so that an initial movement of the gate was required before sustained oscillations would take place.

(d) Aeration, or ventilation of the nappe was found to make the gate less stable. The change brought about by ventilating the nappe was greater when it resulted in a marked change in the trajectory of the nappe.

CONCLUSIONS

52. The main conclusions to be drawn from the model studies are as follows:

(a) A value of maximum hydrodynamic torque (which occurs for the diving jet condition close to the point where it changes to or from the surface jet) for given values of gate crest height (H_c) and head over the crest (H), may be obtained from the expression:

$$T_{max} = (0.28H_c + 2.25)H^{2.28} \times 10^3 \text{ kN m}$$

For gate crests between -1.2 m O.D.N. and +3.0 m O.D.N., the accuracy will be about ± 5,000 kN m.

(b) Low level random excitation of both transverse and torsional modes was observed. In the case of the 1st torsional mode (mass oscillation mode), the amplitude was found to increase as the natural frequency of the simulated gate/machinery system was lowered until finally an unstable condition was reached (see (c) below).

Typical rms amplitudes recorded were:

(i) Transverse modes (radial motion at gate mid span section) — 0.01 to 0.02 inches prototype

(ii) Torsional modes (circumferential motion at rim of arm) — 0.02 in. prototype at 0.7 Hz rising to 0.7 in. prototype at 0.36 Hz

It appears that peak amplitudes would be 4 times the rms amplitudes.

(c) At mass oscillation natural frequencies of 0.30 Hz and 0.36 Hz, regions of instability were found in terms of gate angle, downriver and upriver water levels. It was recommended that the gate natural frequency for mass oscillations should be greater than 0.45 Hz and that at this minimum frequency, the operating conditions should not exceed the limits indicated below:

D/R level ⩽ 2.7 m (9 ft) O.D.N.
or Head Difference ⩽ 2.7 m (9 ft)
        Gate Angle 45°

D/R level ⩽ 4.9 m (16 ft) O.D.N.
or Head Difference ⩽ 2.0 m (6 ft)
        Gate Angle 60°

D/R level ⩽ 6.4 m (21 ft) O.D.N.
or Head Difference ⩽ 3.0 m (10 ft)
        Gate Angle 75°.

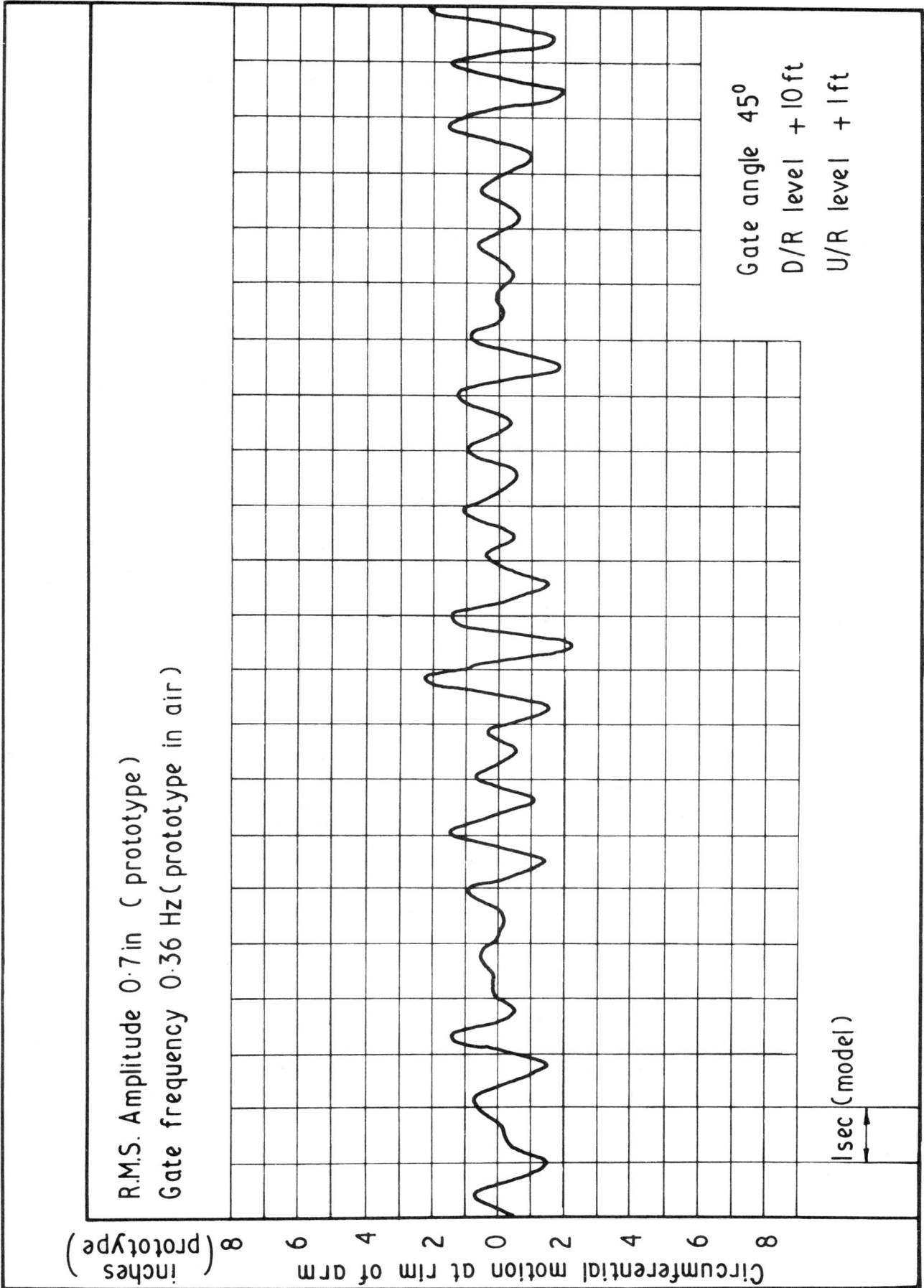

Fig.13. Typical U-V trace of gate response in first torsional mode

139

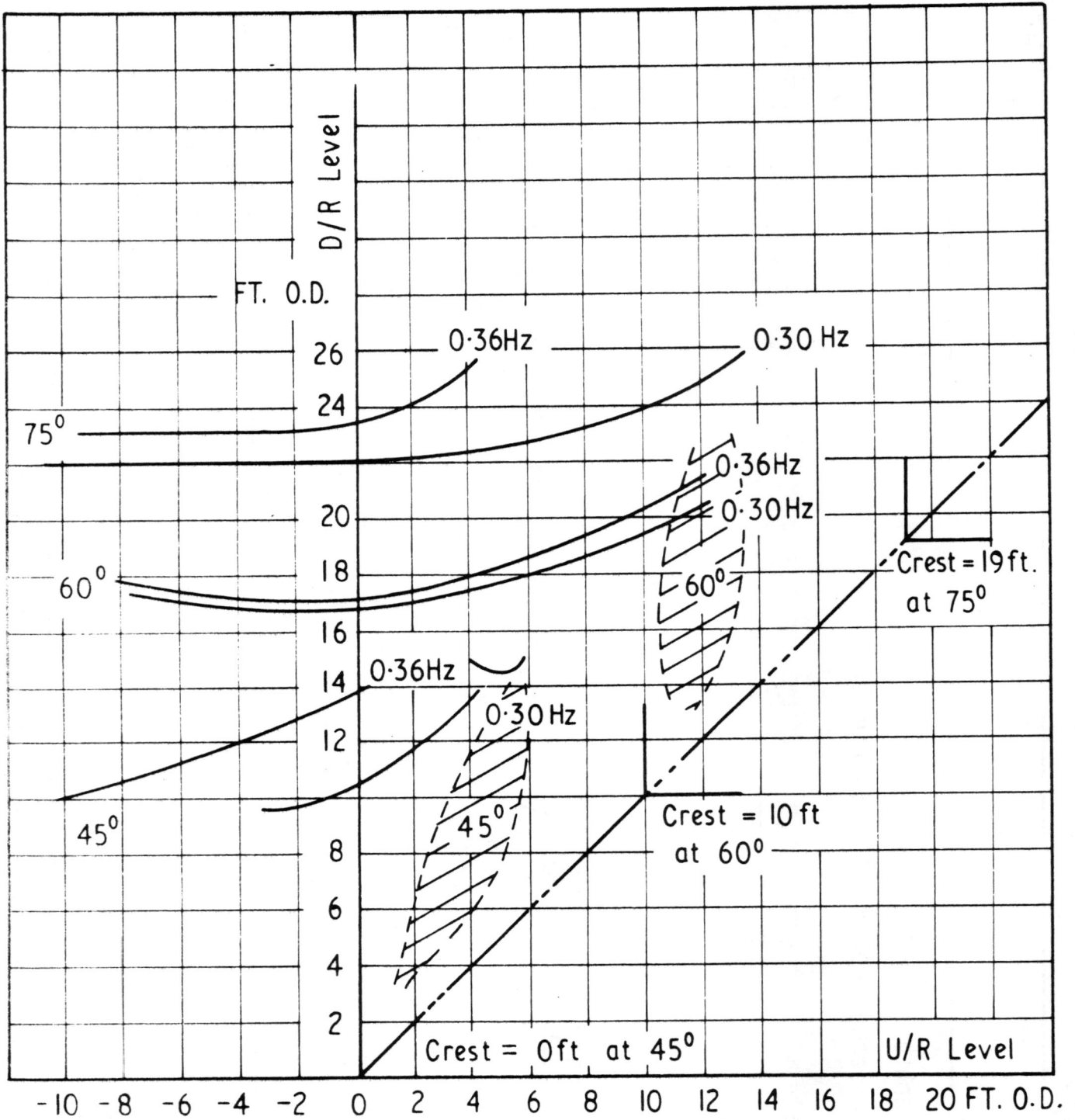

Fig.14. Stability boundaries for mass oscillation mode

REFERENCES

1. CROW, D.A., Thames Barrier Rising Sector Gate - Vibration Studies on a 1/20 scale hydro-elastic model. BHRA Report RR 1225, 1974, Feb.

2. KING, R., Thames Barrier Vibration Studies. BHRA Report RR 1193, 1973, Jun.

3. KING, R. and PROSSER, M.J., Thames Barrier - hydraulic model of the Rising Sector Gate. BHRA Report RR 1195, 1973, Jun.

4. CROW, D.A., Thames Barrier Rising Sector Gate. Report of hydrodynamic force measurement on a 1/20 scale model. BHRA Report RR 1239, 1974, Jun.

5. GELEEDST, M. and KOLKMAN, P.A., Comparison of measurements on the prototype of an elastically similar model of the Hagastein Weir. Proc. 10th IAHR Congress, London, 1963, 3, Sept., 25-32.

12. Hydraulic model studies of the rising sector gate conducted at Imperial College

J. D. HARDWICK, BASc, DIC, PhD*

INTRODUCTION

1. The novel features of the Rising Sector Gates for the Thames Flood Protection Scheme, reflect the numerous and often conflicting requirements which the design had to meet. For example, the sector-shaped gates (Fig. 1a) had to resist attack by salt water while submerged for most of their life over a concrete sill in the river bed. Skin plates were required on all surfaces of the gates to provide important structural as well as hydraulic functions; in addition the skin plates serve to exclude silt from the interiors of the gates. The demand for little impediment to river traffic resulted in the provision of wide, 61 m openings between piers and imposed the flat profiles shown in Fig. 1a. The design had to satisfy a compromise between a low, visually acceptable superstructure and shallow, economically acceptable excavations in the riverbed.

2. At the approach of a potentially dangerous tidal surge, it was originally intended that the gates would be rotated through 90° by a hoist (Fig. 1c); in this closed position the structure would maintain low water-levels on the London-side, and arrest the surge on its down-river side where the riverbanks would be suitably raised. The structural elements of each gate were designed to transmit the hydrostatic thrust of the surge through arms and trunnions to piers at each end; the sector-shape allowed for the stiffest structural members where most required, and in addition, reduced the torque needed to raise the gate, since all thrust on the curved face must pass through the trunnions.

3. Hydrodynamic loading on the gates was a secondary consideration until design of their hoisting mechanism commenced. It then became apparent that if river-closure were delayed while a surge built up, the gates would have to be raised while appreciable flows of water passed over their crests (Fig. 1b). For such conditions, hydrodynamic forces on the flat face of the gates would not pass through the trunnions and the hoists would have to provide additional torque. It was not possible to predict this torque and so model studies were carried out to determine both steady and unsteady hydraulic loadings. Many of these investigations were carried out on a large scale at the

British Hydromechanics Research Association but others were more suited to treatment on a smaller scale in the Civil Engineering Hydraulics Laboratory at Imperial College. Salient features of these latter studies are described in the following sections.

FORCES REQUIRED TO HOIST GATES

4. Studies conducted at the BHRA showed that the hoist mechanism could be exposed to large step-increments of load as a gate was raised (ref. 8). These variations coincide with changes of flow pattern from a "surface jet" to a "diving jet". Imperial College was asked by the Consultants to examine these flow patterns with the aim of discovering ways of smoothing out the variations of load. For this purpose, small models to a geometrical scale of 1:80 were constructed; the models reproduced to scale only a section of the 61 m gate-span such that the assembly of gate and sill could be inserted in a 100 mm wide flow channel. The glass sides of the channel permitted flow visualization, and static pressures could be measured with tappings on the models' surfaces. Fig. 2 illustrates one of the models fabricated in Dural with Perspex false walls. Although the channel was narrow, boundary layers on the walls were not considered to be important over such short distances when the flow was strongly accelerated.

5. Fig. 3 shows estimates of the force which the hoist would have to provide at various gate-angles, θ. The forces were integrated from pressure measurements for conditions representing a "late lift" on a surge tide. For $\theta < 30°$, the hoist-load was statically determinate and is represented here for simplicity as a tension T_{stat}; this tension merely provided the moment to resist the torque-difference between a counterweight located in the gate-arms and the (lighter) submerged gate. In the range $30° < \theta < 40°$, Fig. 3 shows a sudden, four-fold increase of load due to a hydrodynamic contribution, T_{dyn}.

6. The source of T_{dyn} is illustrated in Fig. 4 where three typical conditions are presented:

(i) When $\theta = 30°$, the approaching flow separated from the sharp leading edge of the gate and formed a "surface jet". The lower boundary of the jet rapidly broke down into a

*Lecturer in Hydraulics, Imperial College

Thames Barrier Design. Institution of Civil Engineers, London, 1978, 143-152

143

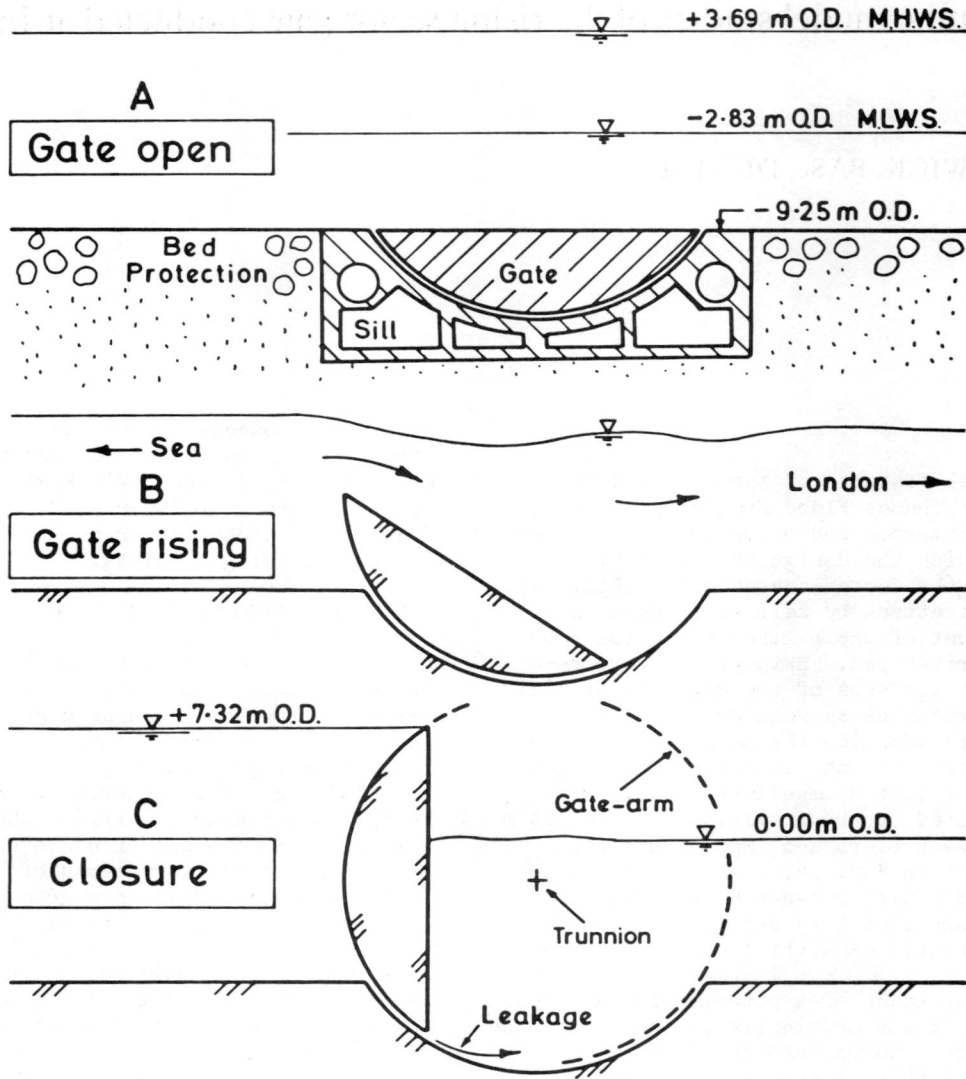

Fig.1. Rising sector gate attitudes

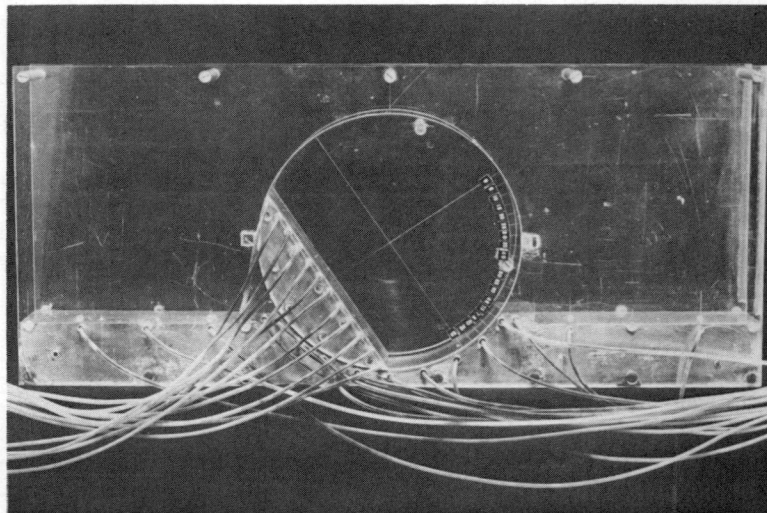

Fig.2. 1:80 model with pressure tappings

series of eddies which progressively mixed the jet with the layers of water beneath. The water entrained and accelerated by the eddies in this way was supplied by flow from a roller beneath the jet. The velocities within the roller were so low that the pressures on the flat, <u>outer</u> face of the gate were virtually hydrostatic. The pressures on the flat, <u>inner</u> face of the water-filled gate were also considered to be hydrostatic, because in the prototype, ports would be incorporated near the gate's trailing edge. Since pressures on either side of the flat face would be equal, no hydrodynamic torque would be applied while the jet remained on the surface.

(ii) As θ increased beyond 30° the jet was increasingly diverted downwards under the action of gravity; as the gap narrowed between the lower boundary of the jet and the flat face of the gate, the supply of water for entrainment by the jet was increasingly throttled. The throttling tended to lower the pressure in the gap and draw the jet even closer to the gate. At some point in the range $30^\circ < \theta < 40^\circ$ the jet re-attached to the flat face, enclosing a zone of separated, recirculating flow. For $\theta = 40^\circ$, in Fig. 4 it can be seen that in the separation zone, the pressures on the gate were lower than hydrostatic in order to support the locally strong curvature of the jet. Near the point of re-attachment, the jet suffered a slight deceleration where a small pressure rise can be seen. The jet thereafter accelerated down the face of the gate and finally decelerated as the flow impinged on the scallop of the sill. The low pressure in the separation zone applied a large hydrodynamic torque on the gate which the hoist would have to counteract. The re-attachment of the "diving jet" was sudden rather than gradual, and this accounted for the discontinuity of loading shown in Fig. 3.

(iii) The hydrodynamic torque for the diving jet increased as θ increased beyond 40°. For $\theta = 60^\circ$ (Fig. 4) the separation zone had so shrunk in size that the pressure on the outer flat face of the gate again tended towards a hydrostatic distribution.

7. Two attempts were made to smooth out the sudden change of load on the gate-hoist. In the first, the leading edge of the gate was rounded with the aim of causing re-attachment for a lower θ when flow-velocities would be lower. Earlier re-attachment was indeed achieved, but it will be seen in Fig. 5 that the rounding had transformed the gate into a hydrofoil. A second trial attempted to delay re-attachment with the aim of lowering the maximum hoist-load; water was ducted in numerous openings through the body of the gate to alleviate the throttling of entrainment-water for the jet. It can be seen that limited success in reducing loads was achieved but the satisfaction of the hoist-designers was not matched by that of the gate-designers.

8. It was concluded from the study that the sector shape was probably the best compromise

which could be achieved. There was apparently no way of avoiding sudden variations of loading on a gate-hoist although the magnitude of the variation could be reduced by lifting the gate at an early stage in the build-up of a surge. In a more positive way the study provided hydraulic design-loads for the plates forming the gate's flat face.

HYDRODYNAMIC FORCES ON THE SILLS

9. It was originally proposed that the gates should completely close the river when opposing a surge, (Fig. 1c), but this method of operation was later questioned. Theoretical studies by the Hydraulics Research Station (ref. 1), suggested that such a closure of the gates would generate a wave which would then travel downriver. The height of this wave would require the level of the embankments on each side of the river to be additionally increased to prevent overtopping; and it was thus proposed that the gates be operated in a manner which would reduce the amplitude of the reflected wave. It was claimed that if the gates were raised to $\theta = 45^\circ$ and held in this attitude while a surge built up, the reflected wave would be of negligible importance while the flow over the gates would not greatly reduce the flood-protection offered. A disadvantage of this mode of operation was that there could be higher flows of water over the gates than had been previously contemplated. A re-examination of hydraulic forces was necessary because the sills were designed to be simply-supported between the piers and were thus sensitive to any additional loading.

10. The original design-study had assumed that the pressure on the upper surfaces of the sills would be hydrostatic. This assumption was investigated with the model of Fig. 2, and typical observations for surface and diving jets are compared in Fig. 6. The pressures over the sill were generally higher for the diving jet than for the surface jet; in the lower half of the figure, pressure differences on a sill's upper and lower surfaces have been plotted, and it can be seen that the vertical component of hydraulic load is four times larger for the diving jet. The reason for these higher forces can be deduced from Fig. 7 which shows how the momentum of the diving jet was completely destroyed as the flow was turned through a 90° angle in the scallop.

11. The loading imposed by the diving jet during surge-regulation had not been envisaged in the design of the sill. As an alternative method of regulation, the Consultants proposed that flow might be allowed to pass <u>beneath</u> the structure as for a tainter (or radial) gate (Fig. 8). For this flow pattern the stream separated from both gate and bed to form a free jet with turbulent, re-circulating flow above and below. The jet suffered a small change of direction as it impinged on the opposite side of the scallop, but large changes of direction were avoided. The vertical forces for under-flow and overflow patterns are compared in Fig. 9, where it will be seen that underflow

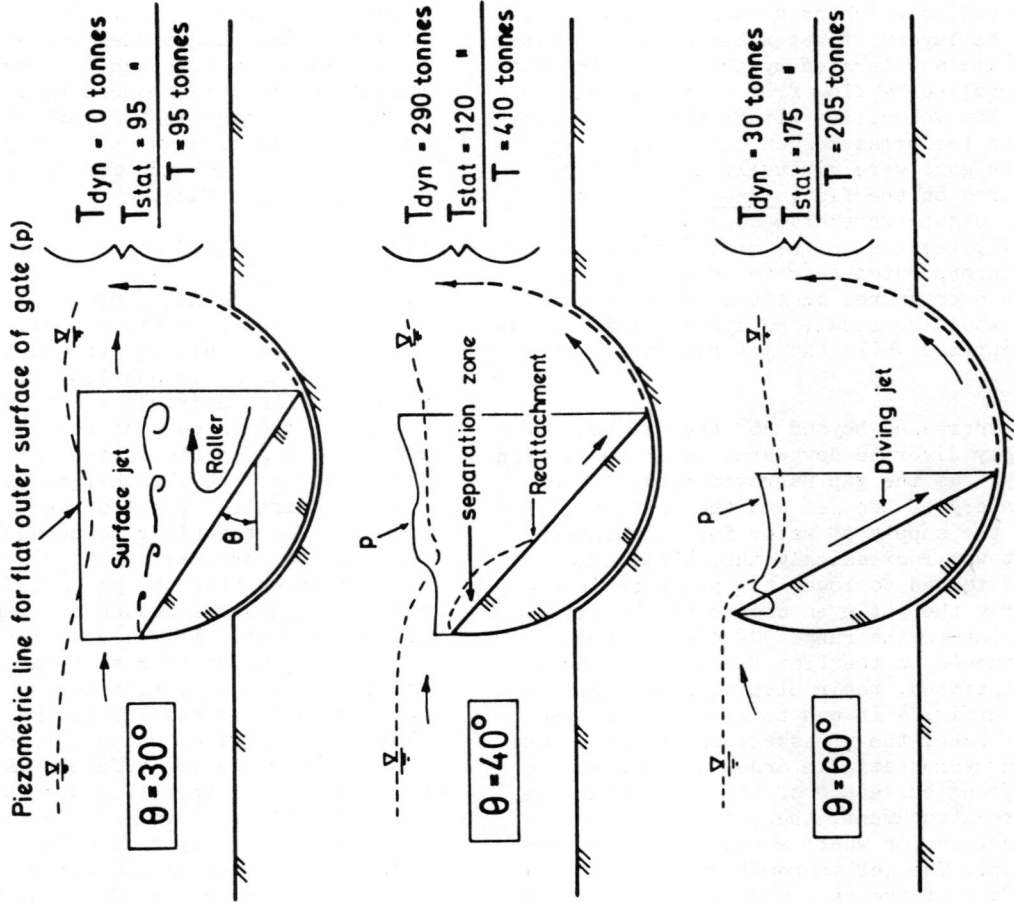

Fig.4. Pressure distributions during closure

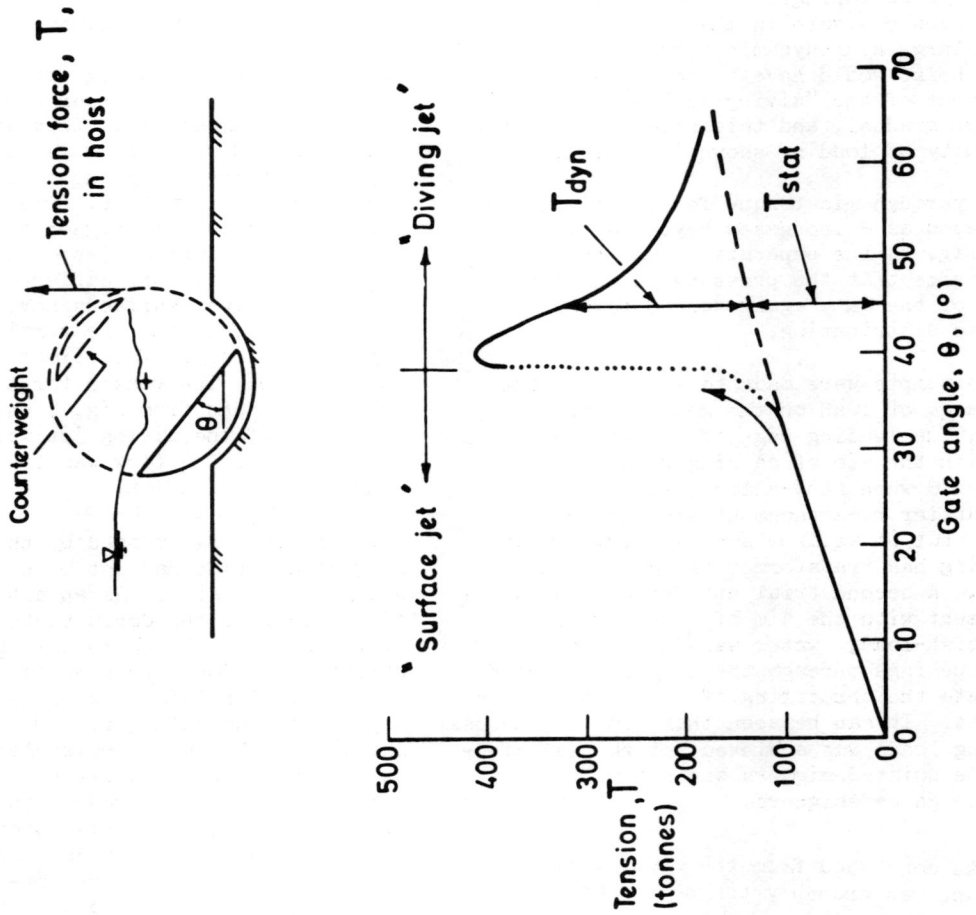

Fig.3. Hydraulic loads during closure

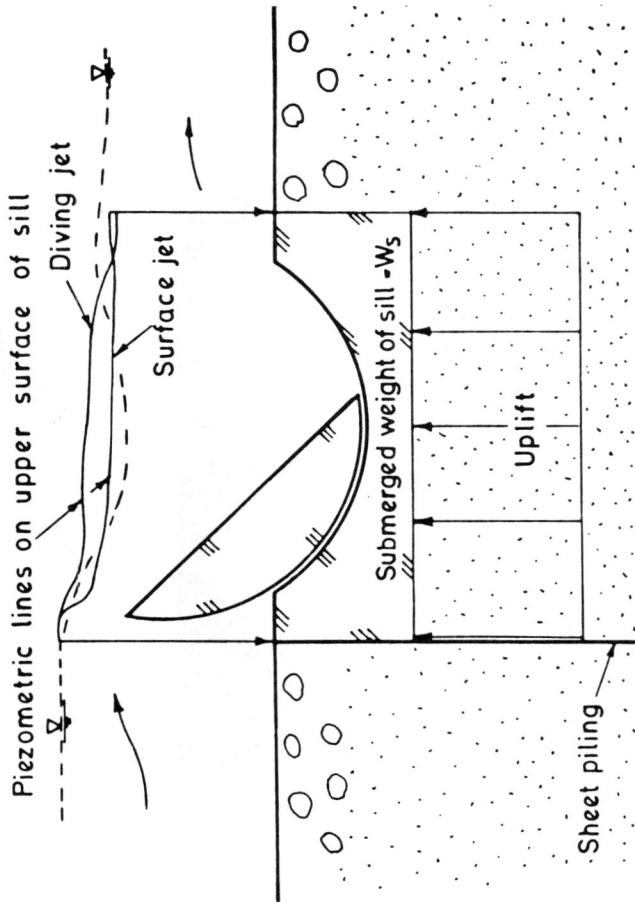

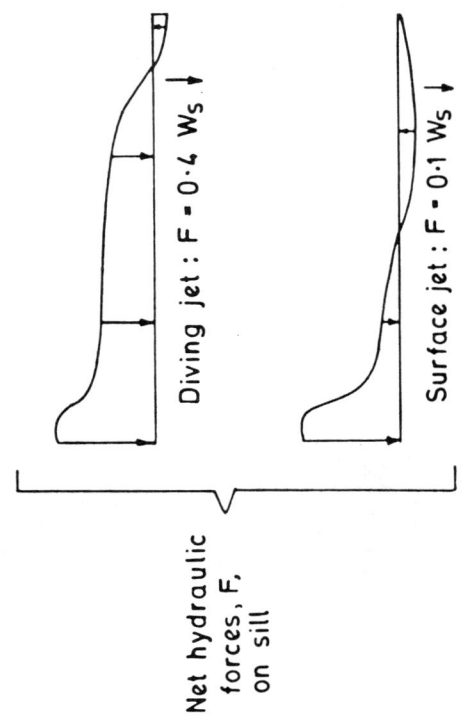

Fig.6. Overflow forces on sill

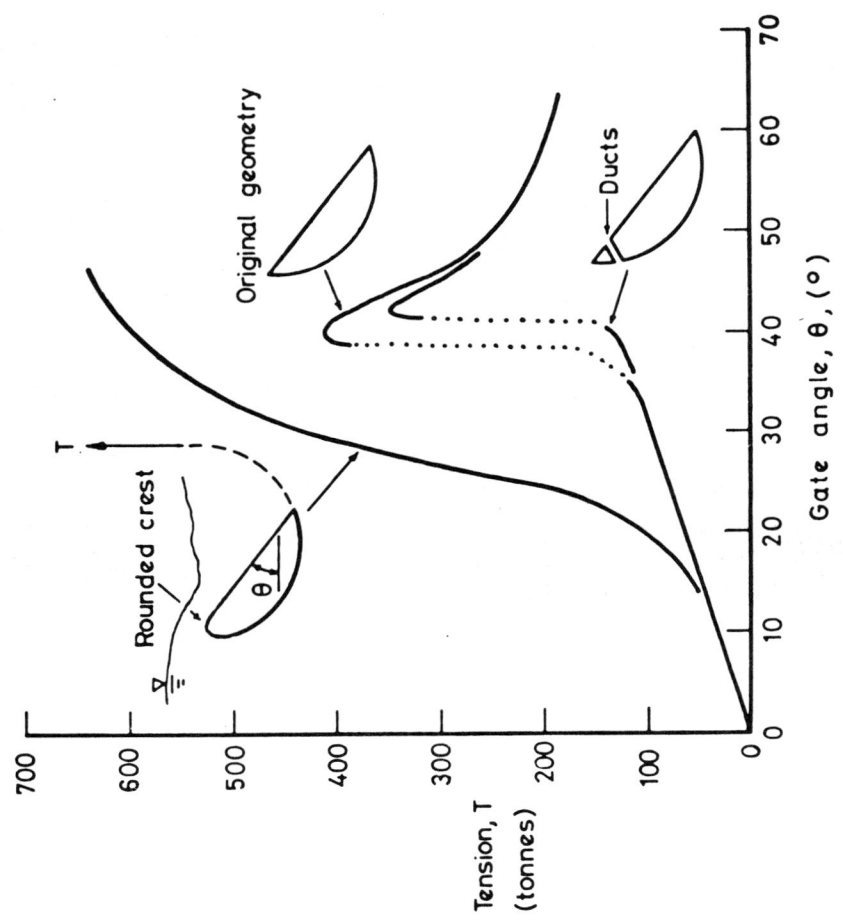

Fig.5. Modified loadings on hoist

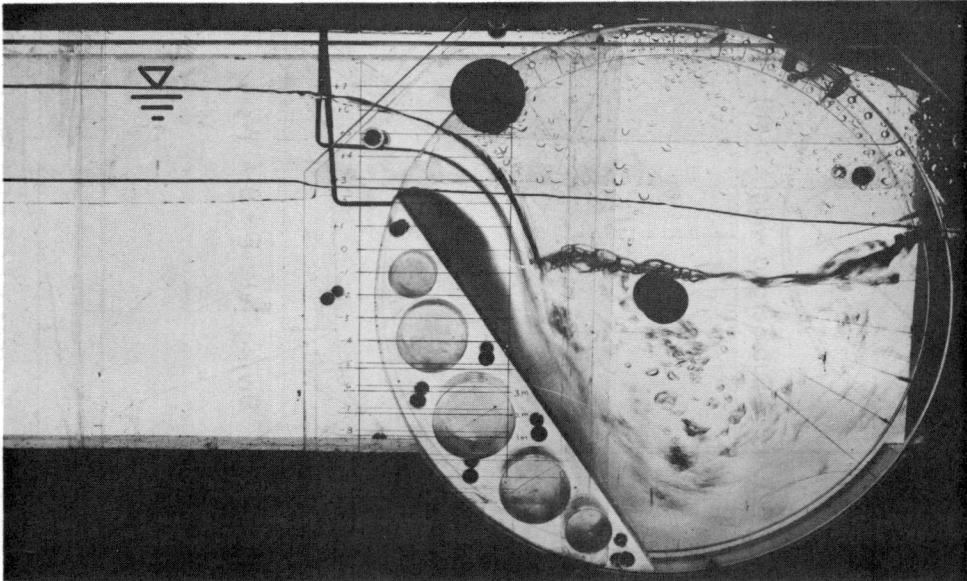

Fig.7. Flow pattern for overflow

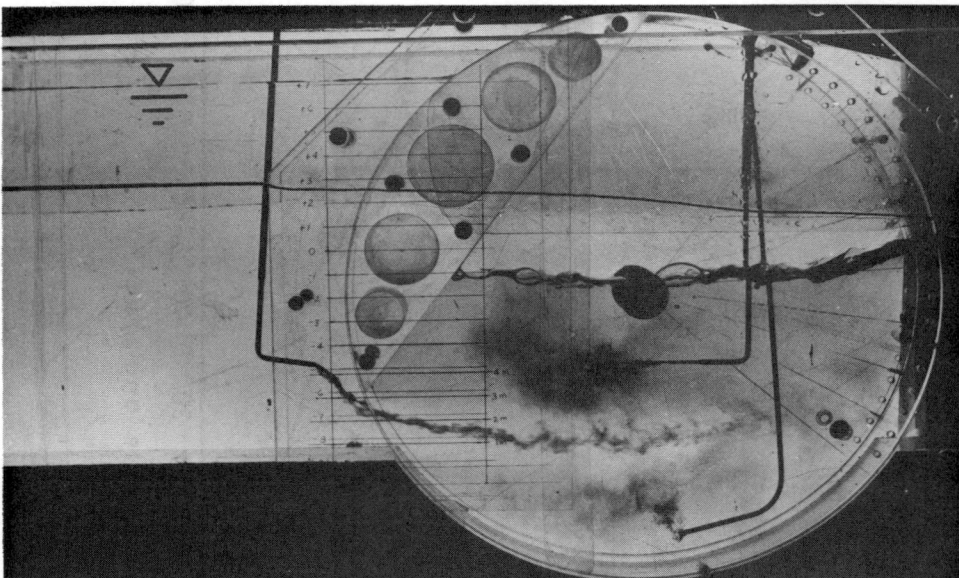

Fig.8. Flow pattern for underflow

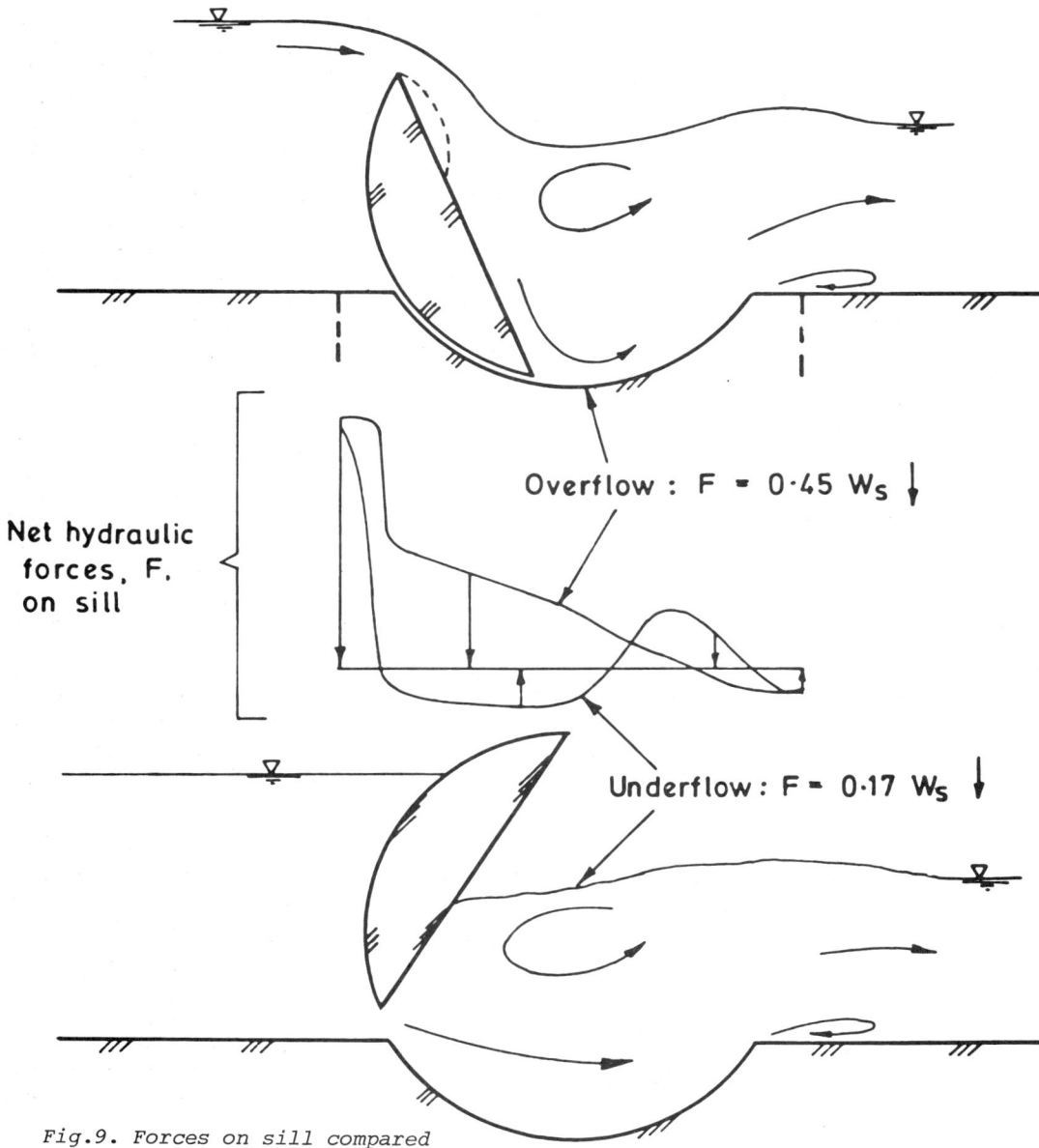

Fig.9. Forces on sill compared

Net hydraulic forces, F, on sill

Overflow : $F = 0.45 \, W_s$ ↓

Underflow : $F = 0.17 \, W_s$ ↓

Fig.10. 1:80 model for vibration sudies

considerably reduced the loading on the sill (refs. 2 & 3).

12. The model Fig. 2, was also used for qualitative observations of bed-scour just beyond the end of the concrete sills. In comparison with overflow, underflow (Fig. 9), appeared to have less scouring potential because of the smaller curvature of the main stream as it passed beyond the scallop of the sill.

13. It was concluded from this study that underflow operation could provide a feasible alternative to overflow and offered advantages, such as reduced sill-loading and scour, which had to be considered in the light of disadvantages such as the finer control required to regulate the surge.

FLOW-INDUCED VIBRATION FOR UNDERFLOW & OVERFLOW

14. In addition to the steady forces of the previous section it was necessary to compare the effects of unsteady forces on the gates when assessing the relative merits of overflow and underflow. The flow-induced vibration of the gates had been studied by the BHRA in great detail, (refs. 9 & 4), at model scales of 1:40 and 1:20, but only overflow had been considered. Two important conclusions from this study were that:

(i) vibration in a direction normal to the flat face of a gate would not be significant; and

(ii) vibration of a gate about its trunnion axis ("mass oscillation") could be reduced to an acceptable level if the gate-hoist were sufficiently stiff.

15. The BHRA investigations had been completed, and the apparatus dismantled, before underflow regulation was contemplated. A correspondingly elaborate investigation for the new condition was not warranted in view of the normally stable characteristics of tainter gates (ref. 5). Since much was already known about overflow vibration, it was considered sufficient to make a direct comparison of overflow and underflow in a small model.

16. The gate used for this comparative study (Fig. 10) was again a 100 mm wide sectional model on a 1:80 geometrical scale. The model reproduced only the important rotational degree of freedom about its trunnion axis in accordance with the conclusions from the BHRA studies. The hoist mechanism was simulated by a leaf-spring which can be seen in the figure engaging the gate near the top of the gate-arm; the upper end of the spring was anchored on the plate shown protruding above the Perspex walls. Fluctuating pressures on the gate's flat face tended to rotate the model on its trunnions, but these motions were restrained in proportion to the stiffness of the leaf-spring. The model's vibration amplitude was deduced from the output of calibrated strain gauges fixed to the spring. Care was taken in the model's design to reduce the friction in the trunnions and at the point of contact between gate-arm and leaf-spring.

17. Fig. 11 shows a typical comparison of flow-induced vibration amplitude for underflow (upper trace) and overflow (lower trace). The scaled natural frequency of the model was here representative of the prototype's frequency. It can be seem immediately from the figure that the excitation in both traces was essentially random, but the levels of vibration for over-flow exceeded those for underflow. Each trace, moreover, exhibited a regular component of frequency which corresponded to the gate's mass oscillation; such motion was dominant for over-flow but for underflow, strong pulsations with periods 5-10 times longer, made a significant contribution.

18. When considering the sources of excitation, it was observed that during underflow (Fig. 8), a portion of the gate's flat face was exposed to a turbulent region of separated flow lying above the jet. Eddying between this region and the jet caused pulsations of the water surface, to which the gate was highly sensitive. It was argued that these pulsations were associated with the longer-period components seen on the vibration trace. The oscillation of the gate was interpreted as a series of transients (with the gate's natural period) excited by impulsive wave-loadings and it was unlikely that the small movements of the model modified the pulsating exciting forced in any significant way.

19. In contrast, the response of the gate for overflow showed strong evidence of an interaction between the gate's oscillatory movement and the exciting force. The motion seemed to comprise intervals of resonance, or "self-excitation", interspersed with "forced" behaviour like that for underflow. This intermittent evidence of self-excitation was probably a residual of the strong excitation process observed in the BHRA research when the gate's mass oscillation frequency was made unrealistically low.

20. In an attempt to explore this phenomenon the 1:80 scale model's frequency was lowered to a value corresponding to only 20% of the prototype's design frequency. Spectacular resonance was observed for this condition and Fig. 12 shows flow patterns for two instants in a vibration cycle. In the upper picture, the tight curvature of the diving jet indicated a low pressure within the dyed separation zone; the low pressure applied to the model in this region was probably responsible for the gate's upward motion. The lower picture shows how the dye in the separation zone appeared to explode towards the gate's trailing edge. At this instant when the curvature of the jet was greatly reduced, the pressure in the separation zone must have risen and the gate accordingly descended. For such a flow pattern, it is hardly surprising that there should be strong interaction between flow and structure when the oscillatory movements of each were so large. But it is remarkable that self-excitation persisted, even when the vibration amplitude was reduced (with a stiffer leaf-spring) to levels imperceptible to the eye, as was the case in Fig. 11b. This excitation

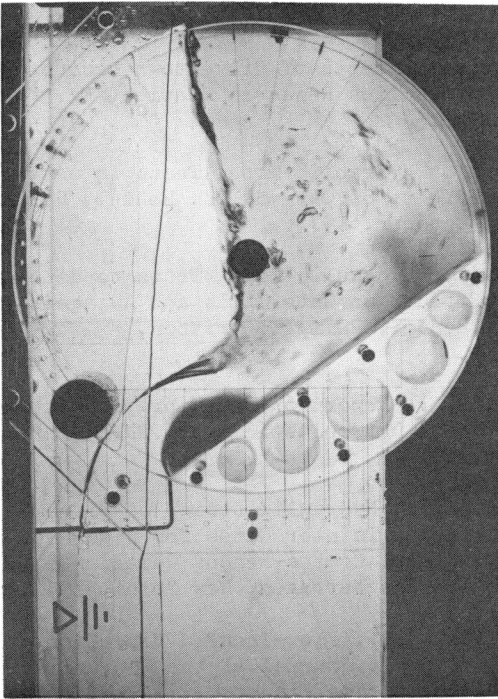

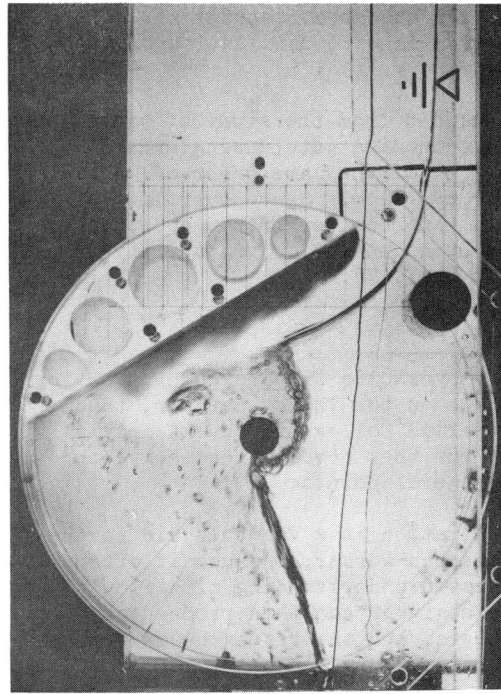

Fig.12. Flow patterns for resonance: gate rising (top); gate falling (bottom)

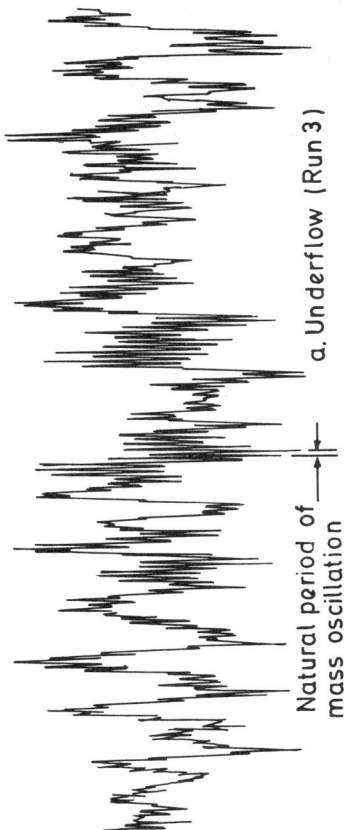

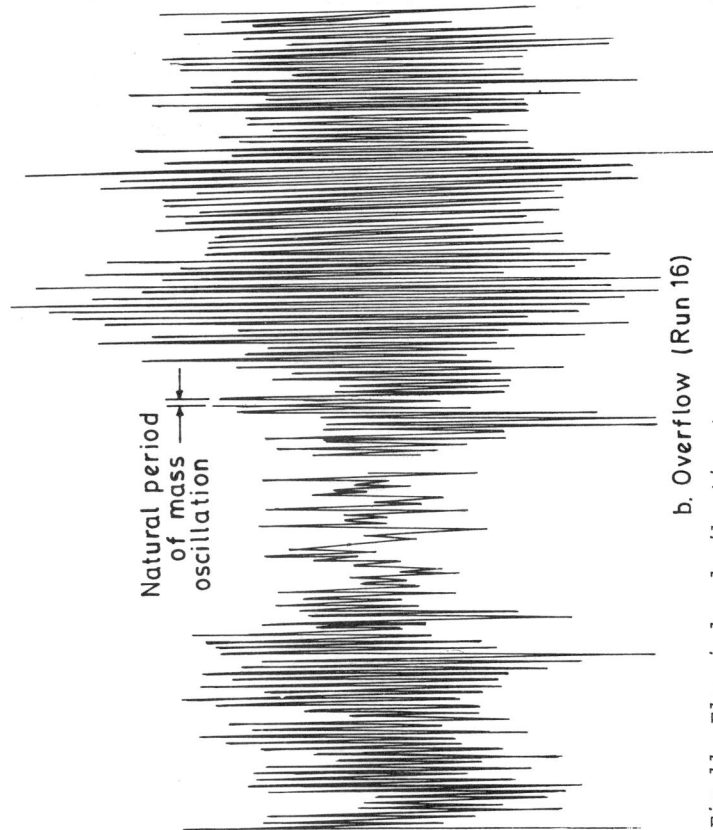

a. Underflow (Run 3)

Natural period of mass oscillation

Natural period of mass oscillation

b. Overflow (Run 16)

Fig.11. Flow-induced vibration traces

process illustrates the great sensitivity of a confined separation zone to oscillation of the separation point (refs. 10 & 6).

21. It was concluded from the study of vibration that the oscillation of a gate during underflow regulation would be of the same order of magnitude as that for overflow. Since overflow vibration levels had been considered acceptable, there seemed to be no threat of instability during underflow (ref. 7).

CONCLUSION

22. Of the many hydraulic studies undertaken for so large a project as the Thames Barrier, those conducted at Imperial College represent only a small fraction, but they nevertheless serve to illustrate two general points:

(i) Flow visualization is a valuable aid to the designer and research-worker, because it often leads to an improved understanding of a flow's behaviour. The designer can then proceed on a more rational basis, and the research-worker can pose more appropriate questions.

(ii) Some classes of problem must be studied with large-scale models and some can be studied on a small scale, but it is often difficult to deduce the class to which a new problem belongs. If a problem requires a hydraulic model study, it is prudent to commence with a small-scale model which can be relatively inexpensive and rapidly tested. If then an investigation on a larger scale is warranted, the experience of the preliminary model can lead to a more suitable design for its successor. Alternatively, the thoughtful interpretation of results from a small, carefully designed model can often be sufficient.

REFERENCES

1. Downriver Effects of Silvertown Barrier Closure. Hydraulics Research Station, Ex. 749, October, 1976.

2. Report on Hydraulic Model Tests to determine Overspill Effects on Gate Sills. Rendel, Palmer & Tritton, October, 1973.

3. Report on Hydraulic Model Tests to determine Undershot Effects on Gate Sills and Adjacent River Bed. Rendel, Palmer & Tritton, March, 1974.

4. CROW, D.A., Report on the Vibration Studies on a 1/20 Scale Hydroelastic Model. BHRA, August, 1973.

5. DOUMA, J.H., Hydraulic Design of Slide, Vertical-Lift and Tainter Gates for High-Head Reservoir Outlets. Comtes-Rendus Quatrieme Congres de Grandes Barrages, New Delhi, 1951.

6. HARDWICK, J.D., Flow-Induced Vibration of Vertical-Lift Gate. Journal of the Hydraulics Division, American Society of Civil Engineers, HY5, May, 1974.

7. HARDWICK, J.D., Studies of Flow-Induced Vibration due to Underflow and Overflow made on a 1:80 Scale Sectional Model. Imperial College of Science and Technology, August, 1974.

7. KING, R., and PROSSER, M.J., Thames Barrier Hydraulic Model Study of the Rising Sector Gate. BHRA, November, 1971.

8. KING, R., Thames Barrier Vibration Studies. BHRA Report RR 1195, June, 1973.

10. NAUDASCHER, E., On the Role of Eddies in Flow-Induced Vibrations. Proceedings of the Tenth Congress of the International Association for Hydraulic Research, 3, London, 1963.

13. Design of bed protection using mobile bed models

J. S. BURGESS, PAIWES*

INTRODUCTION

1.   In the event of the level of a predicted
high water being such that the water level upriver would be unacceptable in relation to the
level of the flood defences, one mode of operating the Barrier would be to close it completely,
at low water or during the rising tide, thus
eliminating the effects of the high water upriver
of the site. However, flow studies had shown
that it was possible to reduce peak levels on
the downriver side by allowing some water to
pass the Barrier and, furthermore, that by
judicious settings of the gates, a useful reduction could be achieved without exceeding
limiting levels upriver.

2.   Of the various alternative partial closure
settings, the favoured arrangement was to operate the four large gates undershot with the remainder of the gates fully closed. The head
difference which would occur across the Barrier
under these conditions would generate high
velocities through the partly open gates and
thus it was imperative that the local scour and
protection situation should be studied.

3.   Thought was also given to the possibility,
however remote, of some malfunction of the gates
causing scouring conditions to develop. In this
context, analysis had indicated that, of the
situations considered, the most unfavourable
would be that associated with the failure of one
of the Rising Sector Gates to close, i.e. all
but one gate fully closed and the remaining gate
in the open position. Should this happen, very
high velocities would occur through the gap and,
although upriver flooding would be averted
except in extreme conditions, the bed would need
quite heavy protection locally to prevent undermining of the structure.

THE INVESTIGATION

4.   The Hydraulics Research Station undertook
to carry out a physical model investigation of
local protection and scour. Tests were conducted in two mobile-bed models to study bed and
bank protection, local scour, flow patterns and
velocities at and near the Barrier. In addition some discharge calibration tests were
carried out for selected flow situations.

*Senior Scientific Officer, Hydraulics Research
Station

5.   The design of the investigation took
account of partial closure and gate failure
situations as follows:

(a) Partial closure, four large gates operating
undershot;

(b) Failure of one small Rising Sector Gate;

(c) Failure of one large Rising Sector Gate.

In addition, the Barrier was tested in its
normal operating mode with all gates open.

6.   Prototype data, supplied by the Consulting
Engineers, were used to design and operate the
two models. The first, constructed in a large
flume, included a pier flanked by part of a
Rising Sector Gate on either side, to a linear
scale of 1:25; it was used for the most part to
make preliminary studies relatively quickly
(Fig. 1). The second, which reproduced the
whole Barrier and a little over 1 km reach of
river channel on either side (scale 1:50), was
required for detailed experiments where faithful
reproduction of the three-dimensional characteristics of the flow patterns was important
(Fig. 2). The scales selected for the models,
resulted in flow conditions which ensured that
stability of the model bed protection could be
scaled up to prototype dimensions with confidence.

7.   An interesting and often important topic
in engineering hydraulics had come to the fore
during the planning stages of the investigation.
It concerned determination of the extent and
rate of local scour beyond the protected area,
and here a complete solution by mobile-bed model
was not possible. The difficulty stemmed from
an inability to reproduce to scale in the model,
the behaviour of the prototype bed at the
Barrier Site. It was known that the bed consisted of materials of greatly differing
particle size including cohesive alluvium overlying gravel and Thanet Sand. The erosion of
cohesive material could be represented only
qualitatively; the presence of sand and gravel,
should scour extend to the lower layers, added
further complications. The net reult was that
it had to be accepted that the extent and rate
of scour were indeterminate in the present
study. All that could be offered was a qualitative assessment of the areas subject to scour.
In the models a comparative measure of the
severity of scour action was indicated by the

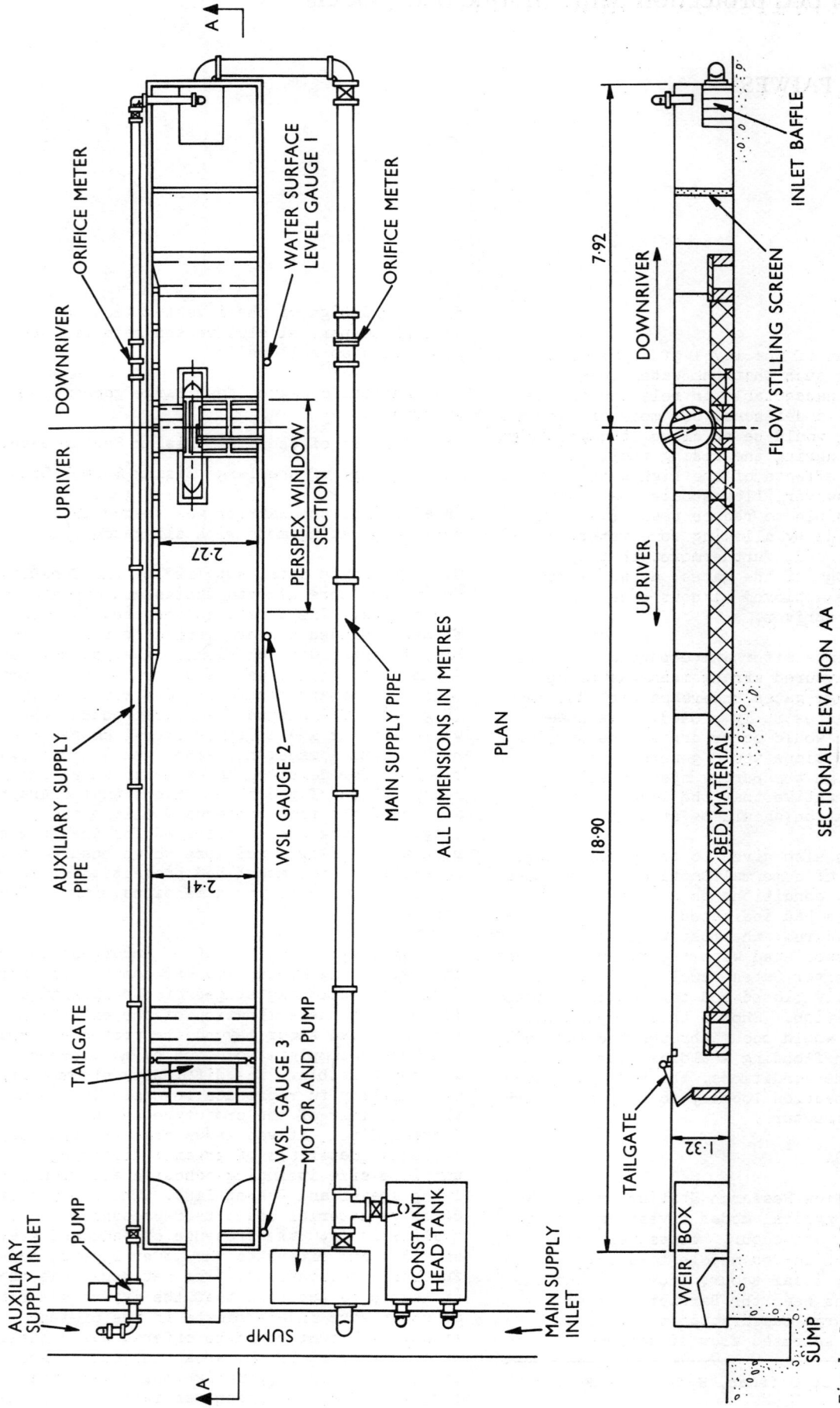

Fig.1. Layout of 1:25 scale model

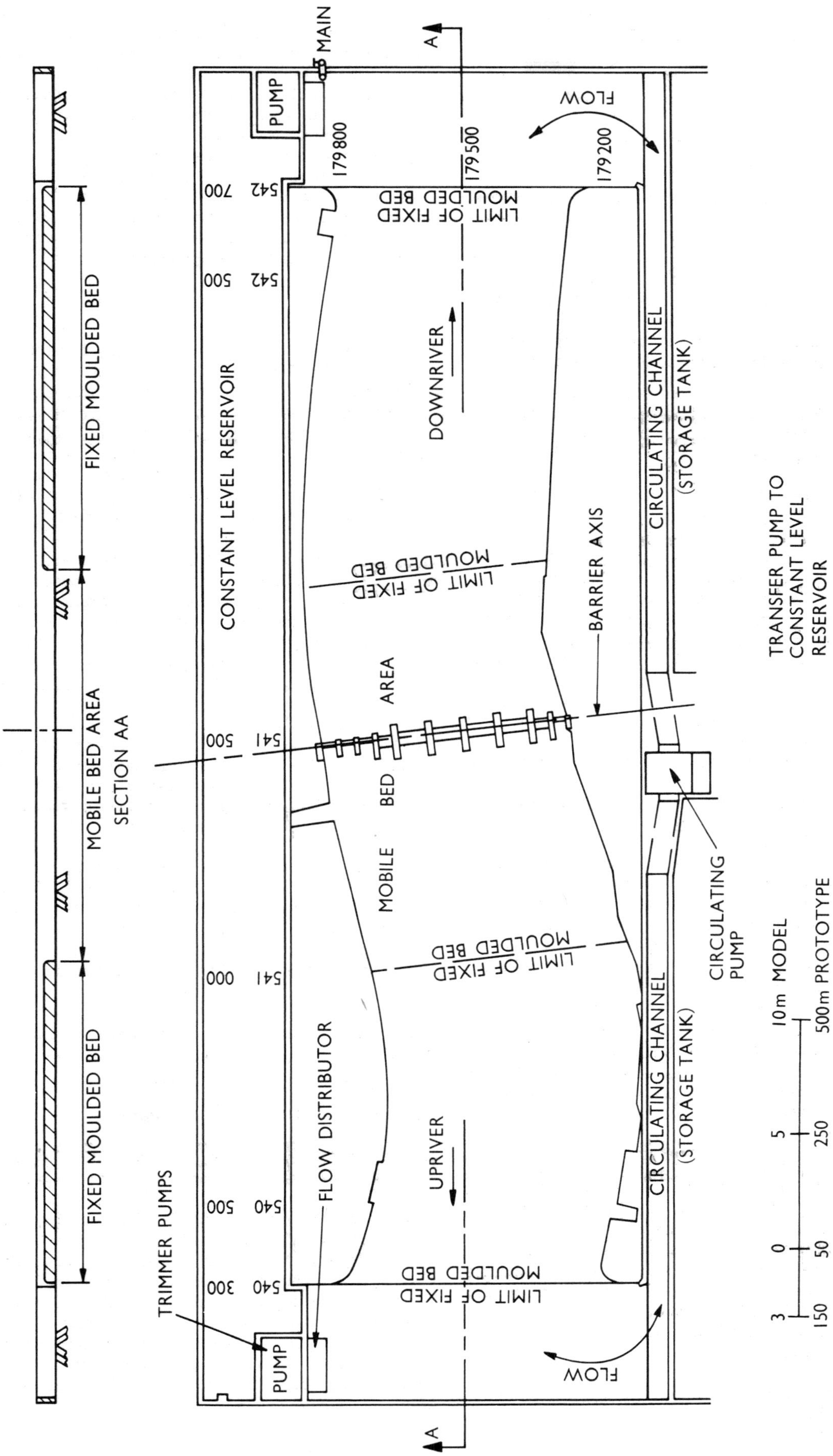

Fig.2. Layout of 1:50 scale model

movement of cohesionless material, granulated coal, used to form the river bed.

8.   The investigation may be considered under three headings:

(a)   Scour of the bed without protection;

(b)   Design of bed protection;

(c)   Scour of the bed beyond the protected area.

MODEL TESTS

9.   Tests representing the following gate situations were carried out under each heading:

(i)   Gate failure, i.e. where one Rising Sector Gate fails to move from its housed position;

(ii) Controlled discharge operation, i.e. the Barrier partially closed with the four large gates set to operate undershot (1.2 m opening) and the remaining gates fully closed.

Local scour of the river bed was checked for the situation with all gates open.

10.   The model tests carried out without bed protection indicated areas subject to local scour action.   The results, together with re-cords of water surface profiles along the piers, aided the Consulting Engineers in an investiga-tion of the possibilities of locally protecting the foundations rather than protecting the bed. The 1:50 scale model, i.e. the one of the whole of the Barrier, was operated over the design surge tide cycle, from low water where levels were the same on either side of the Barrier, to the point when the gates would be opened as levels again equalised on the falling tide.

11.   The test without protection indicated severe scour action for gate failure situations. This was in no way a suprise, because estimates of peak velocities were 9 m/s and 5 m/s for the small and large gates respectively.   However, for the controlled discharge operation, when flow takes place through the large gates opera-ting undershot, remarkably little disturbance of the mobile bed occurred.   The jets under the gates were deflected to the surface by the re-cesses in the sills and, by the time the flow had expanded to fill the section, velocities and turbulence were much reduced and very little scour was observed.

12.   Once the Consulting Engineers had decided to go ahead with model tests on bed protection and had specified conditions, a programme was drawn up designed to determine the stone sizes and extent of riprap blankets, upriver and downriver of the Barrier.   The protection was required to withstand the conditions which would occur should one Rising Sector Gate fail to close, this being judged as the most severe scouring condition.

13.   Including velocity surveys and some gate discharge calibration work, some thirty tests were made in the two models.   The results showed the sizes and disposition of riprap

needed to prevent undermining of the foundation for the gate failure situation.   The protection in the prototype will involve placing a woven man-made-fibre filter cloth, the extremities of which will be held down by concrete anchor blocks.   This was represented in the model by nylon curtain material.   The stones forming the riprap blanket were randomly dumped on the filter cloth.   Stones used in the tests repre-sented sizes ranging from 60 kg to 6 t.   The stones representing the smaller sizes in the model were sieve-graded while the larger stones were individually weight-graded.

THE RESULTS

14.   The final design resulting from the tests is shown in Fig. 3.   Of particular interest and importance was the need for relatively heavier stone in areas subject to turbulence generated along the boundaries of the zones of flow separation close to the piers.   In general the experiments indicated that stones somewhat smaller than anticipated in the original design would provide the protection needed.

15.   In the course of the investigation, velo-cities were recorded at selected positions close to the Barrier by miniature current meter, and the results were used to assist in a first choice of the appropriate size of stone.   The sizes finally selected showed good agreement with the work carried out by Hallmark and Smith (ref. 1).

16.   Overhead photography was used in the 1:50 scale model to record the movement of surface floats and hence to determine flow patterns and velocities.   These measurements were of interest in connection with planning river traffic control and mooring facilities.

17.   Information about the discharge characteris-tics of the Rising Sector Gates was required, to improve the performance of a numerical model being used at the time to compute water levels along the estuary.   The data needed were obtained from calibration tests carried out in both models for the following conditions:

(a)   Gate failure, i.e. one gate in housed position;

(b)   Weir action over a gate.

(c)   Flow undershooting a gate.

Discharge equations were presented covering both modular and non-modular flow situations.

18.   The results of tests of scour beyond the protected area indicated that severe erosion was to be expected upriver should one gate fail to close, but that the Barrier structure would be protected from undermining by the riprap blanket. Only very little scour of the river bed occurred under controlled discharge operation conditions, i.e. with the four large gates set to operate undershot, while for the condition of all gates open, scour was insignificant.

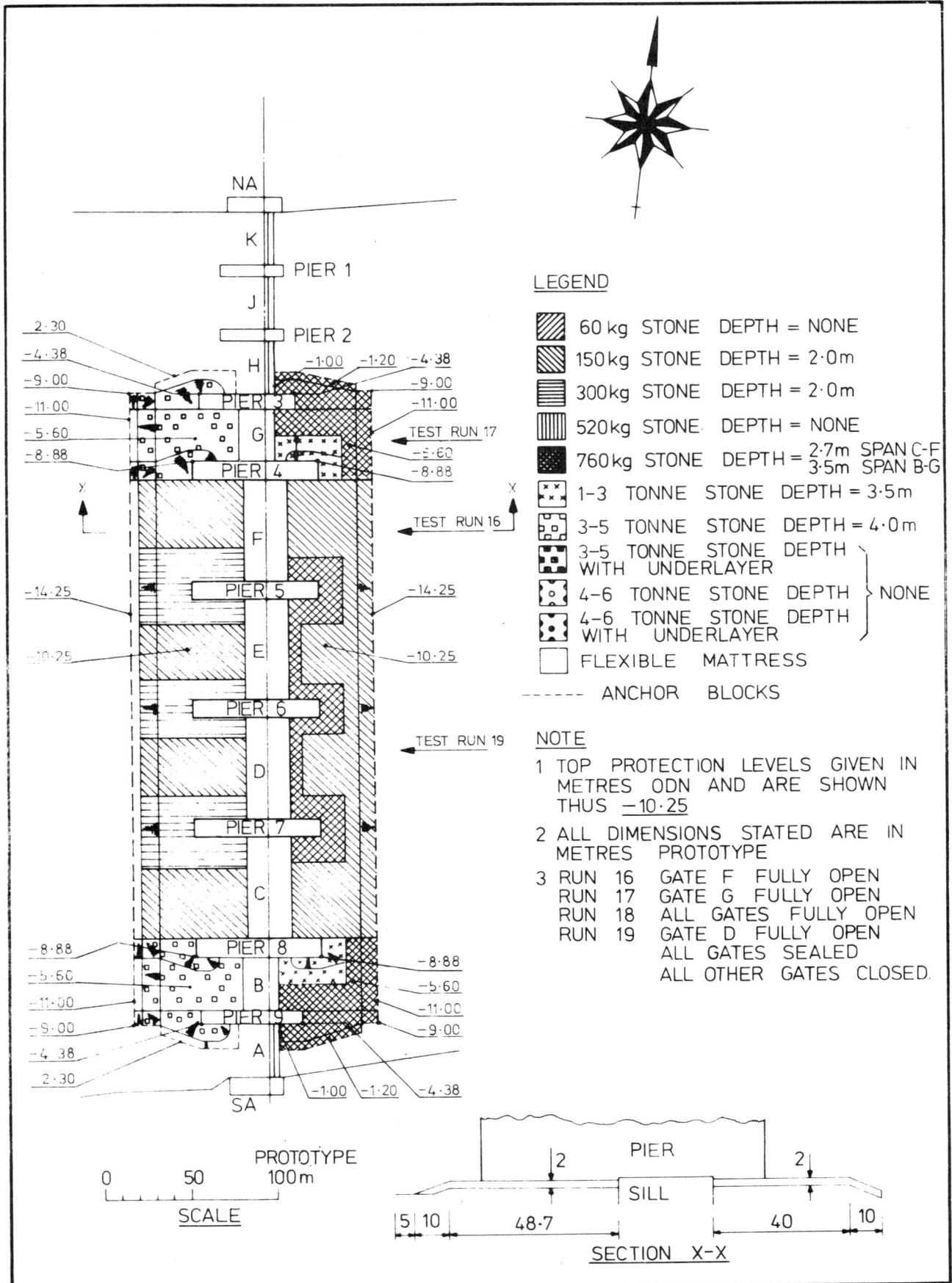

LEGEND

▨	60 kg STONE	DEPTH = NONE	
▨	150 kg STONE	DEPTH = 2·0 m	
▤	300 kg STONE	DEPTH = 2·0 m	
▥	520 kg STONE	DEPTH = NONE	
■	760 kg STONE	DEPTH =	2·7m SPAN C-F / 3·5m SPAN B-G
⊞	1-3 TONNE STONE	DEPTH = 3·5 m	
⊡	3-5 TONNE STONE	DEPTH = 4·0 m	
⊡	3-5 TONNE STONE DEPTH WITH UNDERLAYER		
⊡	4-6 TONNE STONE DEPTH		NONE
⊞	4-6 TONNE STONE DEPTH WITH UNDERLAYER		
☐	FLEXIBLE MATTRESS		
------	ANCHOR BLOCKS		

NOTE

1 TOP PROTECTION LEVELS GIVEN IN METRES ODN AND ARE SHOWN THUS −10·25

2 ALL DIMENSIONS STATED ARE IN METRES PROTOTYPE

3 RUN 16   GATE F FULLY OPEN
  RUN 17   GATE G FULLY OPEN
  RUN 18   ALL GATES FULLY OPEN
  RUN 19   GATE D FULLY OPEN
           ALL GATES SEALED
           ALL OTHER GATES CLOSED.

Fig.3. Layout of protection based on model investigation

ACKNOWLEDGEMENTS

19. This paper is published with the permission of the Director of Hydraulics Research.

Dr. W.R. White's section was responsible for the 1:25 scale model and the Author's section for the 1:50 scale model. Mr. A.D. Crabbe and Mrs. F.M. Farnsworth were respectively in charge of experimental work on the two models.

REFERENCE

1. HALLMARK D.E. and SMITH G.L., Stability of channels by armorplating. Proc. ASCE, Vol 91, WW3, pp 117-135, August 1965.

14. Civil works

G. R. CARR, BSc(Eng), FICE*

A. O. MITCHELL, BSc, MICE†

INTRODUCTION

1. Papers 1, 2 and 3 in this Symposium have described the historical background, and the various investigations which culminated in the decision to provide a flood defence Barrier at the Woolwich Site (located as shown in the key plan, Fig. 1), in association with the raising of the downriver banks to afford the same degree of protection against surge tides, as that to be ensured upriver by the Barrier itself.

2. Following the detailed site investigation carried out during the winter of 1971/72, which was described in Paper 4, a contract was let for the demolition of a number of the disused factory buildings on the south bank. This was followed, in April 1973, by a further preliminary works contract for the construction of a 300 m long section of the permanent new river wall on the south bank immediately downriver of the Barrier location, to provide a more convenient berthing facility for the future construction operations, as well as to reclaim more working area. In addition, other preparatory works were included in this preliminary contract, to facilitate the start of the main construction contract which was subsequently begun in July 1974.

3. A further preliminary to the start of the main works was the widening of the deep channel to enable shipping to be diverted to the northern half of the river during the first stage of Barrier construction. This dredging contract was completed in February 1975. Maintenance dredging contracts were subsequently let, to deal with any significant siltation of the diversion channel.

4. The current construction programme anticipates the completion of the southern half of the works by late 1978, following which the machinery and gates for the related navigation spans will be installed. Shipping will then be diverted through those spans to enable the northern works to be completed. Completion of the project is scheduled for 1982. At present, the cofferdams of two of the main piers in the first half of the works remain to be dewatered, the other structures and a number of the gate sills being in more advanced stages of construction.

*Associate and Project Manager, Rendel, Palmer and Tritton
†Managing Engineer, Rendel, Palmer and Tritton

GENERAL DESCRIPTION

5. The general arrangement of the Barrier is shown in Fig. 2. This arrangement was developed as a logical outcome of the requirements determined by the preliminary studies, the physical characteristics of the site, and the adoption of the Rising Sector type of gate (described in Paper 6) which was made possible by the Port of London Authority's agreement to a reduced width for the main navigation openings (Paper 2).

6. The Barrier consists of nine piers and two abutment structures, providing a total of ten openings, six of which are equipped with Rising Sector Gates, and four with falling radial gates. The bank to bank length of the Barrier is 520 metres. The Rising Sector Gates occupy shaped recesses in the gate sills when in their open position, thus giving unobstructed navigational openings 61 m wide with a mean water depth of 9 m in the case of the four main gates, and 31.5 m wide with a water depth of 4.5 m at the ancillary navigation spans. The falling radial gates, one at the south side and three at the north, are also 31.5 m wide but, with their sills at mean tidal level and headroom restricted by the access bridges and the gates in the open position, these inshore openings will not be used by shipping.

7. The provision of four main navigation openings will ensure safe two-way passage for shipping even when an opening is closed for gate maintenance or other reasons. The width of the deep portion of the natural channel accommodates three of these openings. The fourth therefore necessitated widening the channel, a requirement which was in any case imposed by the navigation diversion during construction. In general, however the sill levels conform closely to the river bed profile.

8. The piers and the abutment structures are generally of mass concrete construction. However, the various chambers, galleries and shafts, as well as the zones of high stress around the gate shaft support structures and elsewhere, have required the provision of large quantities of reinforcing steel, particularly in the upper levels of the structures. The pier widths have been kept to a minimum commensurate with fulfilling their function of supporting the gates and accommodating equipment and operational spaces. Having been fixed early on for the pur-

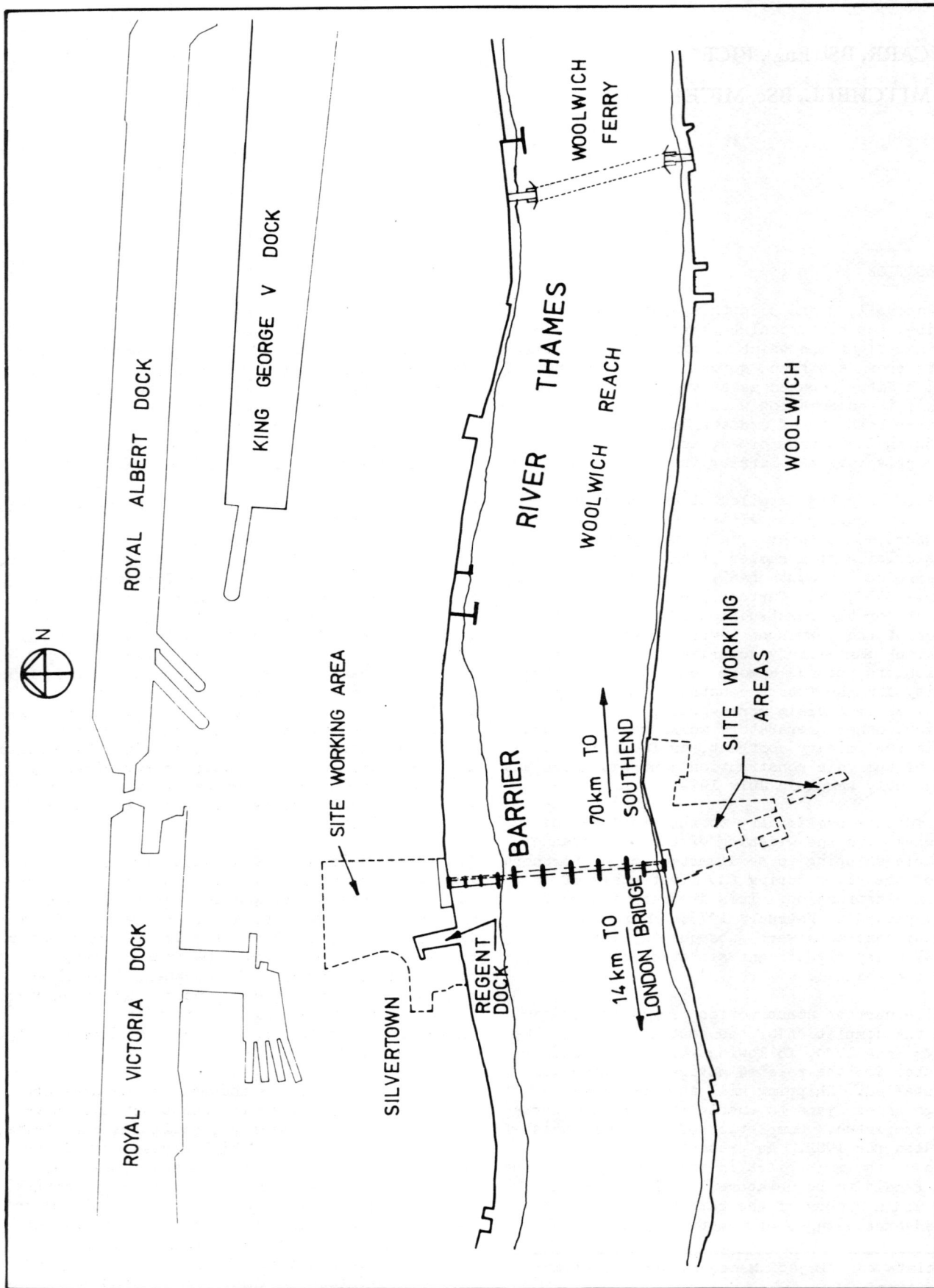

Fig.1. Location of barrier

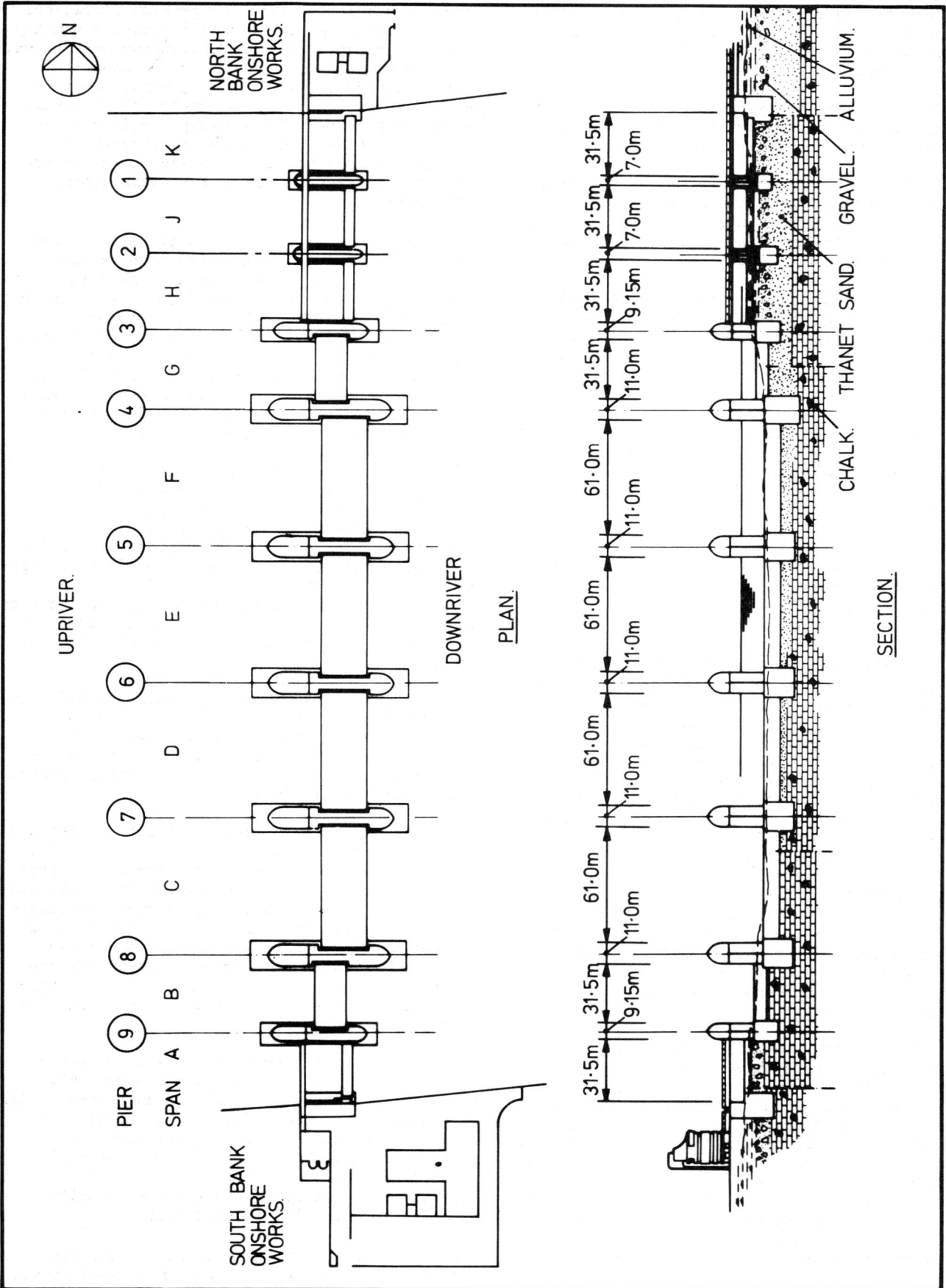

Fig.2. General arrangement of barrier

pose of the enabling legislation (ref. 1), these widths could not be increased, with the result that considerable ingenuity has since been required to accommodate the machinery and other equipment, as fuller details of these items became known.

9. The principal operating requirements determining design loadings are that the gates and structures shall be capable of:

(i) During full closure operation, withstanding (without overspill over the gates) a water level of 6.9 m O.D.N. on the seaward side of the gates when the upriver level is -1.5 m O.D.N., thus giving a differential head of 8.4 m.

(ii) During full closure operation in an emergency, allowing 0.3 m depth of overspill with the crests of the gates at 6.9 m O.D.N. and the upriver levels at -2.7 m O.D.N., thus giving an emergency differential head of 9.9 m.

(iii) During partial closure operations to minimise the downriver effects, withstanding a water level of up to 6.9 m O.D.N. on the seaward side of the gates, while allowing to an agreed extent a controlled flow of water in the upriver direction either over or under the four main gates.

(iv) To cater for a continuous closure over two consecutive surge tides, withstanding a reverse differential head of 6.1 m with the upriver level not exceeding 4.3 m O.D.N.

10. As indicated by the second of the above operating requirements, the Barrier is designed to afford protection against an extreme tidal surge level of 7.2 m O.D.N., which compares with a maximum recorded water level at Woolwich of 5.2 m O.D.N., (31st January 1953), and a mean high water spring tide level of 3.7 m O.D.N. The downriver decks of the piers have a parapet wall level of 8.2 m O.D.N., and the upriver deck parapets are at 5.6 m O.D.N. to meet the existing statutory flood defence requirements. The main and intermediate piers have machinery floors above these levels, the uppermost floor in the case of the main piers being at 15.0 m O.D.N. The enclosing shell roof extends up to 24.6 m O.D.N. With the deepest foundation level at -27.0 m O.D.N., the roof of a main pier is thus, in one case, more than 50 m above the structure base.

11. The sills for the six navigation openings are of cellular precast concrete construction, prestressed in the case of the four large ones, and are shaped to house the Rising Sector Gate in its lowered position. The inshore shallow openings with falling radial gates have in-situ concrete sills. All sills are supported on the piers rather than on the excavated river bed. This ensures that settlements cannot interfere with the sill-pier subway joints or the sill-gate geometry. Problems of uncertain load distribution in the sills are also avoided by this means.

12. Duplicate routes to each pier for services and access from both banks are provided by a pair of subway tubes in the precast sills, and at the inshore spans by a combination of subway tube and bridge. The shafts serving each subway in the piers and abutments have no interconnection below 4.0 m O.D.N., so as to avoid any risk of the duplicated routes being simultaneously flooded. The subway-to-pier connections are designed for some degree of pier settlement and structural movement, and in the case of the precast sills, for installation tolerances.

13. The precast sills will be floated in and sunk on to their pier bearings in a dredged trench. This trench will then be backfilled by pumped sand. A steel sheet pile cut-off at the downriver edge of the sills is designed to limit seepage flows under the sills and prevent piping during surge conditions. At the north and south abutments, which will align with the reconstructed river walls, the lateral seepage path will be controlled by the adjacent sheet piled walls and a drainage system.

14. Heavy stone armouring will be placed locally on the river bed upriver and downriver of the sills and piers as scour protection. Little protection is needed for normal tidal river flow, but operating conditions involving controlled discharges impose protection requirements. The controlling criteria for the bed protection are dictated, however, by the need to safeguard the structures in the unlikely event of inability to operate a gate before or after a surge.

15. The principal operating, administration and maintenance facilities are based on the south bank. The north bank works will be limited to access and security provisions and the equipment installations required for standby electric power supply from the grid.

16. The detailed foundation investigations and their results have been described in Paper 4. They generally confirmed the preliminary borehole information undertaken before the site was finally selected, and gave more precise information for foundation design. No important new factors were revealed, and reassuring indications were obtained regarding certain aspects such as the effects of faulting, and the possible incidence of major fissures and anomalies in the chalk bedrock.

17. At the south shore, the chalk surface is about 10 m below zero datum, but dips gently northwards so that at the north bank where it is overlain by the Thanet Sand it is about 25 m below datum (see Fig. 2). Apart from the area under and near the south bank where weathering has taken place to a depth of 7-10 m, the quality of the chalk was determined as quite suitable for the pier foundations.

18. The chalk at the south bank is directly overlain by Flood Plain gravel and, above that, by recent alluvial deposits (peaty clay and silts with sand and gravel lenses). Similar gravel

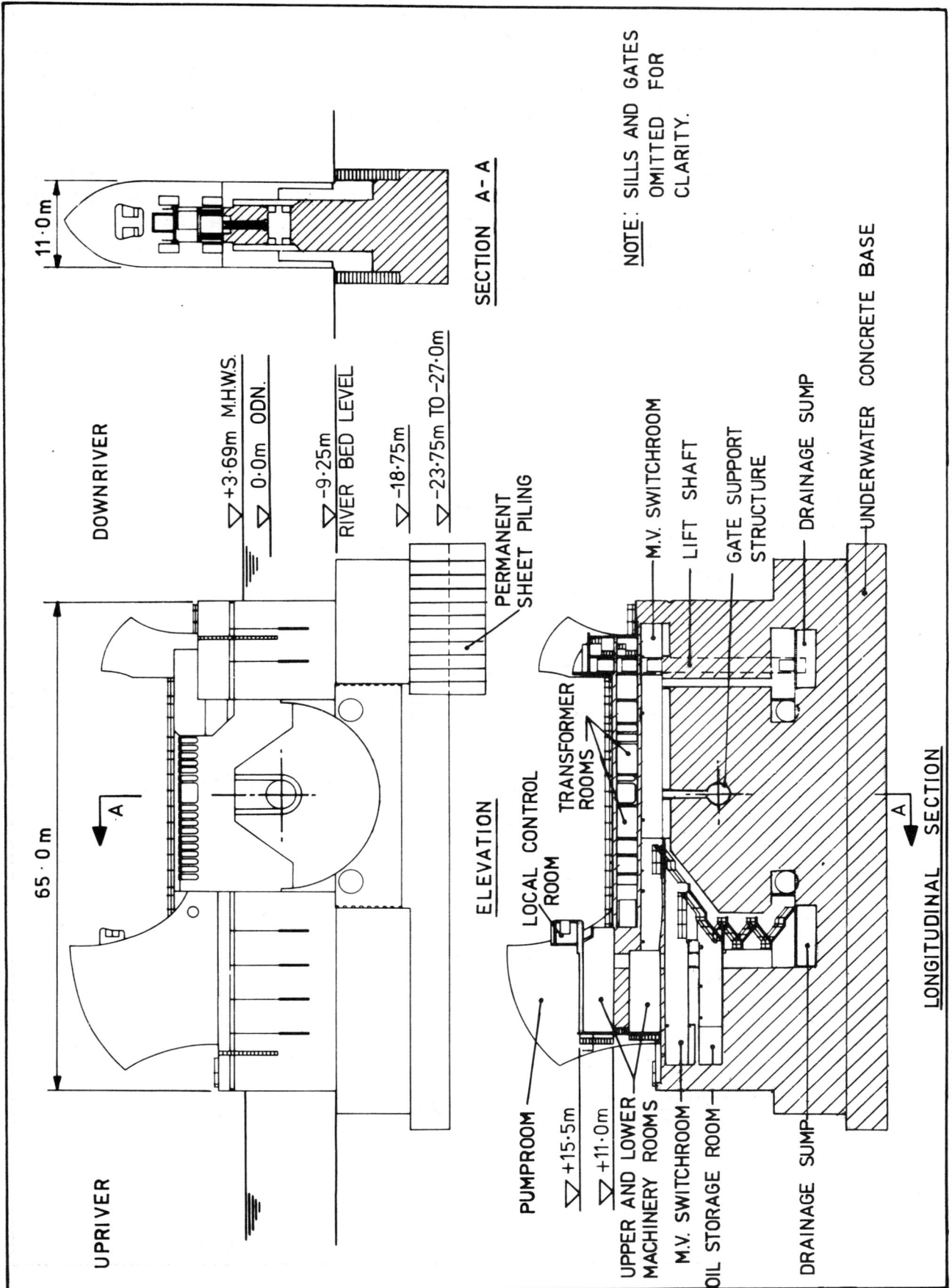

Fig.3. General arrangement of main pier

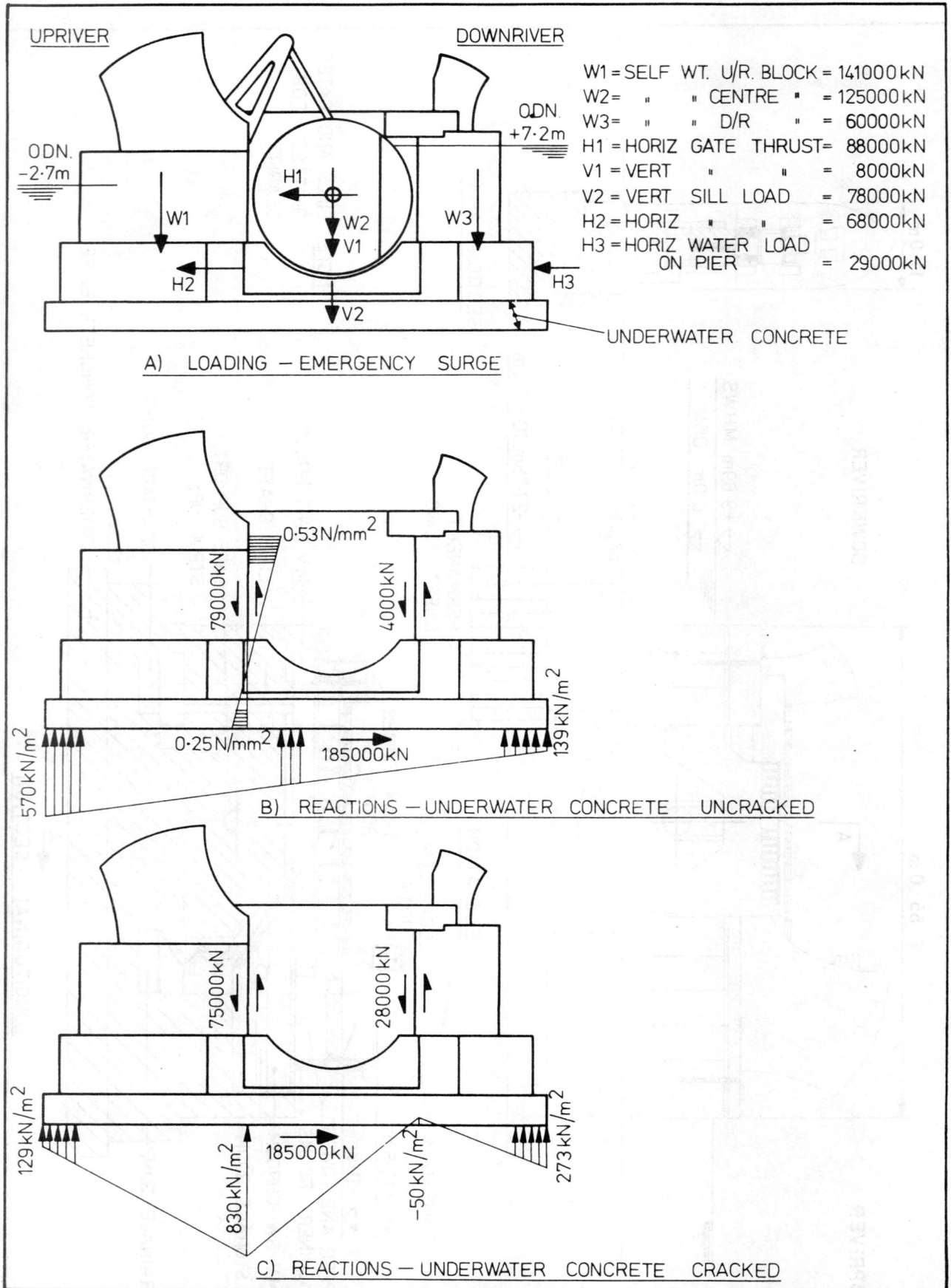

UPRIVER   DOWNRIVER

ODN.
+7·2m

ODN.
-2·7m

W1

H1

W2
V1

W3

H2

H3

V2

W1 = SELF WT. U/R. BLOCK = 141000 kN
W2 =   "   "   CENTRE   " = 125000 kN
W3 =   "   "   D/R   " = 60000 kN
H1 = HORIZ GATE THRUST = 88000 kN
V1 = VERT   "   "   =   8000 kN
V2 = VERT SILL LOAD   = 78000 kN
H2 = HORIZ   "   "   = 68000 kN
H3 = HORIZ WATER LOAD
       ON PIER   = 29000 kN

UNDERWATER CONCRETE

A)   LOADING — EMERGENCY SURGE

0·53 N/mm²

79000 kN   4000 kN

139 kN/m²

570 kN/m²

0·25 N/mm²   185000 kN

B)   REACTIONS — UNDERWATER CONCRETE UNCRACKED

75000 kN   28000 kN

129 kN/m²   273 kN/m²

830 kN/m²   185000 kN   −50 kN/m²

C)   REACTIONS — UNDERWATER CONCRETE   CRACKED

Fig.4. *Main pier foundation loading diagrams*

and alluvial deposits overlie the Thanet Sand at the north side of the river.

19. The chalk has minimal cover in the river bed towards the south bank where the gravel has been eroded. Evidence from the site investigations and during construction, indicates that mean groundwater levels in the chalk and in the gravels tend to be somewhat below mean tide level and have a diminished tidal response, particularly as distance inshore from the river banks increases. Levels in the gravels at the north shore are reported by the Institute of Geological Sciences to be influenced by those in the Royal Docks.

20. In view of the compactness of the Thanet Sand stratum at the locations of the North Abutment and Piers 1, 2 and 3, it is planned to found those structures above the chalk, thus avoiding excessively deep construction.

PIERS - GENERAL DESCRIPTION

21. The function of the piers is to contain the gate support structures and the gate operating machinery, and to support the ends of the gate sills and house the various operating facilities. These include the local control room, switchrooms, oil tank rooms, drainage sumps, navigational aids and cranage. Fig. 3 shows the general arrangement of a main pier supporting two 61 m Rising Sector Gates.

22. The main gate operating machinery is arranged at the upriver end of the pier at three levels, with the hydraulic cylinders on the two lower floors and the pumpsets and local control room on the highest floor. A stainless steel clad timber roof structure encloses the machinery floors. The gate shift mechanism is housed in separate superstructures at the downriver end of the pier. At the upriver end of each main pier the access from the subway to the upper decks is by stairway while at the downriver end a lift serves all levels between the other subway and the upper decks. In addition, at each end of the pier there is a hoist well, a services riser and ladder access.

23. The cutwater shape at the pier noses was selected on the basis of hydraulic model tests to give minimum disturbance to river flow. The dimensions of the piers and abutments are the minimum commensurate with their function. The foundation levels are dictated by the sill seatings overlying a pre-determined depth of tremie concrete base, except where a deeper foundation is required to achieve competent bearing capacity on the basis of the geotechnical assessment.

24. The piers are equipped with bollards, mooring rings, and external access ladders to assist maintenance operations. Timber fenders protect the bottom edges of the roof structures.

PIER FOUNDATION DESIGN

25. The piers are founded on unweathered chalk or undisturbed Thanet Sand. The selection of suitable strata and levels is based on penetra-

tion test values, which give a guide to the chalk quality or the competence of the Thanet Sand. The foundations have to be sufficiently competent to accept the imposed loading whilst limiting differential settlements to acceptable values.

26. Vertical loading on the foundation is due to the dead loads of the piers, gates, machinery and sills, together with superimposed water and uplift loads. In cases where gate spans differ on each side of the pier, these vertical loads are asymmetrical. Transverse loading occurs at the abutments and at Piers 3, 4, 8 and 9, which have different bed levels on each side, with consequent lateral soil loadings being imposed. Horizontal hydrostatic loading on the piers and associated gates and sills under operational conditions have also to be taken by the foundations.

27. Paper 4 describes the extensive geo-technical investigations which were undertaken in order to determine the characteristics of the foundation strata. Design assumptions may be summarised as follows:-

(a) For sound chalk of Grade III or better as classified by Wakeling (ref. 2):

Allowable net bearing capacity   600 kN/m2

Modulus of elasticity   1000 MN/m^2

Concrete/chalk friction angle   40°

(b) For undisturbed Thanet sand:

Allowable net bearing capacity   600 kN/m^2

Concrete/sand friction angle   34°

28. In setting acceptable limits of differential settlements within a single foundation and between adjacent piers, the clearances between the gates and the concrete structures had to be considered. These however were less critical than the limits of movements that could readily be provided for in the sill-pier subway connections and the sill seatings. Unless careful control is exercised during foundation preparation, the settlements caused by consolidation of the disturbed formation and of loose deposits on the excavated surface, are likely to be of greater magnitudes than settlements from all other causes. Although much of these initial settlements will have taken place before forming the sill seatings and other controlling features of the structures, allowances for non-elastic long-term effects had to be made. Settlement plates have been installed at formation level to monitor behaviour.

29. The foundations were designed on the assumption that they would be formed within deep sheetpiled cofferdams around each pier, and this is the method of construction adopted by the Contractor. The base of each pier or abutment is concreted underwater in one large unreinforced pour, which provides weight to aid in resisting uplift after dewatering, as well as transferring the pier loads to the formation.

30. In determining loads on the chalk or Thanet Sand, it is assumed that the underwater concrete base remains uncracked and capable of accepting moderate tensile stresses. However, check calculations were carried out, assuming the underwater concrete base to be cracked on the line of the main keyed contraction joints in the pier superstructure. Comparative results for these two approaches are illustrated in Fig. 4 for a main pier under emergency full closure conditions, the most severe design case. For the "cracked underwater base" in this particular case the maximum bearing pressure regarded as acceptable was increased in view of the severity of the assumptions.

31. Wall friction and passive end resistance was neglected in assessing factors of safety to horizontal movement of the structures. On this basis, the factor of safety for a main pier under the extreme surge loading condition is approximately 2.0.

ALTERNATIVE METHODS OF CONSTRUCTION

32. At the pre-tender stage a number of alternative possible methods of constructing the piers were examined for comparison with the deep cofferdam scheme. Most of these incorporated caissons of various forms, although the use of piled foundations in combination with a shallower cofferdam was also considered.

33. It was felt that pneumatic caissons should be avoided if possible. The use of caissons placed in predredged excavations on prepared foundations at the final required levels was discounted in these studies on the grounds of impracticability, both because of the extensive disruption to the river bed caused by the very deep dredging, and because of difficulties envisaged in ensuring an adequate foundation. Accordingly, the studies concentrated largely on open-well dredged caissons. The scheme most favoured involved a single caisson of reinforced concrete cellular construction, which would be floated into position (with temporary bulkheads closing off the bottom of the compartments to provide the required buoyancy) and sunk either on to the river bed or into a partially predredged excavation. Due to the need to form the complex shapes required at the sill seatings, subway connections and above, the centre portion of the caisson structure would have been restricted in depth of section, thus causing a number of structural and practical problems (see Fig. 5). The most difficult structural problem was to provide longitudinal torsional stiffness to allow for the condition of the caisson being unsupported at one corner.

34. A steel framed sheet pile cofferdam mounted on top of the concrete caisson would have been required in order to extend the caisson walls to above high water and to seal across the sill bearing recesses. Depending on whether the caisson were to be placed on a predredged intermediate level or the river bed, as well as on considerations of draft limitations between the dry dock and the pier location, the steel-walled superstructure would either have been installed in dry dock or at the pier position.

35. It was envisaged that bentonite injection would have been required to facilitate sinking of the caisson during excavation. If compressed air working at pressures of over three atmospheres was to be avoided, the development of special underwater equipment for cutting the chalk would undoubtedly have been necessary. The length of the caisson (approximately 80 m), and the unfavourable geometry at its central portion, would have necessitated stringent measures during float-out, passage to the pier site and sinking, if the risk of overstress were to be avoided.

36. To overcome the problems associated with a long and weak-sectioned unit, the possibility was examined of installing the pier foundation structure as a series of separate smaller caissons, but the problems associated with forming adequate seals between the units, as well as the reduction in structural continuity of the completed base, appeared to present serious disadvantages.

37. All caisson schemes would have required the solution of a number of problems arising from likely inaccuracies of line and level in final positioning, this being particularly the case for caissons sunk into position by excavation.

38. In view of the difficulties indicated by these studies and the range of variants possible, it was not considered worthwhile to present, in the tender enquiry documents, a specific alternative design based on a caisson scheme. As it turned out, one tenderer offered outline proposals for a single-unit caisson scheme (in which the excavation would have been fully completed before positioning the caisson on prepared pads and subsequently under-filling), but at the same time indicated that its cost would exceed that of the deep cofferdam method. The scheme was not therefore examined in any detail although it was evident that some of the problems referred to earlier regarding a fully predredged installation would doubtless have led to grave doubts about its feasibility.

39. The tender enquiry documents contained another alternative foundation scheme, however, namely the use of bored in-situ interlocking concrete piles ("secant" piles), both for the main piers and for the lesser piers and the abutments. The same tenderer proposed the adoption of this scheme for the smaller structures.

40. The secant pile foundation scheme is indicated in Fig. 6. It would have involved the installation of a cellular arrangement of longitudinal and transverse walls of secant piling to support a mass concrete base located at a level generally higher than that needed for the deep cofferdam method, thus permitting large reductions in excavation and other quantities. The in-situ piling would have been cast by tremie concreting to the requisite level within the excavated and bentonite-filled self-boring casing (the "Benoto" system or similar). A shallower sheet piled cofferdam would then have

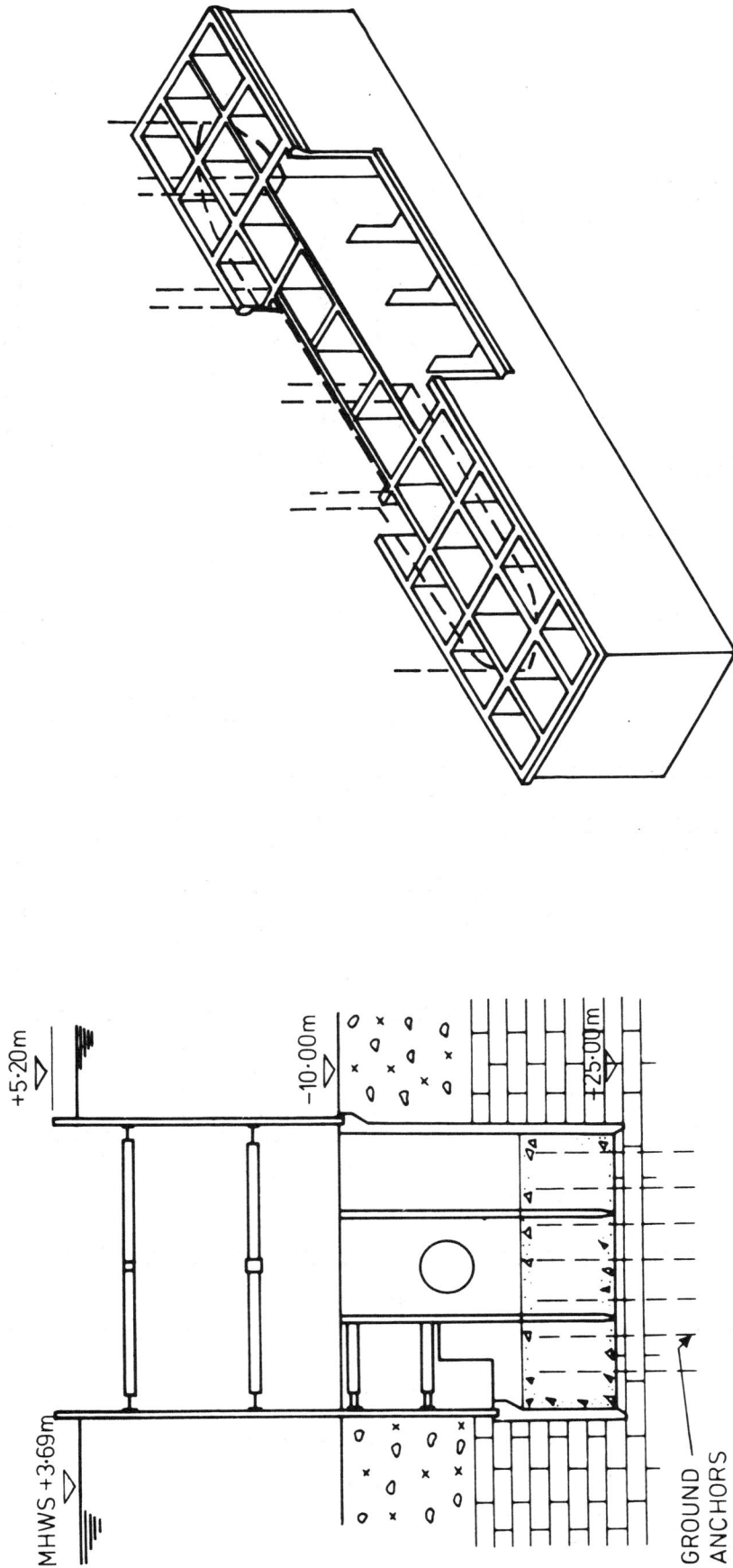

ISOMETRIC OF CAISSON

MHWS +3·69m

+5·20m

−10·00m

+25·00m

CROSS SECTION

CAISSON SUNK, TREMIE CONCRETE
SEAL PLACED AND STABILISED
COFFERDAM COMPLETED AND DEWATERED.

GROUND
ANCHORS

Fig.5. Caisson alternative proposal for pier construction

167

−19·75m TOP OF SECANT
PILES AT SILL RECESSES

−13·15m TOP OF SECANT
PILES AT U/R AND D/R END

SHEET PILED
COFFERDAM

PROPOSED
FOUNDATION
LEVEL.

CONTRACTION
JOINTS

PROPOSED
FOUNDATION
LEVEL

CONCRETE QUALITY EP(40)

CONCRETE QUALITY F(40)

▽ +5·2m

▽ 0·0m
ODN

M.H.W.S.
+3·69m ▽

▽ −9·25m

2·0m
MIN.

▽ −16·0m

▽ −18·75m

▽ −20·0m

▽ −22·0m

▽ −26·0m

▽ −25·0m

PROPOSED FOUNDATION LVL.

−26·0m ▽

▽ −31·0m

▽ −32·0m

2·0m
MIN.

25·0m   36·0m   13·0m   15·0m

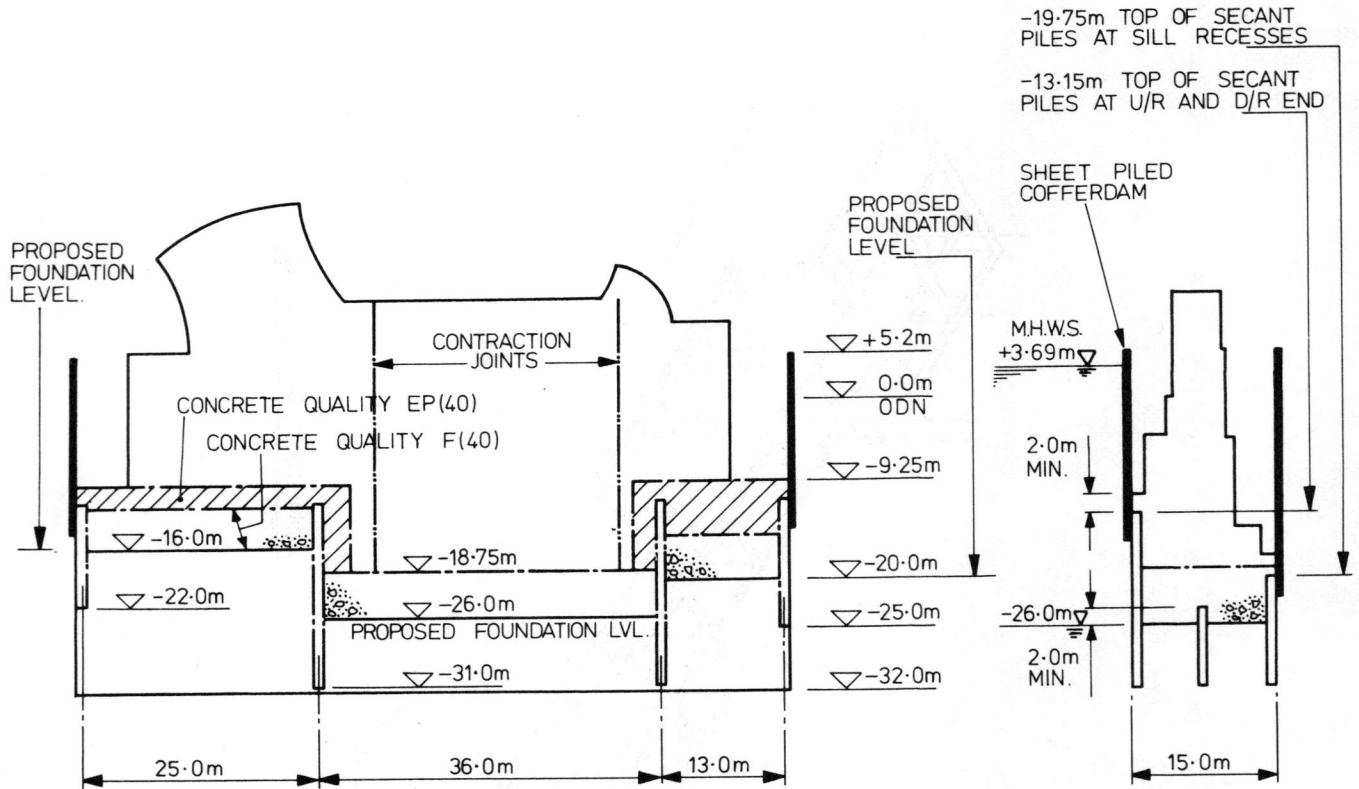

Fig.6. Foundation wall alternative proposal for pier foundations

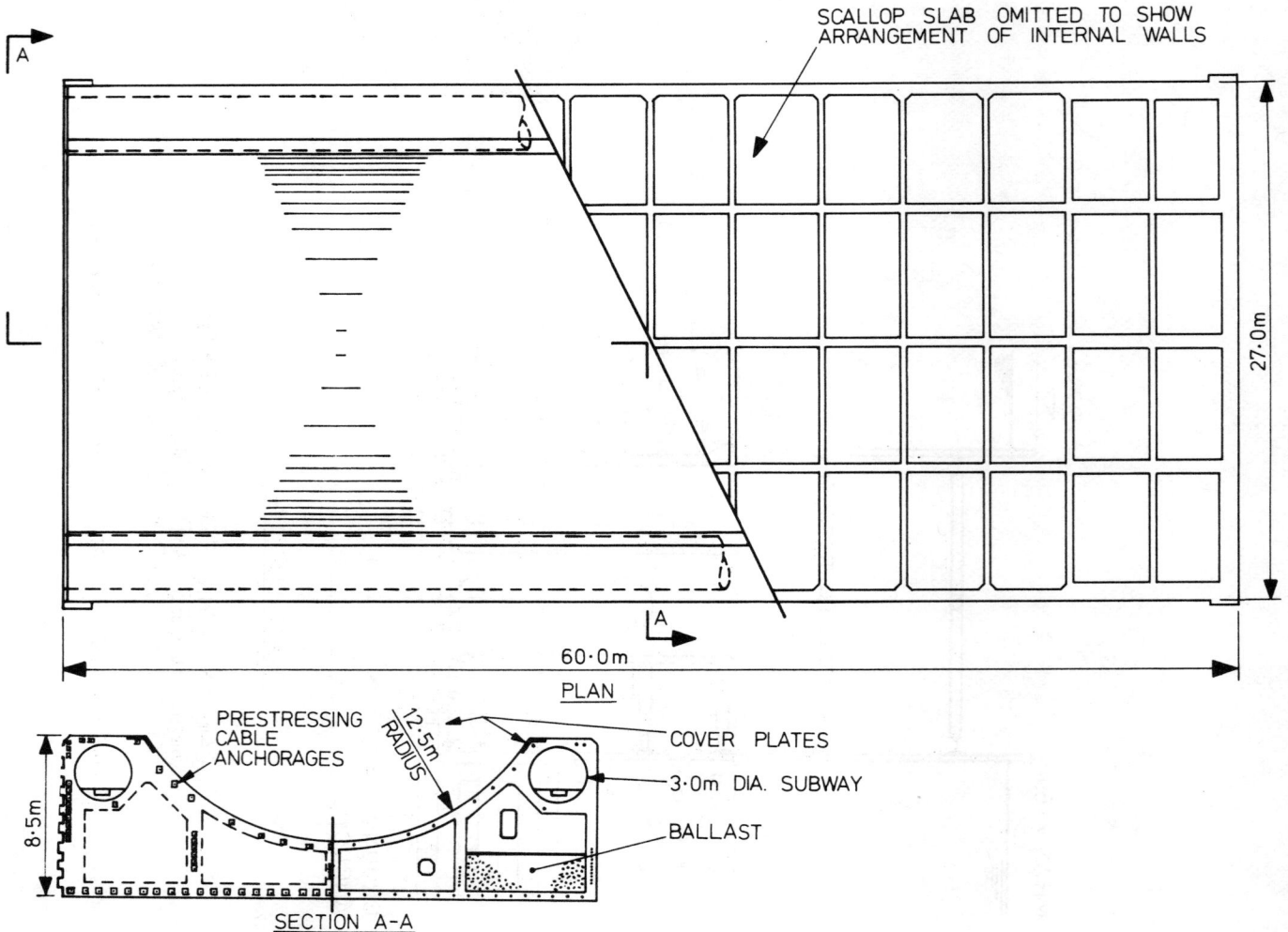

SCALLOP SLAB OMITTED TO SHOW
ARRANGEMENT OF INTERNAL WALLS

A

A

27·0m

60·0m

PLAN

PRESTRESSING
CABLE
ANCHORAGES

12·5m
RADIUS

COVER PLATES

3·0m DIA. SUBWAY

8·5m

BALLAST

SECTION A-A

Fig.7. Precast concrete sill: type 1

been driven, sealed against the secant pile walls, and braced to permit excavation under water. After preparation of the pile heads by diver, and cleaning of the interior faces of the piled walls, the tremie-concreted slab would have been placed. The remaining cofferdam bracing would then have been installed and the cofferdam dewatered. The depth of the piling was designed to be such as to transfer the vertical and horizontal loads to the chalk or Thanet Sand by skin friction on both faces of the piled walls. The piles would have been reinforced sufficiently to anchor the dewatered cofferdam base against uplift during construction, and to transfer any net tensile forces arising during Barrier operation.

41. Minimum interlock dimensions were specified to ensure a high degree of continuity of the secant-piled foundation walls. This was important to ensure cofferdam sealing, avoidance of "piping" risk, and a full "shear wall" action in the completed structure.

STRUCTURAL DESIGN OF PIERS

42. The lower parts of the pier tend to be generally mass concrete while the upper parts are reinforced. The major loads to be transferred to the foundations, in addition to self weight, are from the gate support structures and the sills, which apply maximum design loads of the order of 9,000 tonnnes and 10,500 tonnes respectively to the piers. These loads vary in magnitude and direction for different operating conditions. The resultant levels of stress are checked at all locations, and where necessary the concrete is reinforced accordingly.

43. Because of the lengths of the piers, either one or two vertical contraction joints are provided for the full height of the pier above the underwater concrete base, to control thermal cracking. The contraction joints are keyed and, after a suitable interval to allow the concrete to cool, are grouted to ensure the transfer of compression and shear loads between blocks. The magnitude of the shear stresses requires the keys to have design reinforcement.

44. Structural reinforced concrete design is generally in accordance with C.P. 110. The minimum cover to reinforcement on faces exposed to water is 75 mm and anti-crack reinforcement is provided on all exposed surfaces. All external walls to chambers below water level are checked against the requirements of the water retaining code in regard to cracking at the water face.

45. The operating machinery was described in Paper 8. The items applying the greatest loads to the structure are the gate support structures, the rocking beam support brackets, the gate operating cylinders and the gate shift mechanisms. Extensive use is made of prestressed foundation bolts to limit movement under load. The structural design of the reinforced concrete elements supporting the machinery is generally based on design working loads, but is checked at appropriate factors of safety for extreme conditions which could arise from malfunction or

maloperation, e.g., full hydraulic thrust developed in a falling radial gate cylinder with the gate fully closed.

46. Account was taken of possible ship impact by avoiding external wall thicknesses of less than 600 mm wherever possible, and by generous provision of reinforcement at potentially vulnerable locations.

CONCRETE SPECIFICATION FOR PIERS

47. A characteristic cube strength at 56 days of 15 N/mm^2 was specified for the unreinforced concrete in the underwater bases, together with a maximum cementitious content of 450 kg/m^3. Pulverised fuel ash (PFA) replacement of Ordinary Portland Cement (OPC) for up to 50% of this cementitious content was required, to moderate the rate of heat output and for economic reasons.

48. The lower parts of the pier superstructures comprise mainly massive concrete sections. Except where higher strength concrete is necessary, e.g. 40 N/mm^2 concrete at sill support points, the specification requires these portions of the pier to be of low strength (15 N/mm^2), low heat, unreinforced hearting concrete surrounded by a 600 mm thick skin of facing concrete (20 N/mm^2) containing anti-crack reinforcement. To control heat output, the cementitious content of these concretes is specified to contain 30% PFA and 70% OPC. The maximum cementitious content of the hearting concrete is restricted to 200 kg/m^3.

49. During construction, the Contractor has proposed a "unified" concrete mix, roughly similar to the facing concrete, in place of the specified hearting and facing concretes. This is to avoid the problems of low workability of the hearting concrete and of placing two grades of concrete simultaneously, and has been permitted subject to control of temperature by the use of insulation. Thermocouples have been installed for monitoring purposes. Use has been made of recently published work by the Cement & Concrete Association in adopting practical criteria for temperature control (ref. 3).

50. The upper parts of the piers, which are reinforced and contain large numbers of chambers, are generally in 30 N/mm^2 concrete, but this is varied where there are special requirements, e.g. 40 N/mm^2 concrete is used around the gate support structures.

51. Routine tests are carried out on all aggregates to check their potential reactivity with alkalis. Results of mortar bar tests of the North Sea dredged aggregate being used are well within the limits permitted by A.S.T.M. C227-71.

PIER ROOF STRUCTURES

52. The shape and stainless steel finish of the roofs covering the main machinery rooms were adopted for architectural reasons. Programme constraints required that construction work on

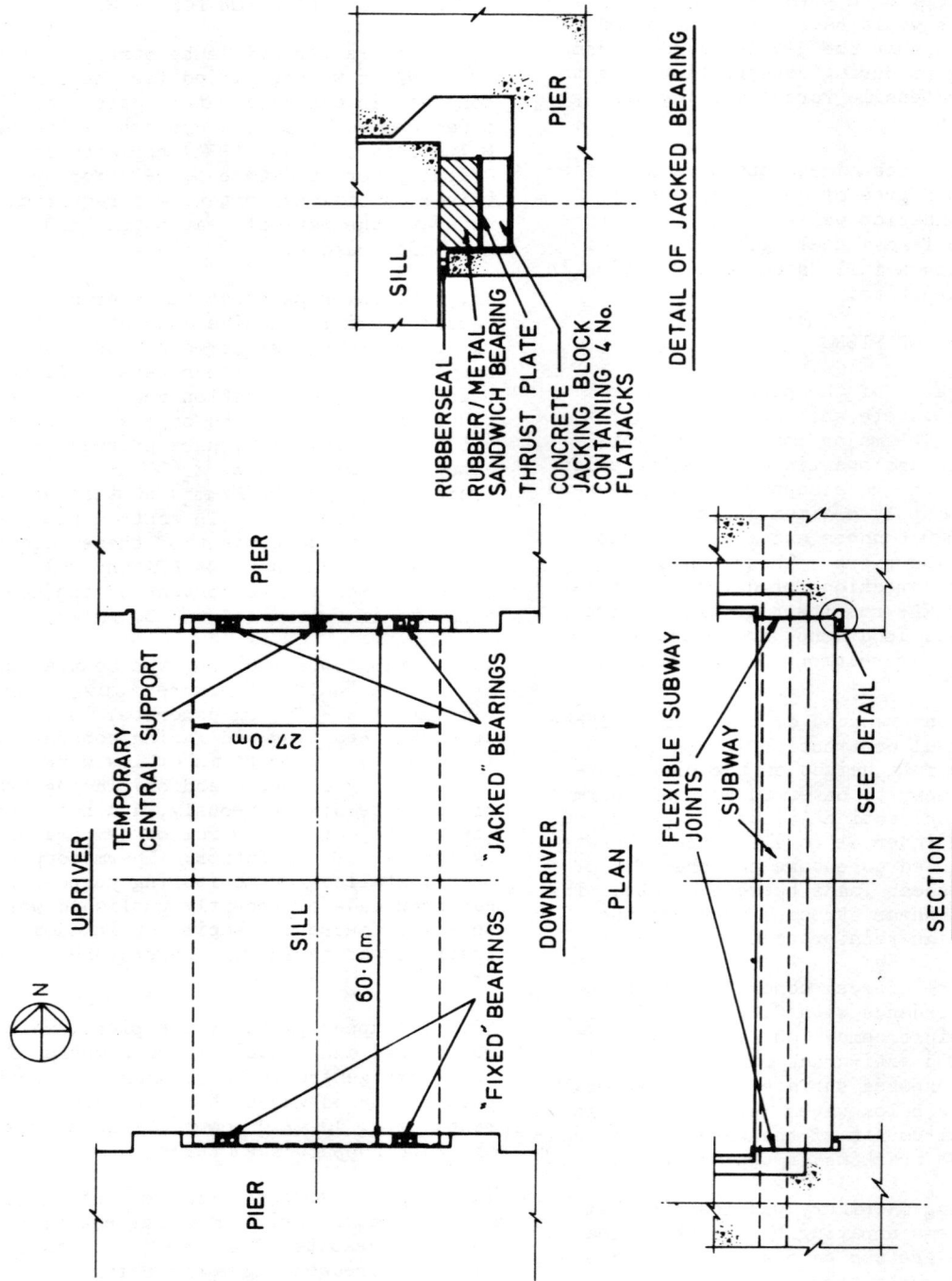

DETAIL OF JACKED BEARING

RUBBERSEAL
RUBBER/METAL SANDWICH BEARING
THRUST PLATE
CONCRETE JACKING BLOCK CONTAINING 4 No. FLATJACKS

UPRIVER

PIER
TEMPORARY CENTRAL SUPPORT
SILL
27.0 m
60.0 m
"JACKED" BEARINGS
"FIXED" BEARINGS
DOWNRIVER
PIER

N

PLAN

FLEXIBLE SUBWAY JOINTS
SUBWAY
SEE DETAIL

SECTION

Fig. 8. Bearings for sill type 1

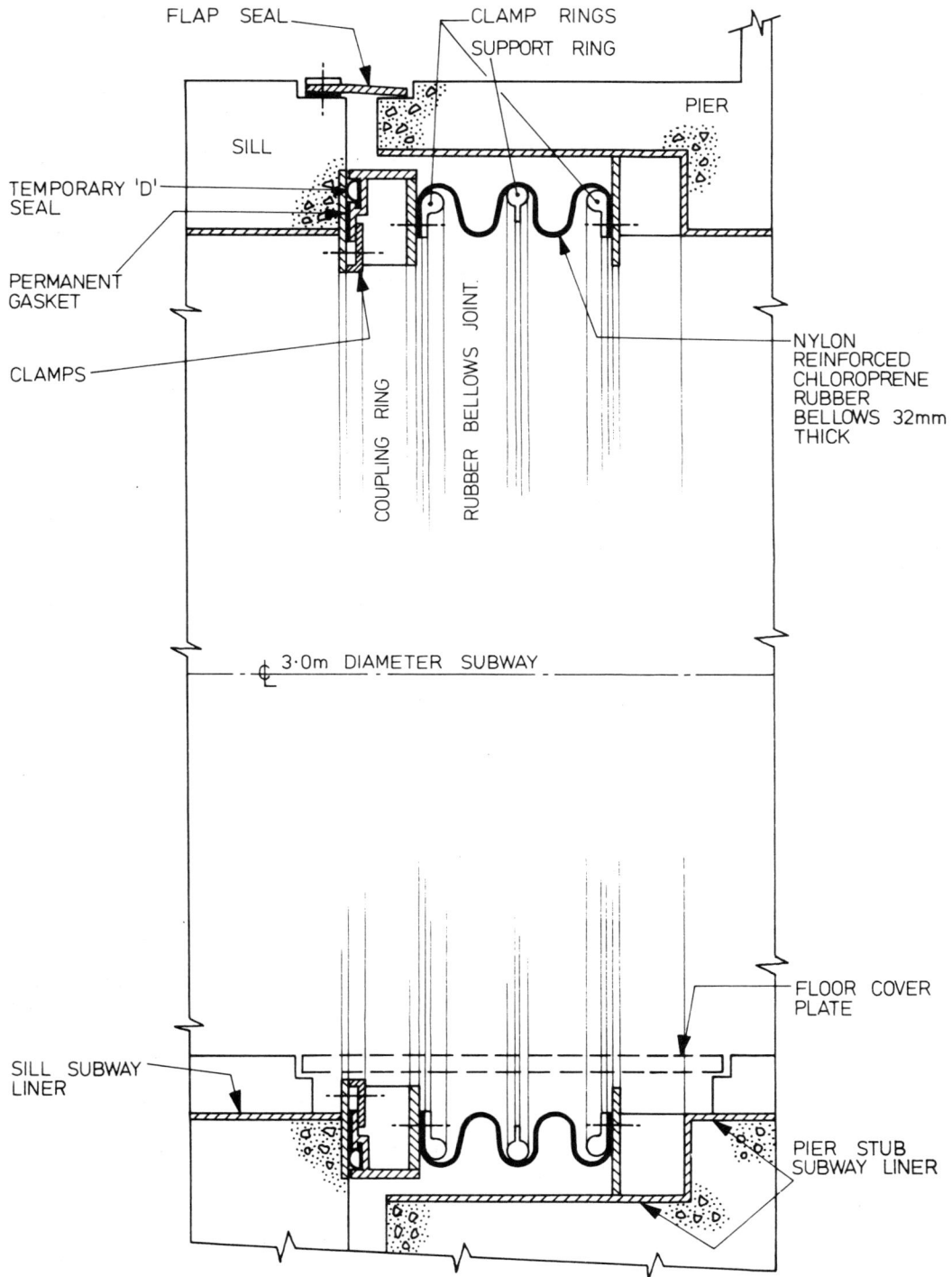

Fig.9. Flexible subway joint

with flexible fillers and sealants to prevent
ingress of silts while allowing the full design
range of movement. The subway joint between
the sill and the pier or abutment is sealed by
conventional rubber waterstops cast into the
surrounding concrete.

73. The design of the reinforcement is to
CP 110. Sulphate-resisting cement is specified
for the concrete. Minimum cover to reinforcement is 75 mm.

74. As in the precast sills, the subways are
steel lined to ensure watertightness, and the
liners are checked against buckling under external water pressures.

BED PROTECTION

75. The bed protection scheme is designed to
protect the Barrier structures from the effects
of normal river flow, operation of the gates
(including partial closure operation), and the
unlikely eventuality of a Rising Sector Gate
failing to close during a surge tide closure
operation.

76. Mobile bed hydraulic model tests were
carried out at a scale of 1:50 to determine the
extent of the bed protection and the stone sizes
required at each location in the works to meet
the design criteria. The tests also proved the
detail of the extreme upriver and downriver
edges of the protection.

77. As part of the test programme, investigations were made of the scour that would develop
under gate-failure conditions if the bed were
not protected, and using the profiles so determined, the stability of the structures was
examined. However, in view of the uncertain-
ties relating to the accurate reproduction of
erosion effects in an idealised model, it was
considered that full bed protection for this
case must be provided.

78. The results of the model tests were
rationalised in the working drawing arrangement to avoid practical difficulties in placing
relatively small areas of different sized
stones adjacent to one another. Adjustments
to the stone sizes were also made to compensate
for differences between the specific gravities
of the model and specified prototype materials.

79. The design comprises a polypropylene
filter cloth to which is attached a mesh of wiepen which float the cloth and keep it in shape
during installation. The wiepen also retain
in position an underlayer of rockfill which
sinks the cloth and protects it from being
torn by the overlying rip-rap. The stone
sizes and layer thicknesses of the rip-rap vary
depending on location, as shown in Fig. 10.
Concrete weights are attached to the extreme
ends of the cloth as anchors, and to help the
protection to conform to the shape of any scour
hole which might form beyond the protected area
of the river bed.

ACCESS BRIDGES

80. The access bridges provided at the falling
radial gate openings carry foot traffic, cables
and pipework. The northern bridges also provide mobile crane access from the shore to
Piers 1 and 2 for the removal of items larger
than can be handled by trolley.

81. A steel truss design was selected as a
light and rapidly-erected construction which
provides a top deck for personnel or vehicular
access and space for an enclosed services route
below. The steelwork design is in accordance
with B.S. 153.

82. For architectural reasons, the services
route is enclosed by profiled stainless steel
cladding located within the bridge framework.

RIVER WALLS

83. The river walls within the site boundary
have been, or are being, realigned to straighten
the banks, and raised to meet the statutory defence levels. The walls are of tied-back sheet
pile design, anchored by sheet piled anchor
walls or by piled relieving platforms, depending
on local ground conditions and the nature of
the adjacent onshore works.

84. Provision is made for the public to have
access to the river walls wherever possible.
On the south bank a public walkway is routed
through the onshore works site below the operating deck level, passing behind the south abutment but without access to the operating areas.

ONSHORE WORKS

85. The onshore works on the south bank comprise the control building, generator building,
workshop, substations, services subways, car
park, public walkway, operating deck and ramps,
gatehouse and boundary walls. Fig. 11 indicates
the layout.

86. Foundation conditions are generally made
ground and alluvium overlying gravel and chalk,
the upper portions of the chalk being weathered.
The earlier industrial developments in the area
have been cleared, leaving existing floor slabs
and piled foundations in place. Nearly all of
the new buildings and structures will be
supported by end bearing piles with toe levels
either in the gravel, which is of variable
thickness and density, or the chalk.

87. The building foundations and basements are
of concrete construction, sulphate-resisting
cement being used to counter aggressive soil
conditions. The building frameworks are of
normal reinforced concrete or structural steel,
clad generally with precast concrete cladding
panels. For architectural reasons, the roofs
are clad in stainless steel on timber boarding.

88. The onshore works on the north bank comprise substations, service subways, gatehouse
and boundary walls.

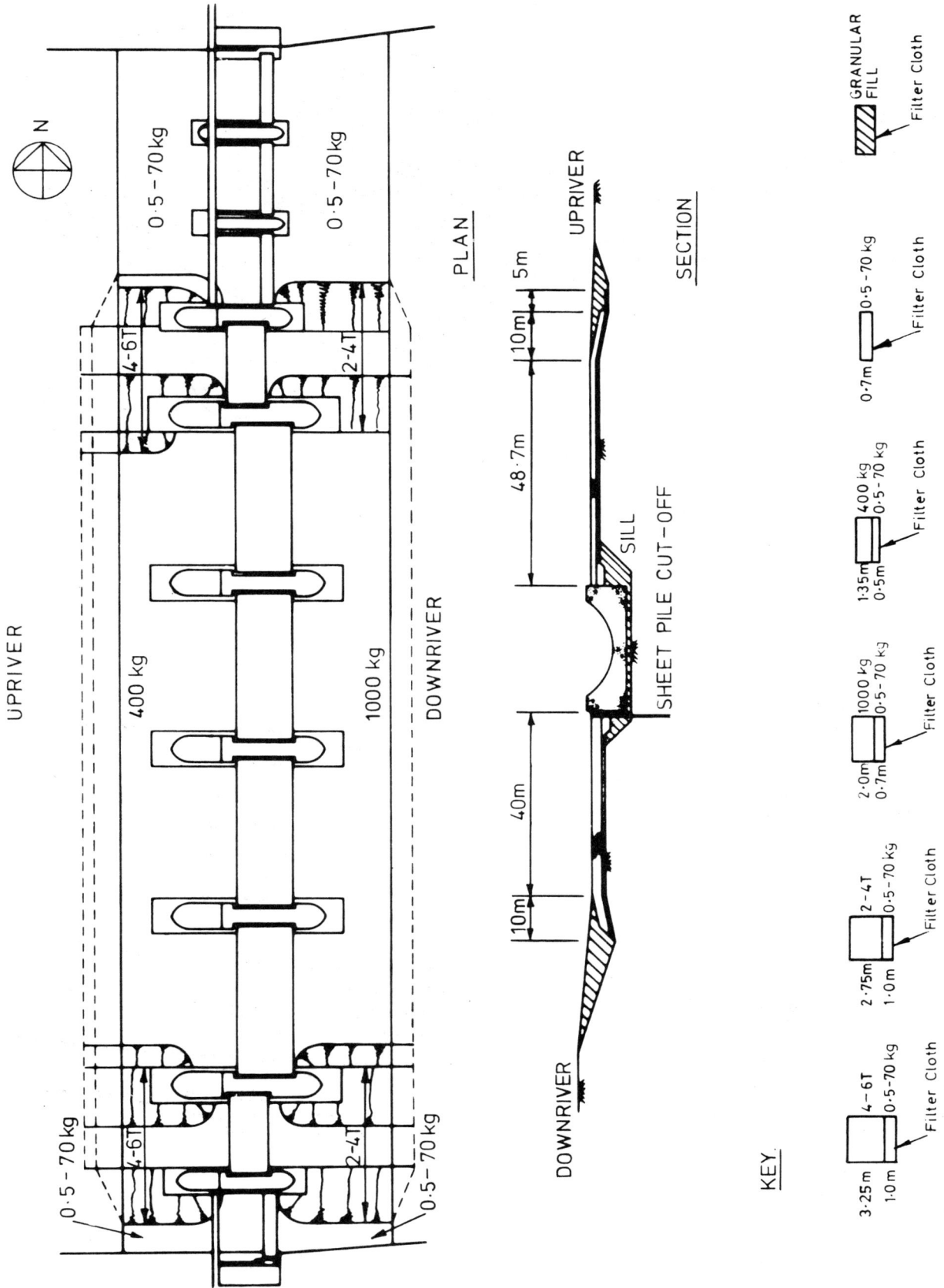

Fig.10. Bed protection

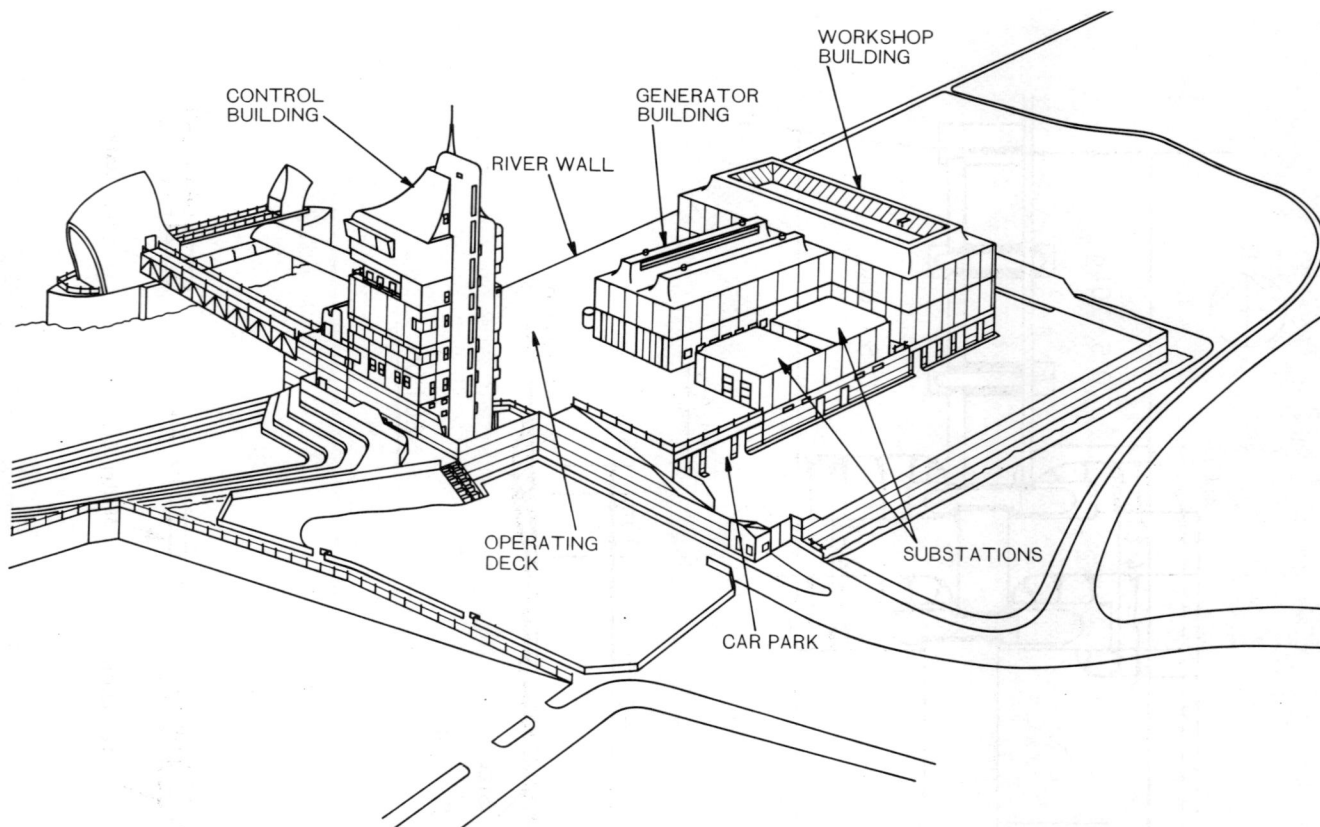

Fig.11. Onshore works: south bank

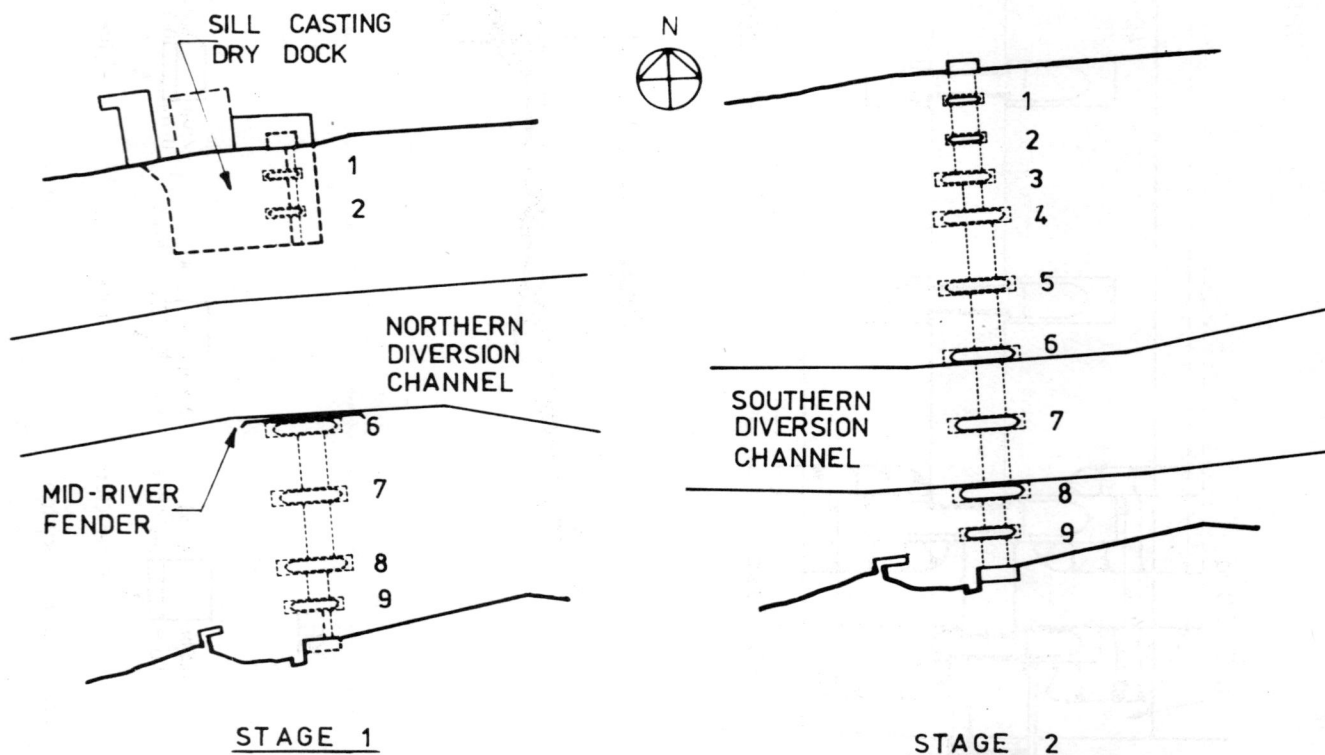

STAGE 1

STAGE 2

PROTECTION DOLPHINS AND ACCESS JETTY NOT SHOWN.

Fig.12. Construction stages and river diversion

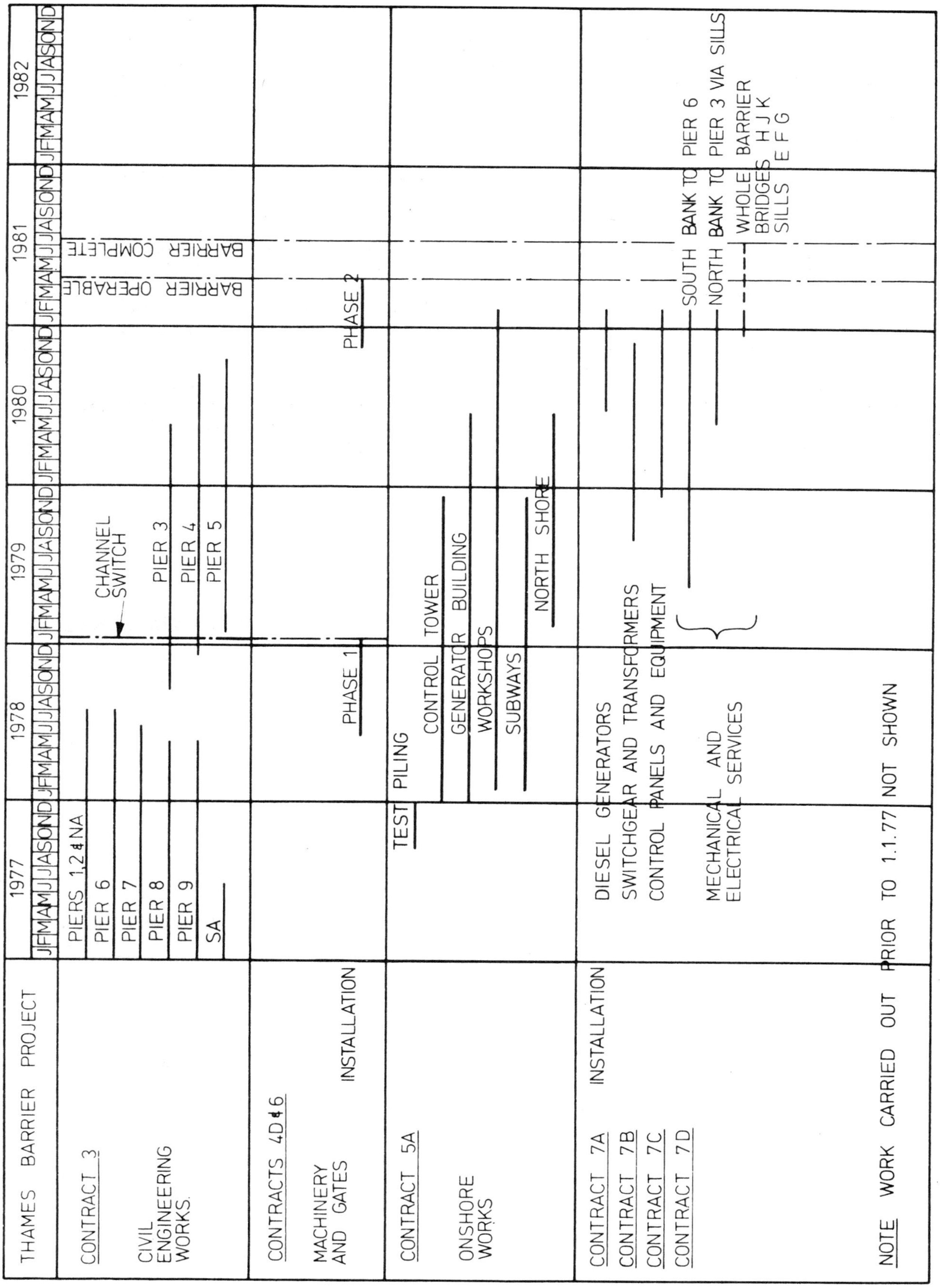

THAMES BARRIER PROJECT

| | 1977 | 1978 | 1979 | 1980 | 1981 | 1982 |

CONTRACT 3
CIVIL ENGINEERING WORKS.

PIERS 1, 2 & NA
PIER 6
PIER 7
PIER 8
PIER 9
SA
CHANNEL SWITCH
PIER 3
PIER 4
PIER 5

BARRIER OPERABLE
BARRIER COMPLETE

CONTRACTS 4D & 6
MACHINERY AND GATES   INSTALLATION

PHASE 1
PHASE 2

CONTRACT 5A
ONSHORE WORKS

TEST PILING
CONTROL TOWER
GENERATOR BUILDING
WORKSHOPS
SUBWAYS
NORTH SHORE

CONTRACT 7A   INSTALLATION
CONTRACT 7B
CONTRACT 7C
CONTRACT 7D

DIESEL GENERATORS
SWITCHGEAR AND TRANSFORMERS
CONTROL PANELS AND EQUIPMENT
MECHANICAL AND ELECTRICAL SERVICES

SOUTH BANK TO PIER 6
NORTH BANK TO PIER 3 VIA SILLS
WHOLE BARRIER
BRIDGES H J K
SILLS E F G

NOTE WORK CARRIED OUT PRIOR TO 1.1.77 NOT SHOWN

Fig.13. Programme of site works

89. The works are partly behind the previous river wall, where made ground overlies alluvium, gravel and Thanet Sand, and partly in new fill to be placed behind the North Abutment. To control settlements, the structures are piled. The north bank area was the site of a chemical factory, and sulphate-resisting cement is used in concrete below ground.

90. The piling works on either bank have not yet been let and hence the types of piles have not been selected.

CONSTRUCTION SEQUENCE

91. The contract documents laid down that the works were to be constructed in two stages. Stage 1, which is now under construction, comprises the piers and sills from Pier 6 southwards, leaving the north side of the river channel for shipping. A temporary piled fender structure protects the Pier 6 cofferdam. When the Stage I works are complete (including the installation of the Rising Sector Gates and machinery), shipping will be routed through the southerly two main openings to enable Piers 4

and 5 and the related works to be built as Stage 2.

92. Fig. 12 illustrates the two construction stages, while Fig. 13 indicates the works programme as presently envisaged.

REFERENCES

1. Thames Barrier and Flood Prevention Act 1972.

2. WAKELING, T.R.M., A comparison of the results of standard site investigation methods against the results of a detailed geotechnical investigation in Middle Chalk at Mundford, Norfolk. B.G.S. Conference on in-situ investigations in soil and rocks, Paper 2, London, May 1969.

3. FITZGIBBON, M.E., Joint-free Construction in Rich Concretes; Paper 2, Symposium: Large Pours for R.C. Structures, University of Birmingham, September 1973.

15. Architectural aspects of the Thames Barrier

A. M. PETTY, AADipl, SPDip*

F. O. BROWN, BA, ARIBA†

INTRODUCTION

1. Collaborations between engineers and architects are not always blessed with success. Failures, when they occur, often seem attributable to a simple oversight - that no thought was given at the inception stage to the ultimate appearance of the project. Too often the architect is called in at the last moment to add some measure of artistic respectability, perhaps after the engineers' design has been rejected by the Royal Fine Art Commission. Too often it is felt that the purity of the engineers' concept has been ruined by the architect's attempts to give it a veneer of style. The rare exceptions to this generally occur when the engineer and the architect are combined in a single person or organisation, and when working methods favour, rather than impede, the assignment of appropriate values to all design factors, including those concerned with appearance.

2. The discipline imposed by a large and complex project, drawing upon many branches of engineering science, should itself ensure the satisfactory resolution of all the conflicting requirements which inevitably occur, and enable due consideration to be given at an early stage, to environmental and visual aspects of design.

3. In the case of the Thames Barrier, the concern of the client body, The Greater London Council, to set high standards of design and public amenity was of great importance. The Council's aims, as the strategic Planning Authority for Greater London, are set out in the Greater London Development Plan, where attention is drawn to the unique importance of the River Thames in the life of London, both for recreation and commerce. The Greater London Development Plan states that the Council regards the condition and future of the river as matters of strategic concern to the whole of London, and has decided that when new developments occur the opportunity shall be taken to make thorough and effective improvements to the riverside scene.

4. As well as requiring attention to such matters as improvement of public access to the river, the provision of riverside walks and open spaces, and facilities for river sport, the Greater London Development Plan emphasises the importance of high quality for riverside design and landscaping and calls upon the Port of London Authority, the Thames Water Authority and the riparian London Boroughs to bring a special quality to everything connected with the River Thames. The Greater London Council itself undertakes to observe a specially high standard in its own works affecting bridges, approaches, embankments, roads and any other riverside developments.

The Greater London Development Plan refers to the great benefit which a Thames Barrier would confer upon the whole of the riverside London.

5. The Council announced its approach to the design of the Barrier in a statement to the Press in March 1973 which emphasised that, as well as providing the most effective means of protecting London against flooding, the Barrier must be of attractive appearance. It promised that the appearance of the Barrier would reflect its setting in the riverside scene, and that its form, colour and material would be carefully chosen to reduce the scale and apparent bulk of the river pier structures. It concluded: "The Barrier will be one of Britain's finest engineering achievements as well as a tourist attraction. So in visual as well as practical terms it will be something of which every Londoner can be proud".

6. These challenging statements established beyond all doubt the Greater London Council's concern for design quality, and placed a clear responsibility on all those concerned with the development of the project to ensure that the Barrier would look good in its setting and that it would enhance rather than detract from the public's enjoyment of the Riverside.

7. The Local Planning Authorities (the London Borough of Newham on the north side of the River, and Greenwich on the south) have also shown concern for these matters, and in particular for the preservation of existing riverside walks and accesses and for the future use by the public of the lands which will become surplus when the construction of the Barrier is completed.

THE DESIGN APPROACH

8. The design of the Barrier is unique not

*Group Architect, GLC Department of Architecture and Civil Design
†Assistant Divisional Architect, GLC Department of Architecture and Civil Design

only in engineering terms; it presents unique visual problems. It spans the River Thames, a distance of 520 metres, and occupies sizeable areas of land on each shore. Its main elements, the piers, the riverwalls and the shore installations, (which together provide the essential means of flood prevention), consist of a number of separated buildings and engineering structures serving a wide range of supporting functions.

9. The GLC Architects Department was nominated to work with the Consulting Engineers responsible for the detailed design and contract works. The foremost objective identified by the architects was to unify these disparate parts in a clearly recognisable form, and give to the Barrier a visual identity which would express the great significance of the project. The second was to ensure that this important new feature of the Thames would take its place harmoniously in the river scene. Further, that the design treatment would reflect the diversity of the engineering resources involved and enhance the unique qualities inherent in the project.

THE DESIGN SOLUTION

10. The seven main piers are the most visually significant parts of the Barrier and they will largely determine its scale and character. They are massive structures, the five largest measuring 65 metres by 11 metres, and their superstructures must be high enough to conform to the functional geometry of the gate-operating machinery. This presents some peculiar visual problems to the designers. When the Barrier is viewed obliquely, the navigational channels between the piers might not be apparent, and without appropriate architectural treatment, this could give the illusion of a single structure, a featureless wall, spanning the River from shore to shore, (Fig. 1). Seen from mid-stream, the Barrier might have an unfinished appearance resembling a partly completed bridge. Thus a very false and misleading impression could be given, whereas to be successful in visual terms the form of the Barrier should be immediately comprehensible from any viewpoint. The most promising approach in avoiding these visual ambiguities is to give each pier a clear and separate identity so that the presence of the navigational channels will always be implied even when they are not actually visible; further, to ensure that the form of each pier will contribute to the overall visual effect of the Barrier as a whole.

11. The design of the piers has passed through many stages, reflecting the development of the mechanical engineering concept, and at each point it has been studied in the light of the above criteria. A factor common to every stage has been been the requirement for superstructures to house the gate-operating machinery. The primary machinery, four hydraulic rams which operate the linkages to the gate-arms, is located at the up-river end of pier and needs superstructures of considerable volume extending to the full width of the piers (Fig. 2). At the down-river end of each pier the shift-mechanisms are located, requiring smaller housings on both sides of the

piers. At a late stage a lift shaft was introduced at the down-river end.

12. Attempts were made to modify the obvious rectilinear shapes of these superstructures to more sympathetic forms, but none proved satisfactory. Eventually it was decided to adopt curved arching forms which satisfy both engineering and visual criteria (Fig. 3). By introducing a recognisable theme which can be incorporated into the shore installations, it is hoped that the appearance of the Barrier will be unified and established in a memorable form. The changing patterns in the silhouette of the piers will provide a variety of interest when viewed by river users and pedestrians on the shore. The large linkage mechanisms which operate the gate arms require no protective enclosures, and these dynamic engineering forms will provide a dramatic contrast to the static forms of the superstructures, particularly when the gates are seen in motion during the operation of the Barrier (Fig. 4).

13. The model (Figs. 5 and 6) gives some idea of the close relationship which has been achieved between the forms of the superstructures and the machinery which they enclose, and the provision which has been made for the protection, maintenance and replacement of the machinery. Of the many proposals investigated, no other arrangement succeeded in reconciling the often conflicting demands of function and appearance.

DEVELOPMENT OF THE PIER SUPERSTRUCTURE DESIGN

14. The forms adopted for the superstructures are based upon a flexible system of geometric relationships, which has permitted the design and arrangement of the machinery to develop without imposing too great a restriction on space. It has given considerable scope for change in the internal arrangements, and has enabled the internal volumes to be increased by more than one third during this stage, without significant alteration to the external appearance. Because of this flexibility, it has been found possible to accommodate new engineering requirements as they developed, without overriding considerations of appearance. Consequently, in the final design now being built, the integrity of the original architectural concept has been maintained.

15. A number of structural methods and materials were considered for the construction of the superstructures, including gunned concrete, glass reinforced concrete, timber, steel, glass reinforced plastic and metal sheeting in various combinations. Finally, a timber structure clad in stainless steel sheeting with profiled aluminium gable ends, was chosen as offering the best prospects for reliabilty, durability and appearance.

16. The design of the timber structures was entrusted to the Timber Research and Development Association, and was based upon information provided by wind-tunnel tests carried out at Bristol University. The ribs, which are constructed on the 'Glulam' system, are of laminated

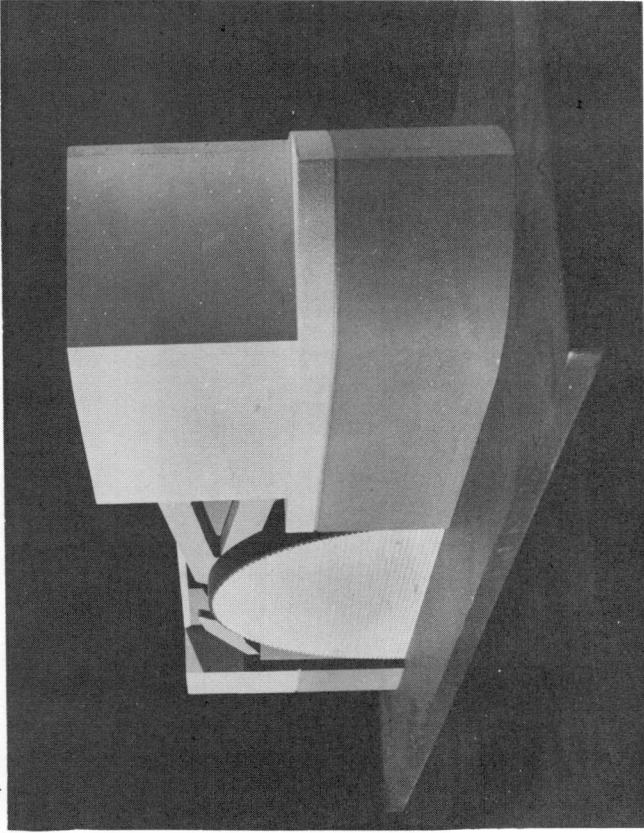

Fig.1

Fig.2

Fig.3

Fig.4

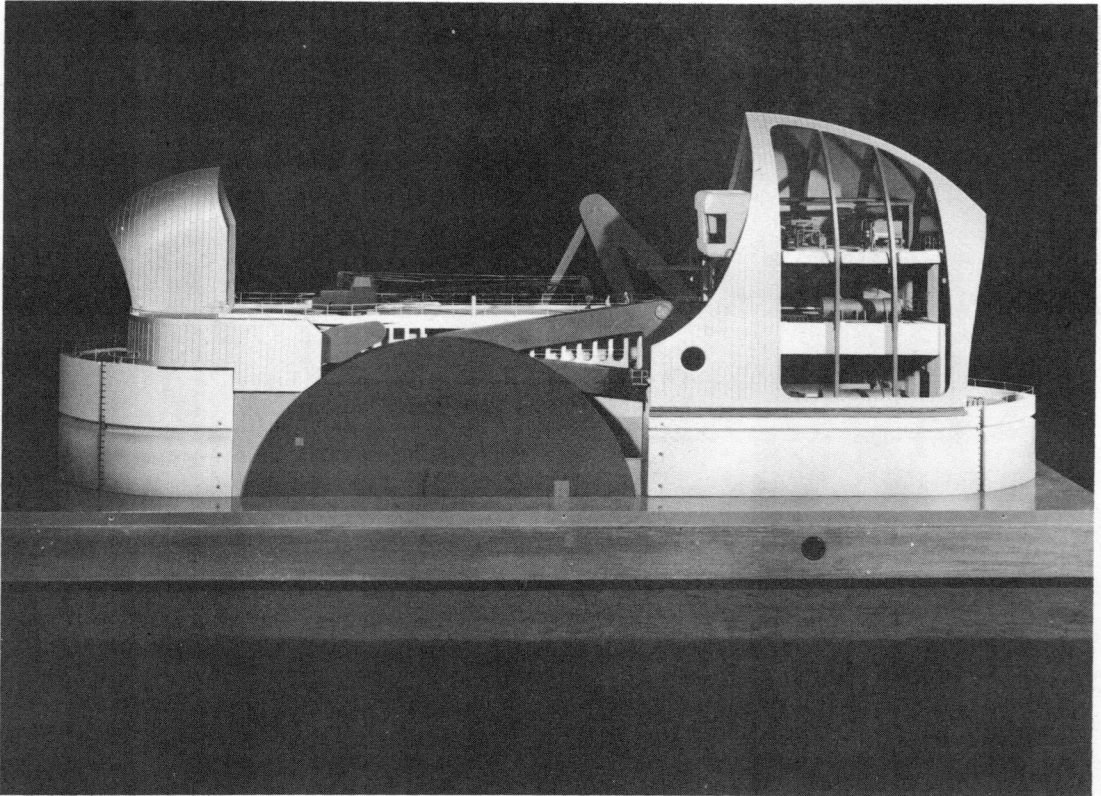

Fig.5

Fig.6

SOUTH ELEVATION.

PLAN AT LEVEL +5000

WEST ELEVATION.

DETAIL OF
STANDING SEAM

EAST ELEVATION.

+5000

DETAIL AT F.

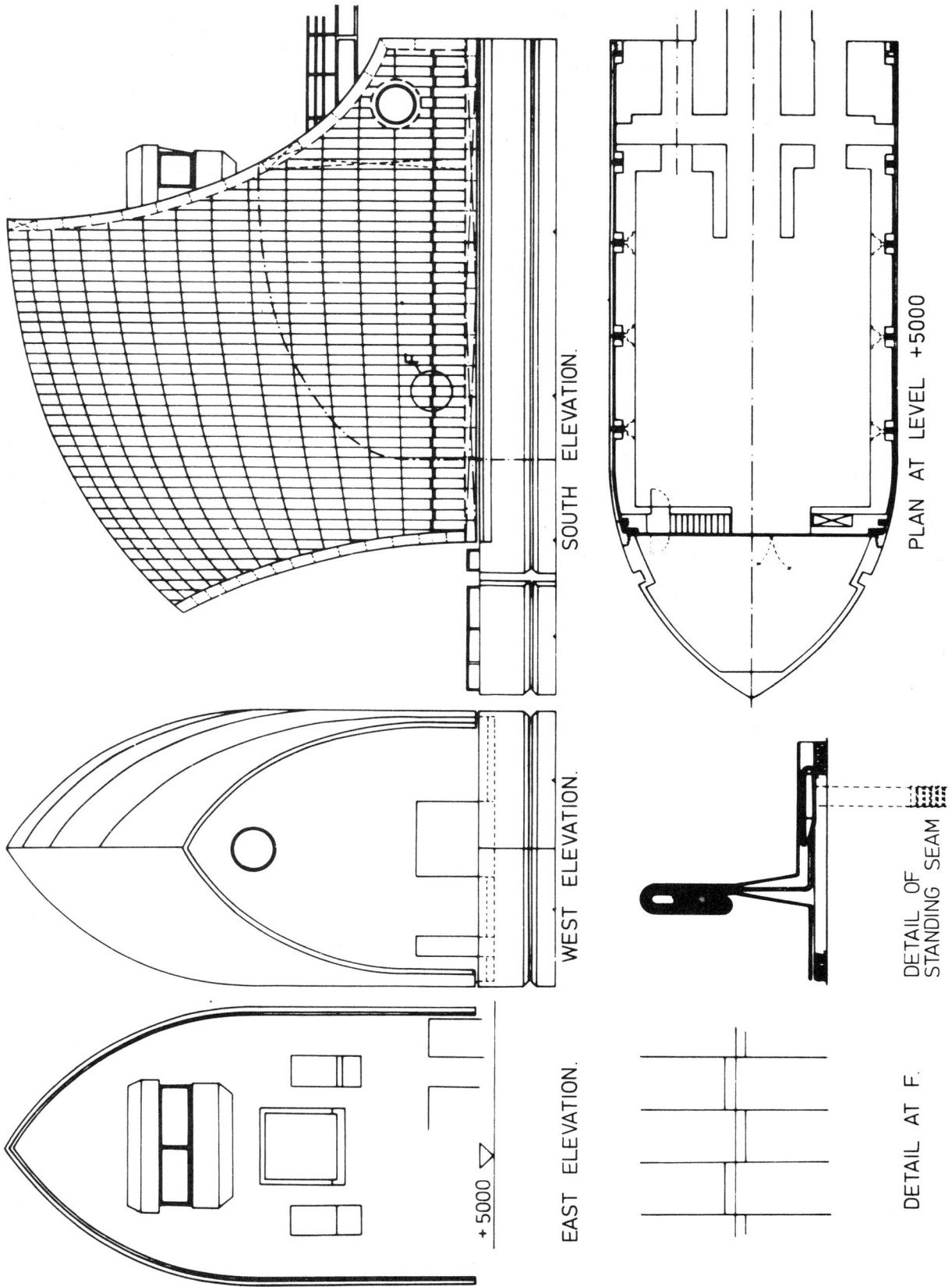

Fig.7. Arrangement of shell roof

iroko, bonded with phenol-resorcinol glue. Three
layers of European redwood boarding are fixed
to ribs using glue and annular-ring nails.
The stainless steel is of 28 gauge, Grade 316
S16 with a 2B mill finish, samples of which are
subjected to regular metallurgical examination.
The cladding method follows roofing practice
traditional in the building industry, except
that with stainless steel the craftsmen have to
deal with a metal unfamiliar in this context in
the United Kingdom. Trial panels were made to
establish optimum spacing of seams and welts,
and to confirm the trade's ability to carry out
satisfactory double-curved work in this
material, (Fig. 7).

17. Stainless steel is the only roofing metal
which combines durability with a light, reflec-
tive appearance; used in conjunction with the
form of superstructures described it will do
much to reduce the apparent bulk of the very
large pier structures in the River.

DESIGN OF OTHER ELEMENTS

18. Having developed a distinctive theme for
the piers, it was logical to extend it wherever
appropriate to the forms or the materials of
other important elements of the Barrier. However,
this had to be done with discrimination, and the
essential differences in the character of buildings based firmly on the land, and piers isolated
from the shore, had to be observed. Without re-
sorting to nautical pastiche, conventional building forms have been avoided wherever possible
on the piers. Conversely, the buildings do not
copy the forms of the piers but reflect them in-
directly, as in the roof shapes.

19. Finally a brief reference to public
amenities. It is certainly the Council's wish
that the public should have access to the riverside adjacent to the site and that they should be
able to enjoy whatever spectacle the Barrier may
provide, as well as gain reassurance from its
presence. At the very least it will add interest
to the riverside walks which exist in the area
and will stimulate the creation of others. On
the other hand the Barrier may well prove to be
an important attraction to tourists as well as
those with a specialist interest in flood prevention. The latter category will be reasonably
well catered for within the operational area of
the Barrier, where provision has been made for
receiving limited numbers of visitors. However,
it is a matter of regret that the national
economic policy has precluded, at this stage, all
but the very minimum facilities for the general
public when they visit the site. For security
reasons they are excluded from the operational
areas and there could well be a demand for
better parking and viewing areas, some catering
facilities, and ultimately some provision for
river passenger craft.

16. Protective treatment

K. JULYAN DAY, FICorrT, FTSC*

* Corrosion consultant

APPOINTMENT

1. The Author was appointed by the Consulting Engineers in 1970 to identify suitable protective systems for the Barrier gates based on information at that time, and subsequently to devise further testing and to develop special testing methods appropriate to special conditions under which the Barrier is to operate.

GENERAL

2. From the foregoing papers, it should be clear that long-term steel protection would have to be provided for the different environments of the various parts of the project. The Rising Sector Gates, gate arms, falling radial gates, box internals and sill subways would need separate consideration at least. The remaining major surfaces such as exposed machinery and access bridge steelwork might be brought in with one of the other schemes, but special consideration was still necessary for some minor areas such as trunnion support structures and bellows-connections although these are not discussed here.

3. The initial protective systems would not be expected to last the life of the Barrier but necessary maintenance should be at the longest possible intervals. Although provided for in the design of the submerged steelwork, any remedial work on it would inevitably necessitate closure of part of the Barrier to shipping. From operational and ship-movement viewpoints, closure sufficiently long to meet recoating requirements should be at intervals possibly as great as ten years.

INITIAL TRIALS

4. As soon as the basic design for the gates had been fixed, but before its concomitant environmental hazards to the possible protective systems for the steelwork could be fully appreciated, the GLC had authorized the Consulting Engineers to erect two test racks in tidal immersion conditions near the construction site. Here were exposed 300 mm square steel panels, blasted and coated by various paint manufacturers, mostly with their proprietary coal tar epoxy systems, but including some interesting alternatives.

5. Very useful information has come from these trials and, in the event, one of the original systems exposed at Woolwich is now being used on this project (para 27 below).

ENVIRONMENTAL HAZARDS

6. It soon became apparent that the various parts of the project were subject to different environments which were also at varying degrees of aggressiveness.

I Rising Sector Gate Spans (R.S.G.)
   (External Surfaces)

(a) Continuous immersion in Thames mud in the open position.

(b) Flowing water at 12 or more metres per second down the slot between concrete sill and steel gate when raised.

(c) Impact and abrasion from river bed materials and jetsam carried down by (b).

(d) Impact and abrasion of the raised gate by flotsam.

II Gate Arms (External Surfaces)

(a) Tidal immersion, partially in mud.

(b) Normal river atmosphere for upper surfaces with occasional immersion.

(c) Some abrasion of front surfaces by flotsam but mostly protected by timber facing.

III Box Internals (of I and II)

(a) Immersion in Thames water and mud.

(b) Flowing water and mud when gates in operation.

IV Falling Radial Gates (F.R.G.)
    (External Surfaces)

(a) Atmospheric environment.

(b) Temporary partial immersion when lowered.

(c) Some impact and abrasion when lowered.

Thames Barrier Design. Institution of Civil Engineers, London, 1978, 185-190

185

V   Access Bridges and Exposed Machinery

Exposed externally only to the normal river atmosphere.

VI   Sill Subway Lining

Subject only to condensation internally.

The above have been arranged in descending order of general aggressiveness. It will be appreciated that the requirement for any coating system to meet all the conditions in I is a formidable one.

CATHODIC PROTECTION

7.   This well known method of applying just enough negative potential to stop electrolytic corrosion currents from flowing, is being provided for fully immersed surfaces.

8.   Sacrificial zinc anodes fixed inside compartments, were part of the basic concept of protection for box internals of arms and spans of Rising Sector Gates, and an impressed current system is being provided to cover their flat top surfaces. Gate arms below, say, the half-tide level would also receive some cover from this system, at least on the available surfaces of their front faces.

9.   However, it was thought unlikely that this latter method of cathodic protection would penetrate very far into the mud between sill scallop and gate span. In addition, the mud at Woolwich had been reported to contain one of the more aggressive U.K. strains of sulphate - reducing bacteria (ref. 1), with all the connotations of attack on unprotected steel which that implies. These two factors of the environment taken together with those listed in I above, focused anti-corrosion thought on providing a paint scheme with the best chance of resisting impact and abrasion, so that bare steel would not be available for these microbes to attack.

COATING SYSTEMS

10.   To meet the various demands of the differing environments outlined above, the following four main protective systems were devised and specified, subject to the results of any tests or trials which might be instituted.

11.   750 Microns of Two-pack Epoxy. A straight, i.e. not coal tar, aromatic amine-cured epoxy was specified for the total immersion conditions of the external surfaces of R.S.G. spans. The thickness was derived from Admiralty work for surfaces subject to cavitation and was to be applied in not less than two coats, (ref. 2). It was likely that the material would be formulated specially, since very little had been offered in the way of impact and abrasion resistant paints of this type. Tests would have to be carried out to find the right product.

12.   200 Microns of Arc-sprayed Aluminium. This system was specifically chosen to meet the tidal immersion conditions for R.S.G. arms, based on the results of the only long term information available: the 19-year test series of the American Welding Society (ref. 3). These showed flame-sprayed aluminium to have superior performance in comparison with sprayed zinc, in full and tidal immersion and in a marine atmospheric environment. In addition, wash-primer and vinyl sealer gave some improvement over bare aluminium spray, and for this project, would also provide a decorative and maintainable surface. Arc spraying rather than conventional gas spraying was selected because of the remarkable improvements in adhesion reported (ref. 4).

13.   500 Microns of Hot-applied Solvent-free (HASF) Epoxy. Although more field experience had been gained with cold-applied solvent-containing coal tar epoxies, there were good reaons for choosing this comparatively new straight epoxy for the internal compartments of gate arms and the cavernous interiors of the R.S.G. span structures:

(a)   The technique of hot-spraying solvent-free epoxies had been well proven in the protection of about 90,000 square metres of steel lining to the 2nd Mersey Tunnel (ref. 5).

(b)   The two great advantages on that project of lack of solvent and ease of high-build application in one coating operation would be the same for these compartments.

(c)   Although the environment would be different, full immersion here would mean that cathodic protection would be possible to look after any deficiencies in the coating which might slip past the inspector.

(d)   A lighter colour, say eau-de-nil, would be easier for maintenance inspection than the black of coal tar epoxy.

The same choice of coating for steel sill liners was made on the basis of (a), (b) and (d) above.

14.   300 Microns of Zinc Silicate/Chlorinated Rubber. Those parts of the Barrier not subject to tidal or permanent immersion, and therefore more in the public eye, would benefit from a paint system that would be easy to maintain yet having a first class rust-inhibitive primer. Falling Radial Gates, Access Bridges and Exposed Machinery would all fall into this category; the relatively short periods of immersion for the lower parts of the gates would be well catered for by such a system.

TESTS AND TRIALS

15.   Apart from the trials mentioned in the section on 'Initial Trials', a number of other tests and trials were envisaged to confirm the choice of paints, to set laboratory acceptance criteria for them and to try out their application properties and site reparability.

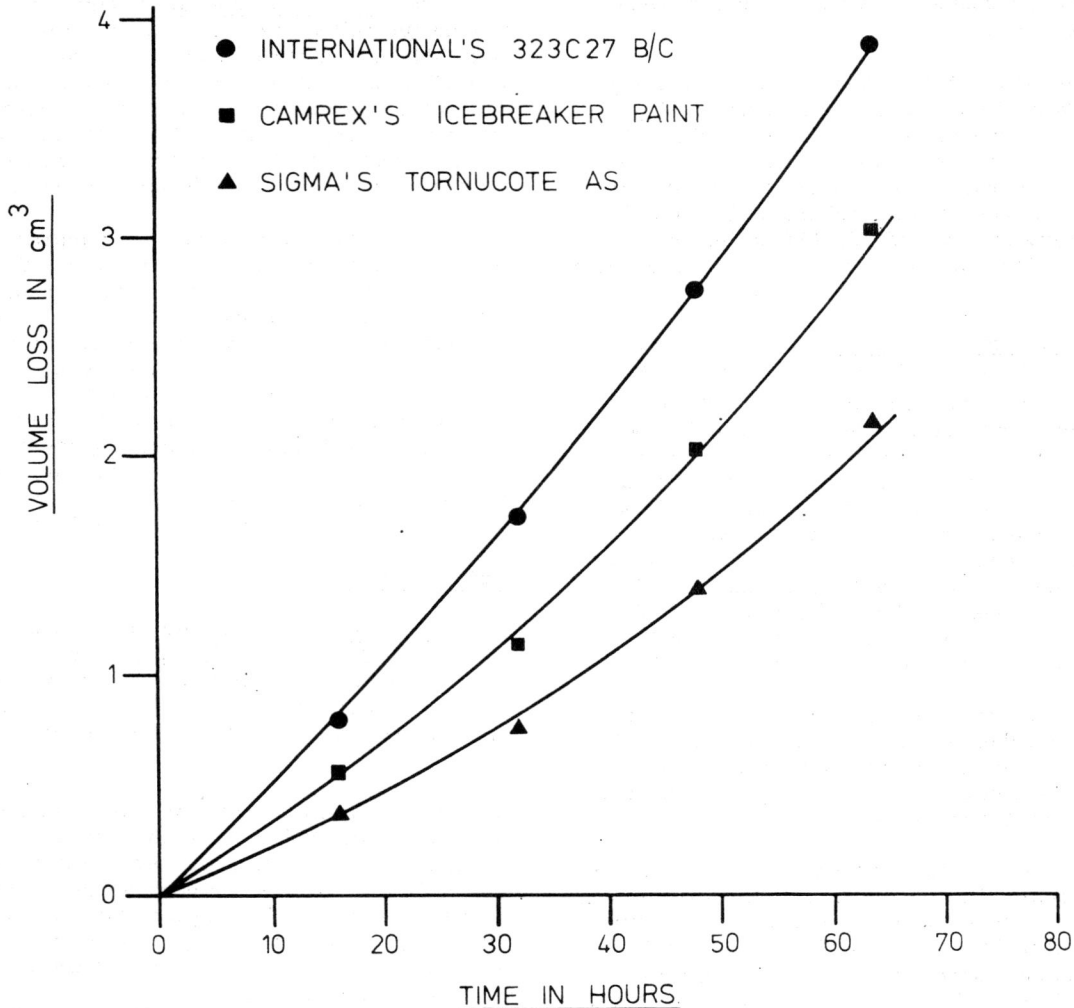

Fig.1. Roller tests for abrasion

Laboratory Tests

16. <u>Rotating Disc Test</u>. 90 mm steel discs of 10 g steel were blast-cleaned and coated with 16 different systems. John Laing R. & D. Ltd. carried out the test in equipment they had themselves developed where the discs were all rotated together vertically in the medium. The latter consisted in this case of Thames water and mud and a proportion of BS 1200 sharp sand. Speed of rotation was 1420 r/min and two periods of 100 hours were sufficient to show that, during this particular test: (a) a straight epoxy, specially developed for Icebreaker hulls, lost half as much volume as the next best system, multicoat neoprene; (b) vinyl sealed aluminium spray was of the same order of abrasion resistance as the neoprene; (c) glass-flake polyesters were no better than a two-fold improvement; (d) the latter could have its abrasion resistance improved four-fold by reformulation, including hard fillers.

From these test results, then, it appeared that a purpose formulated, straight epoxy the 'Icebreaker' paint, could provide the high level of abrasion resistance required for the external surfaces of the Rising Sector Gates.

17. <u>Roller Tests</u>. Abrasion within the sill recess was considered to be more normal to the surface than that provided by the disc test above and likely to have a element of impact as well. So, as part of the development of acceptance criteria, a roller test was devised and developed into a viable method. 114 mm sections of galvanised 63 mm barrel were machined to remove external galvanizing and to true-up the ends. The internals and ends were painted with an epoxy primer. Temporary caps were fitted so that the rollers could be mounted in a stand to rotate them for coating with the test paints. Very smooth uniform coatings were applied in this way after blast-cleaning. When dry, the coated specimens were carefully weighed and special Tufnol end-caps inserted after filling the barrel sections with water. Each specimen was placed in a plastic 'pot', made of sections of 127 mm diameter pipe which were then half-filled with Thames water, mud and a carefully graded and weighed amount of sand. Plastic end-caps were shrunk on to the pots which were then turned on their sides and rotated at about 60 r/min for 24 hours. As the pots rotated, the rollers were cascaded in the abrasive medium. On removal from the pots, the specimens had their caps withdrawn, were washed clean, dried

and weighed. Losses in volume of paint were then calculated.

18. The results of this test on two paints put up by the Main Contractor, are shown in Fig. 1 using the Icebreaker paint as a control. It appeared from the repeatable results obtained that a much shorter method than the Rotating Disc Test had been developed and this has since become a regular test at the Paint Research Association for abrasion resistance in immersion conditions.

19. Other Acceptance Criteria. 200 x 75 mm, 10 g steel plates were blast-cleaned and painted in duplicate with all the proposed external and internal gate systems, (paras 11 and 13 above). After a minimum of 7 days curing, all samples were immersed in trays of Thames water at laboratory ambient temperature for 6 weeks. They were the removed, impact tested to B.S.3900 Part E3, and returned to water-soak. This was repeated after 14 days refrigeration at 0°C followed by 7 weeks of half-immersion of the panels in water at 40°C (ISO test No.1521). Then the panels were impact tested as before.

The lack of outstanding differences in the results may be taken as an earnest of the excellence of these epoxy proprietary paints, but the specially reformulated coal tar epoxy came out the best at all temperatures.

20. The ISO 40°C test showed $\frac{1}{2}$ - 1 mm blisters below the waterline on one aliphatic amine HASF epoxy and on one of the experimental coal tar epoxies, which had been modified with a flexibilizing agent, apparently water-sensitive. The rest of the HASF epoxies offered had stood up well to the variations of temperature in these tests, and among the external systems tried, the Icebreaker paint had maintained its superiority.

Natural Environment Exposure

21. Tidal Immersion. As soon as it was possible, exposure to tidal immersion at Woolwich for internal and external gate span and gate arm systems was arranged, using one of the original exposure racks. (Tidal immersion would be ideal for Aluminium sprayed test plates for the gate arm system, but would be tougher than the full immersion conditions for the paint schemes and therefore act as an accelerated test).

22. Although it is still early days, some results are already in evidence, confirming previous experience with the various materials and systems.

(a) No signs of rusting other than in the score mark simulating damage, to test rust-spread.

(b) Greater rust spread with HASF epoxy than with more highly pigmented epoxies (Icebreaker paint) and least with coal tar epoxies.

(c) The thinner the sealer coats, the better the aluminium spray behaves.

23. Atmospheric Exposure. Fourteen manufacturers provided test plates of their Chlorinated Rubber paints, (para 14 above), of stepcoated undercoat and gloss, in both the golden yellow and black required by the architects. These were exposed on a 45° rack facing approximately south, on the roof of a 4-storey building on the river bank at the Barrier Site.

24. After two years only two manufacturers' panels had stood up sufficiently well to the site environment to obtain unequivocal recommendation. The remaining panels had either checked in undercoat or finishing, or had so chalked or changed colour as to be unacceptable. The Main Contractors chose paint manufacturers with sound undercoats, but some of their new gloss finishing panels are still exposed on the rack. Since the finishing coats are for site application, there should be enough time for these experimental re-formulations to be well tried.

Suitability Trials

25. HASF Epoxy Seven manufacturers were found able to offer these hot-applied solvent-free materials, but the ability to apply these at 500 microns had yet to be tried, as had the associated stripe-coats specified for edges and welds.

A Graco 'Hydracat' hot airless spraying equipment was commissioned, with operator and sprayer who handled all the proprietary materials under the guidance of the respective manufacturer's representatives. The method of use has been described by Brown (ref. 6), and by Newell and Watson (ref. 7), who have had some large scale practical experience.

26. During the 'Hydracat' application, thickness was monitored by a wet-thickness comb between passes of the spray gun. 'Stripe-primer' was applied by brush to half the weld area and edges of the 300 mm square panels.

One supplier made a mistake in the colour, and with another's, the brush applied primer was too heat-sensitive and the whole coating sagged in this area. Another paint was not nearly thixotropic enough, with unacceptable sagging. One other material was not in sufficient quantity for the full thickness to be applied. The other three paints were satisfactorily applied but wide variations in curing time were found, some not being tack free until the 3rd day.

With such results, no doubt remained of the value of these trials but in addition, the small plates, painted at the same time, i.e. by this actual method to be used on the job, provided the laboratory (see paragraph 19) with much more reliable specimens to test than could have been otherwise produced.

27. Arc-Sprayed Aluminium Similar application trials of hot-spraying aluminium by the arc process were successfully carried out. However when the first full-scale trial took place on the construction site, underneath the flat horizontal face of a large gate-arm, such condi-

tions developed, of air-borne aluminium and alumina dust, as to endanger both health and safety.

As a result, this system of protection for the gate-arms had to be abandoned in favour of proprietary coal tar epoxy paint, which has given good service on a number of projects during the last 15 years in similar tidal conditions. However, only service experience on the Barrier gate-arms will show whether this change will result in tne need for more frequent maintenance than should have been the case with aluminium spray, or for further cathodic protection measures to be provided.

28. **Site Reparability** One of the important facilities of the Rising Sector Gate design was 180° rotation from the normal position to an upside down one for maintenance. Areas of damaged coating and/or rusted steel would have to be **blast-cleaned for remedial painting**, but openblasting with grit might contaminate operating machinery. Since high pressure water-jetting would probably be employed for cleaning the surfaces first for inspection, the idea of using the same equipment but including grit in the water-stream for preparing the surfaces for painting was an attractive one, particularly in respect of removal of residual contaminants of salty, Thames-rusted steel.

29. The panels used for these trials were those from para 19 above, long term testing and watersoak. Centre strips were blast-cleaned on one side of each and the bared-steel thoroughly rusted in Thames water to simulate site damage. Then one panel of each duplicate pair was dry blast-cleaned, and the other water abrasive blasted, before the systems were made good again in the laboratory using the respective paint manufacturer's recommendations. After seven days curing at ambient and four weeks further immersion, pull-off adhesion tests were carried out.

The results of these tests showed no significant difference between the two methods of cleaning prior to making good and that, even after the 6 months watersoak these panels had experienced, adhesion was of the order of 10 N/mm^2 or better. It would seem that lasting repairs can be made to the epoxy systems tried, using the proper techniques.

30. Resistance to Impact The B.S. impact test referred to above (12 mm indenter of 4.8 kg) was too severe a test to correlate as well with likely impact on the gates as the Falling Ball tests used by the Admiralty.

4 g mild steel plates were blast cleaned, painted with two-pack epoxy and experimental coal tar systems and the better of the glassflake polyesters, and Thames-watersoaked for 150 days. The steel ball for this test weighs 1 kg but is

dropped from 3 metres; two 'drops' very close to each other constitute a test. The results showed the specially formulated coal tar epoxy, the 'Icebreaker' epoxy, and the glassflake systems to be well able to stand up to such impact.

CONCLUSIONS

31. The programme of tests and trials, although limited in scope in scientific terms, nevertheless achieved its main objective of confirming or amending the specification which had had to be produced in advance. The results are summarized in the Appendix (see page 190).

32. Some contractual and operational lessons were learned as well as technical, but the programme as a whole confirmed that, for major projects, it is just as necessary to assess the site environment at the design stage and decide what tests should be put in hand on protective coatings, as it is to do physical tests on engineering models and mockups. Given longer notice, there would have been a dredger-bucket trial of the external R.S.G. coatings, 'as designed' C.P./coating tests and a better assessment of the efficacy of biocides for inclusion in epoxy paints than has so far been recorded.

ACKNOWLEDGEMENTS

33. Due acknowledgement is thankfully made by the Author to the G.L.C. for permission to publish the work, to Mr. G.A. Newell for the original conception of the roller test, and to the staff of the Paint Research Association who developed the idea and conducted most of the other tests as well.

REFERENCES

1. BOOTH G.H. and TILLER A.K., Proc. 2nd I.C. Marine Corrosion and Fouling, Athens, Greece, 1968.

2. Paint Epoxy Solvent-free, to Draft Specification CDL.P86/72.

3. American Welding Society: Corrosion Tests of Flame-sprayed coated steel, 19 year Report, 1974.

4. BAYLISS D.A., Paper No.6, Ass. of Metal Sprayers Symp. on The Protection of Steel Structures, London, April 1972.

5. MEGAW T.M. and BROWN C.D., Mersey Kingsway Tunnel: Planning and Design. Proc. I.C.E., 1972, 51, 479-501.

6. BROWN P.T., Paint Manufacture, December 1973, 18.

7. NEWELL G.A. and WATSON E.A., Heavy Duty Protective System for the 2nd Mersey Tunnel. Industrial Fin. Conf., London, June 1972.

Appendix: results of tests and trials

PART OF BARRIER	PROTECTIVE COATING SYSTEM	ABRASIVE RESISTANCE		ANCILLARY TESTS		SITE ENVIRONMENT		SUITABILITY & REPARABILITY	REMARKS
		ROTATING DISC	ROLLER	ISO 40°C	BS IMPACT	TIDAL IMMERSION	ATMOSPHERIC		
RISING SECTOR GATES & GATE ARMS EXTERNAL	ICEBREAKER PT	1st	2nd	SATISF'Y	SATISF'Y	V. GOOD	-	SATISF'Y	2 COATS
	C.T. EPOXY	16th	-	-	-	-	-	-	-
	NEOPRENE	2nd	-	-	-	-	-	-	MULTI COAT
	SEALED AI SPRAY	4th	-	-	SATISF'Y	V. GOOD	-	FAILED ON APPLICATION	ARC SPRAY *
	SPECIAL C.T. EPOXY	3rd	-	SATISF'Y	BEST	BEST	-	SATISF'Y	2 COATS
	GLASSFLAKE POLYESTER	11th	-	-	POOR	-	-	-	-
	PROP STRAIGHT EPOXY1	-	1st	SATISF'Y	SATISF'Y	SATISF'Y	-	SATISF'Y	(2 COATS) +
	PROP STRAIGHT EPOXY2	-	3rd	SATISF'Y	SATISF'Y	SATISF'Y	-	SATISF'Y	2 COATS
INTERNAL (ALSO SILL SUBWAYS)	HOT APPLIED SOLVENT FREE EPOXY (7)	-	-	1 FAILED BY BLISTERING	1 FAILED ON LONG WATER SOAK	SATISF'Y	-	3 FAILED SUITABILITY 4 PASSED REPARABILITY	STRIPE COAT, ONE FULL COAT PLUS C.P.
FALLING RADIAL GATES MACHINERY & ACCESS BRIDGES	ZINC SILICATE & CHLORINATED RUBBER 14 BLACKS & 14 YELLOWS	15th	-	-	-	-	AFTER 2 YEARS ONLY 2 OUT OF 14 MANUFACTURERS u/c & GLOSS WERE SATISF'Y	V. GOOD	6 SATISF'Y BLACK SYSTEMS. 2 MANUFACTURERS YELLOWS NOW ON RE-TEST

* GATE ARM EXTERNALS: REPLACED BY 600 MICRONS OF ISO-CYANATE CURED COAL TAR EPOXY COATING. (SEE PARA.27)
+ GATE SPAN EXTERNALS: FIRST COAT 500 MICRONS 'HASF' EPOXY (AS FOR INTERNALS); 2nd COAT AS INDICATED. (250 MICRONS)

Discussion on Papers 6–16

MR CARR

1. Although the main features of the project were defined when the Parliamentary Act was formulated in 1971, a number of major aspects of the design remained to be decided. There was never any doubt that the sills for the rising sector gates would be prefabricated, but their design underwent significant changes. The initial tender design was revised to introduce prestressing, necessitated largely by changed sill loadings arising from the operating methods which were employed to minimize the downriver reflected wave effects. In addition, the revised sill design substituted embedded subway tubes for the earlier freely supported ones which, although achieving minimum weight to assist sill flotation, presented an extremely difficult problem regarding corrosion protection.

2. Prior to the tender stage, design of a suitable double-ended support structure, embedded in the pier concrete to carry the gate pivots, enabled the previously skewed line of the barrier across the river to be straightened and the pier lengths to be shortened.

3. The decision to design the piers as basically mass concrete structures necessitated the incorporation of contraction joints, which in some respects caused design problems and placed limitations on internal layouts. Nevertheless, had a reinforced concrete cellular pier design been attempted, it is evident that the flexibility of layout required to accommodate the evolving demands of the equipment, interior access, and so on, would have been much less.

4. The diversion channel at the north side of the river has not experienced siltation, contrary to some predictions. It is hoped that this short-term evidence may be indicative of freedom from future siltation of the permanent navigation openings. Should it not be so, the navigation openings and the upper surface of the gates can be periodically cleared of silt by use of water jets, pumps, and similar mobile means.

MR MITCHELL

5. The design of the sills took account of a large number of variables, particularly the hydraulic loadings. During surge closures high water levels downriver caused high vertical loads on part of the top of the sill and high horizontal loads on the side of the sill. Uplift pressure was assumed to be uniform and equal to the upriver water level. Design cases including other uplift assumptions were also examined. One of these resulted in net upward loading on the sill, and ballast was introduced to give a factor of safety of 1.25 under the worst conditions. Reliance on the keyed joints at the ends of the sill to resist uplift would have overstressed the sill in upward bending, and resulting deflections would have reduced gate clearances.

6. In the reverse head condition the higher upriver level extends over most of the width of the sill. A low uplift assumption was prudent in this case, thus giving a large net downward load which represented a further critical design case.

7. When the sill is sunk it is dry internally with inward hydraulic loads on all the external panels. When it is ballasted and only just filled with water venting to atmosphere a critical design case results from the sill spanning under maximum weight together with unbalanced hydraulic pressures on external panels.

8. Figure 1 clarifies the detail at the flexible subway joint/sill subway liner interface. When the coupling ring is pushed forward and the "D" seal makes the initial connection, spacers prevent complete metal-to-metal contact. This leaves an annular gap for the injection of the permanent gasket material which takes up any irregularities in the two faces. Bolted clamps complete the connection. The spacers can be removed if re-tightening is considered necessary.

MR DAY

9. Dr Shreir, Rendel, Palmer and Tritton's cathodic protection adviser, and myself were asked what we would do to protect the outside of the subway tubes in the waterfilled part of the sills. Access for inspection and maintenance was impossible and so we devised a complex and costly scheme; there was to be an external coating as well as big sacrificial anodes. Furthermore there was to be a specialist monitoring system from inside to see that the water in the sills was not showing signs of excessive steel corrosion or microbial activity. Special measures would have to be taken, perhaps for as long as 100 years, if these tubes were to remain unperforated in such anaerobic conditions. Rendel, Palmer and Tritton then changed their minds; the steel tubes are now sited in the solid concrete part of the sills and the corrosion situation has been reduced to negligible proportions.

MR J.B. DWIGHT (University of Cambridge)

10. A study was carried out, at the request of the designers, on residual stresses in the rising sector gates caused by weld shrinkage. These structures were more massive than any on which residual stresses had been measured before, and the designers needed data to feed into their calculations.

11. A specimen was made representing, to full scale, a portion of one of the gates. This was 6.5 m long, containing two cross-frames and six longitudinal stiffeners, with a multi-pass buttweld down the middle. Full thickness material was used (40 mm for the plating, 35 mm for the stiffeners). Fabrication took place at the works of Messrs Horsley Bridge, and practical

Thames Barrier Design. Institution of Civil Engineers, London, 1978, 191-202

191

welding procedures were followed.

12. Strains were measured using a Demec extensometer engaging in 1.0 mm holes drilled in the steel, readings being taken after each stage of fabrication. In this wasy the build-up of longitudinal residual stress was followed as welding proceeded.

13. In the plating it was found that a mean compressive stress of about 20 N/mm² was introduced by the initial flame cutting, rising to about 40 N/mm² after completion of the plate-to-stiffener fillet welds. However, the final operation of butt welding the two halves of the specimen together hardly increased the plate stress at all.

14. Surprising results were obtained for the stresses in the stiffeners. One would expect the shrinkage at the plate-to-stiffener joints, being located below the centroid of the section, to produce a stress gradient with tension at the stiffener tips (according to the ordinary rules of bending). In fact a reverse gradient was observed, with the tips reaching 80 N/mm² compression in some cases. This was due to slip occurring in the tacks between the plating and the stiffeners as the welds cooled.

MR A.R. YOUNG (Robert Cuthbertson and Partners)

15. The great torsional rigidity of the gate, referred to in Paper 6, has a valuable purpose in that the gate can be operated from one end by machinery, thereby preventing a big hazard if the other end fails. With reference to the torsional stresses on the gate arising from hydrostatic loading, can the Authors explain why a differential head would not have given uniform loading on the gate?

16. With regard to Paper 9, does the Author think that condensation is possibly a problem in the subways and, if so, what is the solution?

17. Another aspect to do with the operation of the gate is the evacuating of water when the gate is raised and the letting in of water when it is lowered. The faster the gate is lifted the more carried water there is because there is only a limited exit for the water. There must be some limit as to how much carried water the machinery can take and therefore there must be some limit, with the openings that have been provided, to the rate of lifting of the gate. The water has a rather tortuous path getting in and out. Have any model tests been done to check the connection between carried water and rate of opening?

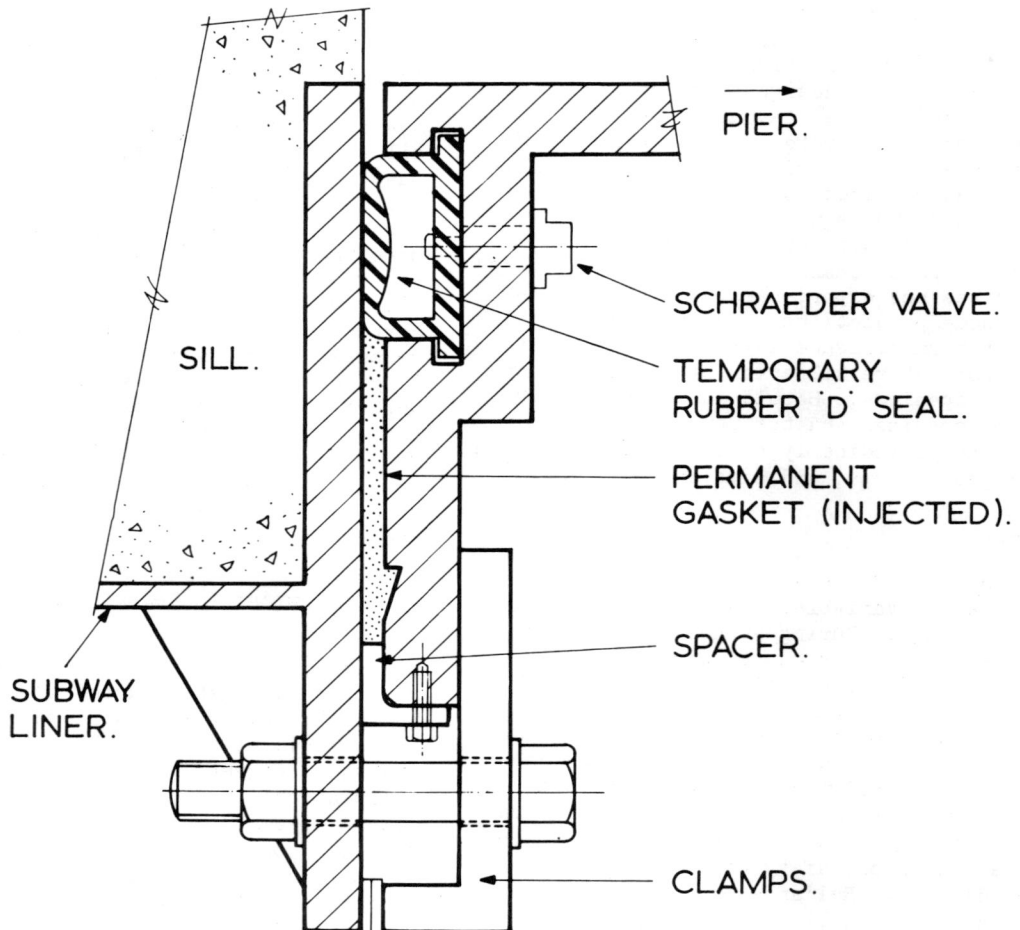

Fig.1. Seal between flexible joint unit and sill

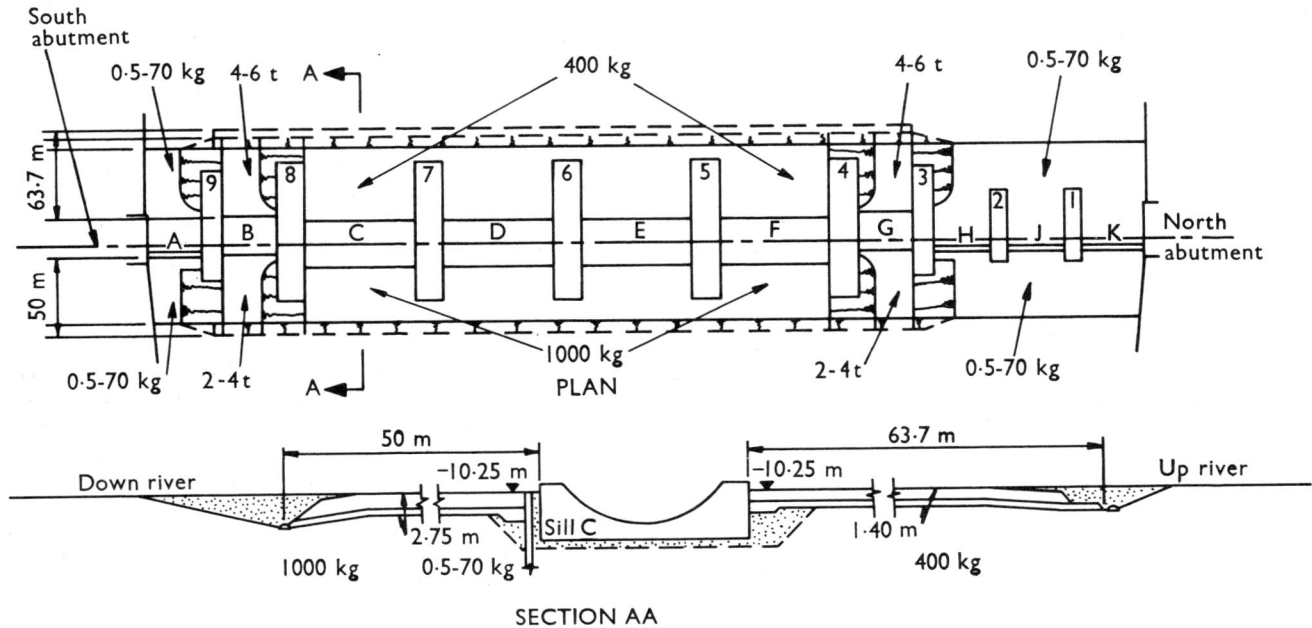

Fig.2. Thames Barrier bed protection design layout

MR P.H. McMILLAN (Sir M. MacDonald and Partners)

18. Will there be only a post office telephone link from the barrier to the storm tide warning service at the Meteorological Office in Bracknell, or will there be a telex link?
19. How much warning will shipping require when the barrier is to close? Will there be sufficient time after such a warning to satisfy shipping when there is a high freshwater flow in the Thames?

MR H.L. WAKELING (Rendel Palmer and Tritton)

20. Despite all the safety devices, described in Paper 9, that were provided to ensure satisfactory operation of the gates, it was still considered necessary to provide for the remote possibility of a gate failing to operate during a surge. This was to ensure that there should be no structural failure which would be almost impossible to repair.
21. When investigating this, it was initially thought that the underlying chalk foundation would be resistent to scour. However, no information on the erodability of chalk could be found and so it was decided to carry out full scale tests. A disused chalk pit was located at Thurrock where the chalk was of similar quality to that in the bed of the Thames at the barrier site. A flume was constructed in the chalk and water pumped across it at the velocity which was estimated to occur during gate failure conditions. The chalk rapidly eroded and within about ten minutes the flume was no longer usable. Hence it was concluded that the chalk in which the piers were to be founded had to be protected against erosion.
22. Figure 3 of Paper 9 shows the minimum sizes of stone which were necessary in various areas to prevent erosion. In order to simplify placing the stone under water a rationalization was carried out and the final design is shown in Fig.2.

MR P.L. MALOY (Sir Alexander Gibb and Partners)

23. Normally the gates will be in the fully open position with a gap between the concrete cill and the gate skin plate. Are the designers confident that no problem is likely to arise with silt deposition? Although, initially, the gates would probably be operated regularly, which would assist in keeping the cill clear of silt, such operations could tend to become infrequent. Thus silt compaction in the gap could add appreciably to design friction forces. Also since the gate skin plate is not a true curve but made up of a series of flats this could worsen the problem by introducing a wedging action. Was consideration given to the provision of a jetting system using either high pressure water or air?
24. If water is allowed to jet under the gate to assist silt removal has consideration been given to the possible problem of the 'wire drawing' effects on the trailing or bottom of the gate?
25. In view of the type of steel used for the gate structure and the plate thickness it is assumed that the fabricated sections have been stress relieved for shipment. Are the site welded joints stress relieved and, if so, what method is being used?
26. With reference to Paper 8, what provision has been made in the linkage for deflection of the gate structure? In the case of the main hydraulic cylinders were enquiries restricted to British manufacturers, and were there any difficulties in obtaining an adequate number of quotations?

MR COX

27. It may be necessary to pass some water through the barrier before it is finally closed. Tests have established that the under-shot mode is the best method of doing this. Bed protec-

tion was studied in relation to this as well as
for the gate failure condition.

MR J. WATERSTONE (Polytechnic of the South Bank)

28. The flood gates are to be raised and
lowered by hydraulic rams. A situation could
occur when additional force is needed to move
the gates after an obstruction has lodged between the gate and the sill, or because the gate
has been distorted by damage, perhaps after a
vessel has collided with it. The gate is in the
form of a hollow steel hull. It would be simple
to blow air into it which would provide a very
large buoyancy force to help the rams raise the
gate.

29. To lower the gate air could be pumped out
of the gate and water drawn into it at a level
above that of the river. This would only be
effective as long as some part of the gate was
above river level. Buoyancy would be more effective for raising the gate, and less so for lowering it. But in times of crisis it would be
important to raise the gate and complete the barrier. If one bay remained closed and out of
action for river traffic and for passing river
flow because the gate stuck in the raised position, this would be a minor matter which could
be put right when the danger of surge tide had
passed.

MR CARR

30. It is not considered possible that the inshore spans can fail to be closed during a
surge. The bed protection being provided at the
inshore span is purely to cater for the flows
through the adjacent rising sector gate spans
should that fail to close. The gates can be
manually pumped and lowered to the sill, should
all standby machinery fail. They will not be
subject to any significant risk of collision, as
major vessels cannot get in there and smaller
vessels are unlikely to do any damage. The gate
is robust, and there will be strict navigational
controls to prevent ship movements in good time
before gate closures are executed.

MR CLARK

31. One justification for model work is being
able at the end of the day to make some improvement as a result of the test data. Although the
holes were obviously the weakest parts of the
webs, the engineer who had specified the size of
the hole was asked if they were necessary in
this high shear part of the structure. After
further consideration it was determined that
they could be removed from the end bays, and the
structure became stronger in consequence.

MR D.P. SHIPMAN (Anglian Water Authority)

32. With reference to Paper 12 it was shown
that a large hydrodynamic torque can be applied
to the gate by low pressure in the separation
zone when a diving jet re-attaches to the flat
face of the gate. It would appear that unless
the prototype gate is operated at or near slack
water, then it could well be subject to the
phenomenon of sudden increase in the gate hoist
loading when a sector gate has been rotated by

30° to 40° from its rest position in the sill.
Has any investigation been made into the possibility of allowing water to be drawn into the
low pressure area from the hollow body of the
gate; the replacement water being driven through
the gate arms by the difference in hydrostatic
pressures? There is the possibility that a damping effect may be achieved on any oscillation
that may be present if the water in the gate
arms were to be guided along the radius of the
gate.

MR H. HENKIN (Rendel Palmer and Tritton)

33. It was necessary when laying down the tolerances allowable for the civil engineering works
to ensure that they were of an accuracy which
would allow the gates and machinery to be fitted
and to function properly. A spherical bearing
was used to support the gates and a line hinge
was provided at one end of the gate leaf for connection to the gate arm. The gate arm could
therefore move in service to the pier walls.
The amount of movement on the spherical bearing
was strictly limited to $\pm 1\frac{1}{2}°$. Moreover, there
had to be proper clearance between the gates and
the sills and provision for the sill subway to
be properly sealed to the pier subway by means
of the flexible rubber joint.

34. An analysis was made of the various clearances which would effect the operation of the
gate. This analysis included consideration of
the tolerances in fabrication and erection of
the gate, the effects of differential settlement
of adjoining piers, including the effect of any
out-of-plumb settlement, and inaccuracies in
the civil engineering construction of the piers
and sills. It was concluded that the movement
of $1\frac{1}{2}°$ on the spherical bearing could be sub-
divided into three equal parts. A half-degree
could be allowed to take up tolerances in gate
construction and erection, and in gate movements such as temperature movements and movements of the ends of the gates. Another half-
degree could be allowed for relative settlement
of the piers, leaving a half-degree for the
allowance for civil engineering construction
tolerances.

35. Special tolerances had to be set in the
civil engineering works for those areas which
affected the fixing and movement of the gates,
and the connection of the sills to the piers.
Both absolute and relative tolerances had to be
considered. A tolerance of ± 20 mm was specified
between the faces of gate support structures in
adjoining piers, and an angular deviation at the
face of the structures of ± 10 min. The tolerance
in positioning the centreline of the gate support structure (which in the case of the large
gates weighs 150 t) was set at ± 25 mm of the
theoretical position. In addition, tolerances
were specified on the relative positions of
installed sill and gate support structure.

36. As regards the fitting of the sills, these
have to be floated out and sunk accurately into
position, so that the rubber bellows joints can
be connected to them within the allowable movement of the joints. In addition the sill scallop surface must be positioned so that it does
not interfere with the free movement of the gate.

37. The space available between the end of the

sill and the pier is required to be within 55 mm and 105 mm, which is extremely small in the context of placing a 61 m long sill. This meant that a tolerance of only 25 mm had to be maintained on the length of the sill, which is extremely difficult to achieve constructionally.

MR W.W. MILTON (Engineering and Power Development Consultants)

38. I think that this work is costing more than originally envisaged and is running a little late on the original schedule. Modifications have been made, leading to substantial savings on cost. Looking back, would the authors like to have given more or different information to the contractors at the time of tender, and would they have made any changes in their design three or four years ago?

MR PRATT

39. Ideally both the flat top surface and the curved underside of the rising sector gates would be cathodically protected, but the structural design prohibits the fixing of anodes directly on the steel skin. Fixing anodes to the cill scallops would be impracticable and could lead to possible damage to the gate should anodes become displaced.
40. The possibility of low profile anodes of low current magnitude has been investigated but a survey throughout the industry produced only one possible solution. This was a strip anode incorporating permanent magnets and other specialized fixings. Investigation is still continuing and if a satisfactory solution is found protection of half a gate may be recommended, say for comparison with the other unprotected half.

MR D. PALMER (Cement and Concrete Association)

41. There is, or will be, provision for cathodic protection for some of the steelwork. One of the worries with North Sea structures is the effect of cathodic protection to the steelwork producing currents in reinforcement and embedded steelwork in the concrete and causing electrochemical corrosion. How much study has been made of this? How far have people satisfied themselves that this will not happen here? Will the reinforcement of the pre-stressing be satisfactory? Are there any plans for monitoring the reinforcement to make sure that no undesirable effects occur?

MR PRATT

42. There is limited published data giving satisfactory conclusions on the effects of cathodic protection currents flowing through reinforcement. We are aware of the hazards and hope to be able to incorporate cable connections at selected areas, from the reinforcement below the waterline to high level positions, for monitoring purposes.

MR B.L. CLARK (Bernard L. Clark and Partners)

43. This is a wonderful project; the gates are

similar to some in Holland which have been up for 10 years.
44. The Thames has been the basis for this country's success commercially for three centuries. The idea of a barrage across the Thames first grew in 1928 when Pimlico was badly flooded, and later in 1953 when we suffered in western Europe and to some extent in the Thames. Immediately after the war I could see that if something was not done with London docks we would very soon lose our place as the leading maritime country in the world. Until that time virtually every port in the world had been able to handle the ships which were plying the oceans. However, the ships then began to get bigger.
45. I worked on the idea of a Thames barrage with sea locks with a view to maintaining constant high water in the Thames, at least above Woolwich, so that the docks could be used all day. The present engineering works at Woolwich do nothing for the country commercially. There used to be 40% of the national commercial production in the South East, around the Thames, but about 90 miles of Thames waterside frontage is now dead because the Thames has been allowed to become a tidal creek. I am convinced that from figures studied, the economic problems we suffer today to some extent result from the decline of London as a maritime port and the fact that Rotterdam, or Europort, is today handling 300 million tons of cargo a year. London is now down to about 40 million tons. London is about the fifteenth port in the world whereas 25 years ago it was the biggest.
46. My idea was simply to have a dam with sea locks and a roadbridge to keep the Thames above Woolwich at a constant high water level, maintaining all the water frontage as it then was with the sea locks enabling the ships to go through at any time of the day or night. The docks have been tied until very recently to the tides and much of the dock activity is geared to the tidal movements whereas most other countries are non-tidal so far as shipping activity is concerned.
47. Rotterdam was 35 kilometres up river, rather like London. So the authorities brought the docks down river to deep water. There are now 35 000 ships a year going into Rotterdam because London cannot handle the big ships. London serves a population of well over 50 million and justifies a major port, but in the last 25 years all has been lost to Europe because of a lack of facilities.
48. This is a fine engineering job. It does not do anything commercially but it will protect us from flooding. A great deal of bank raising has been done up and down the river. Is it now necessary for the barrier to be there at all because we have raised the flood level virtually to the barrier level? Is the barrier level as effective as it was?

MR COX

49. The bank raising and other subsidiary barriers downstream of the Thames Barrier are necessary in dealing with the areas down river. If the barrier had not been built the bank raising upstream would have needed to have been considerably higher than the present interim bank rais-

ing. The cost might have been of the same order but the environmental impact would have been severe. This is one of the reasons why the barrier was constructed rather than some other scheme.

MR PALMER

50. I feel that the continued stability of the concrete piers is as important as the correct functioning of the gates they support. The use of PFA as a partial replacement for cement is reasonably well documented but there is still much to be learned. Can the Authors of Paper 14 explain more fully the basis for the decision to specify PFA to control heat output and say how much, both technically and economically, they expected to be gained from this. In the underwater bases up to 50% PFA was specified rather than the normal 20-30%. Were any tests conducted, before work started, on the relative strength gains and heat outputs of these mixes when compared with lower or zero replacement rates?

51. The piers are said to have a 600 mm thick skin of 20 N/mm^2 concrete. Was this the sole specification requirement apart from that for the materials to be used? If so, are the Authors satisfied about the durability of this facing? With modern materials and techniques of quality control a 20 N/mm^2 concrete can be produced with a low cement content and high water cement ratio. Its durability therefore must be doubtful in anything other than the mildest exposure conditions.

MR MALOY

52. In Paper 14 the Authors, referring to the cill concrete, stated that concrete with a high quality finish was preferred to granolithic concrete or any epoxy mortar. Experience has shown that the latter two types of concrete usually have better erosion resistance. What type of

concrete was, in fact, specified? The scallop edges of the cill are protected by mild steel cover plates. It is surprising that this material was chosen in preference to stainless steel especially since these items are permanently immersed and cannot be easily replaced.

53. With reference to Paper 16, can the Author describe the protective coating eventually chosen for the gate skin plate. Any coating system must stand up to continuous immersion, abrasion or erosion and possible mechanical damage, and from past experience this has usually meant an epoxy based system. Why does the Author not agree with the use of epoxy coal tar paints for this particular application? There are scores of gates of every type in dams, barrages, and so on, many of them operating in far more arduous conditions than the Thames Barrier and they have been protected by epoxy coal tar paint systems. These systems appear to have given no serious trouble, and useful working lives in excess of ten years are becoming common. The effectiveness of any protective coating is dependent on correct surface preparation, on the correct application of the paint in strict accordance with the manufacturer's recommendations, and on checking that the paint chosen has the required chemical constituents for the job. From my own experience most paint failures are caused by non-compliance with one or more of the above basic requirements.

54. Can the Author amplify his statement on the paint tests, i.e. the comparison between high build epoxy coal tar, vinyl and polyurethene? A series of paint trials in South Africa has just been completed using various paint systems put forward by paint manufacturers. These trials over a two year period were on large diameter valve hoods to resist high velocity water with high silt concentration; the jets impinged on the hoods at an angle. The three most successful systems were epoxy based, two with abrasion resistant fillers.

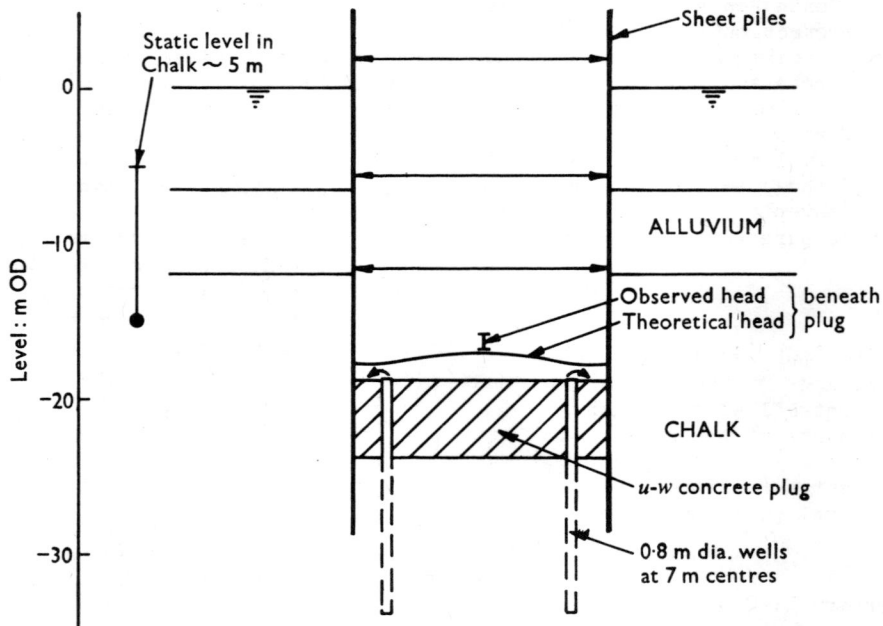

Fig.3. Pier 8

MR D.F.T. NASH (Ove Arup and Partners)

55. The design of foundations for piers cannot be considered in isolation to the temporary works needed to construct them. An interesting temporary works design problem has arisen as a result of the decision to found the northern piers in the Thanet Sand.

56. During the construction of each pier the sheet piled cofferdams have to be dewatered. In the case of the southern piers founded in Chalk this occurs after placing a concrete plug under-water at the base of each cofferdam. However, for the northern piers founded in the Thanet Sand, it is proposed to excavate in the dry in order to minimize disturbance at formation level. In both cases it is necessary to reduce the water pressures in the Chalk; this is to avoid heave of the tremie-concrete plug for the southern piers and to avoid heaving and piping of the Thanet sand for the northern piers.

57. The successful dewatering of the Chalk is essential to ensure that the foundations perform as designed. Design of the relief well systems started from the assumption that seepage through Chalk could be analysed using Darcy's Law. In the case of Pier 8 for example a finite element seepage analysis was carried out which indicated that 0.8 m dia. wells bored 10 m into Chalk were needed at 7 m centres (see Fig. 3); the average head reduction beneath the concrete plug would be 12.4 m with the wells discharging onto the surface of the concrete - a level in the wells of 14 m below the static level in the Chalk. Observations of piezometers in the Chalk showed that the actual head reduction achieved was 10 m to 11 m when the level in the wells was 13 m below static. This measure of agreement between the design and observations suggested that this approach to design was justified.

58. A similar approach was adopted in designing the relief wells in the Chalk below Pier 2. In this instance the water level would be reduced in each well by a submersible pump. Again the assumption was made that Darcy's Law would apply,

and so standard well theory was used to estimate the drawdown. In this case 0.6 m dia. wells were augered 10 m into the Chalk at 13 m centres (see Fig. 4). The drawdown beneath the Thanet Sand in the centre of the cofferdam was estimated to be 11 m, with the water drawn down 28 m in each well. However only 6 m drawdown was observed in the piezometers in the Chalk which would not have been enough (see Fig. 4).

59. To investigate the behaviour of the wells, a stepped drawdown test was carried out in one well (see Fig. 5) which showed that the flow in-to the well was proportional to the square root of the drawdown. Piezometers nearby showed that with the water drawn down 28 m in the well there was less than 2 m drawdown 1.5 m from the well (see Fig. 5a); evidently the water was finding it hard to get into the wells, and locally Darcy's Law was not applicable. From the slope of the drawdown curve the mass permeability of the Chalk was estimated to be 7×10^{-5} m/sec (c.f. maximum of 5×10^{-5} m/sec reported in Paper 4). There was no apparent reason why the comparitively unweathered Chalk at Pier 2 should be markedly different in jointing to the Chalk below the concrete plug at Pier 8. It was con-cluded that flow in vicinity of the wells was turbulent and was probably concentrated in a few narrow joints.

60. Unsuccessful attempts were made to improve the yield by scouring the wells with a hydrojet to remove any smear, and by deepening them. Then one well was acidized to enlarge the joints in its vicinity in a further attempt to improve its yield. This is a common technique used in drilling in carbonate rocks for water supply, but is not widely used in general civil engineer-ing. George Stow and Sons injected about $4\frac{1}{2}$ tonnes of 35% aqueous solution of hydrochloric acid down the well which increased the yield and the resulting drawdown (see Figs. 5a and 5b). When all the wells were pumped, the resulting drawdown was similar to the original estimate (see Fig. 4), and was sufficient to allow the excavation to proceed.

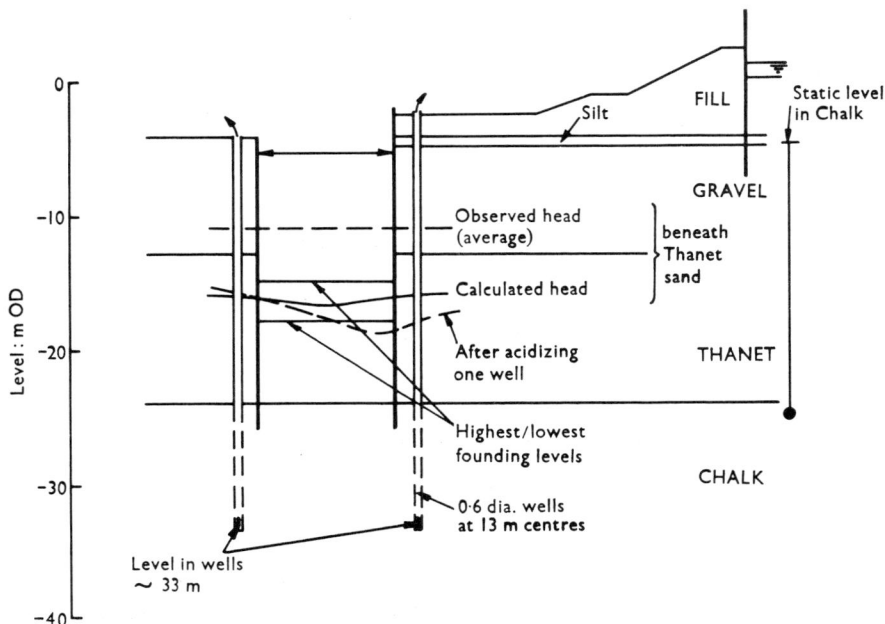

Fig.4. Pier 2

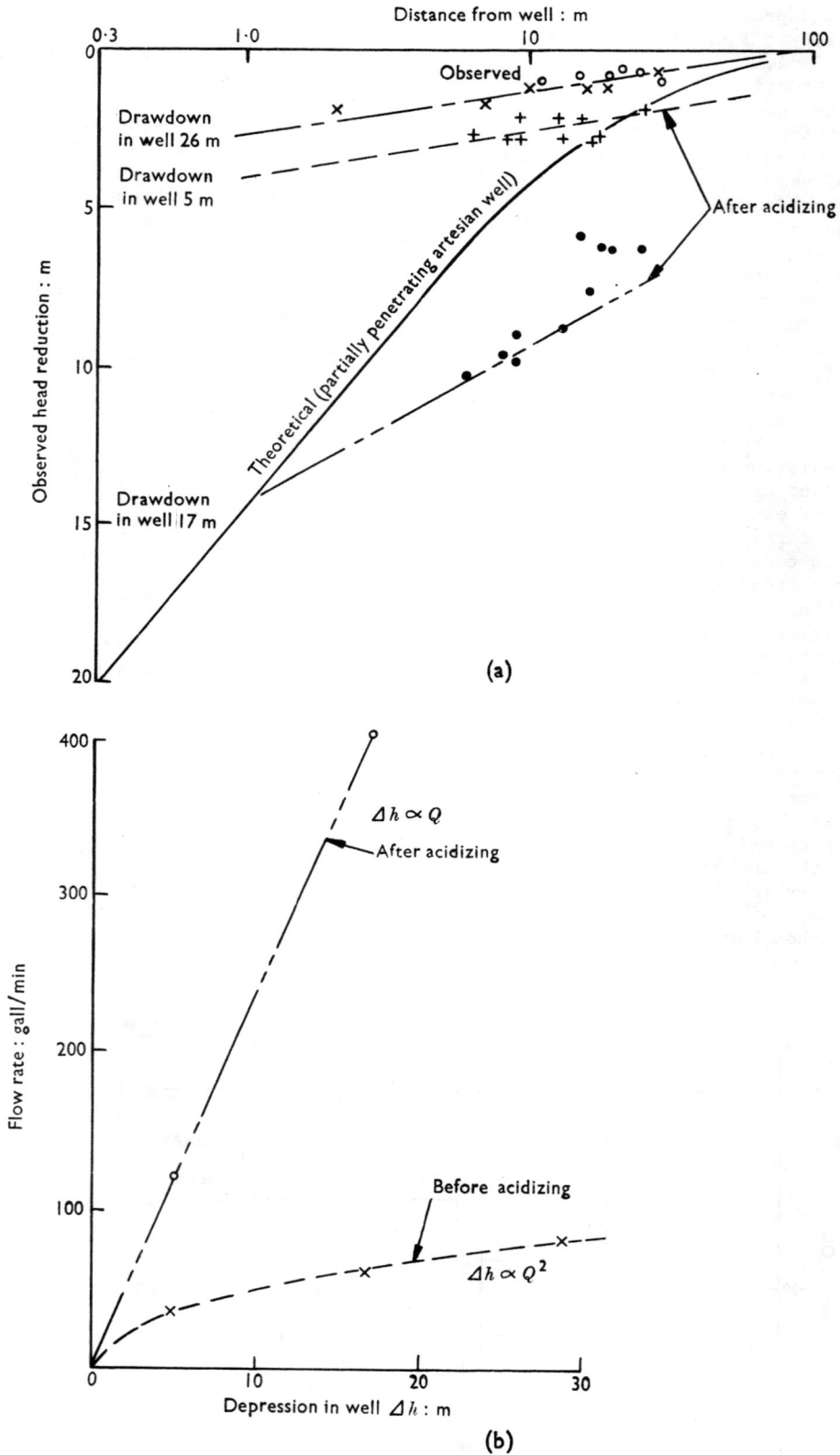

Fig.5. Pier 2 - Pumping from single well before and after acidizing; (a) Drawdown in vicinity of well; (b) Yield-depression curves

61. The success of these relief wells, though a temporary works problem, was an integral part of the foundation design and construction process. They presented an interesting design problem, and unlike many other aspects of the design, this one could not of course be model tested.

MR HOLLOWAY

62. Mr Dwight is too modest about his work. We felt that with a structure of this type we had to go back to the basic considerations of making welds. The experiment for which he was responsible was conducted on that philosophy. At that time there was considerable thought about the importance of residual stresses. There were two things we wanted to find out: what distortion takes place and what are the resulting residual stresses during fabrication. We would then have some idea of what we should specify for that stage.

MR CLARK

63. With regard to Mr Dwight's comments the structure was being designed to its final form during the period shortly after the first deliberations of the Merrison Committee had become public. The spotlight was firmly on residual stress, deformations and so on. We therefore took the precaution of considering very carefully the implications of this work. Fortunately the structure is not sensitive to either residual stresses or initial deformations in the way that many other structures are. The loads on these hydraulic structures are so large that the plate thicknesses are large and result in stocky panels and stiffeners. This means that we have structural elements which yield before they buckle in most cases.

64. We were lucky here because in some cases the results were the opposite of those predicted, which indicates what difficult territory we are in with regard to residual stress.

MR MILLER RICHARDS

65. One of the features of this project has been the large amount of model testing carried out. There have been papers written throughout the world discussing problems of vibration of the gate. One series of gates in the States vibrated themselves to pieces only two years after installation. The welds cracked and subsequently a great deal of remedial work had to be carried out. In the early design stages of the Thames Barrier the decision was made to thoroughly investigate, theoretically and by model testing, the question of vibration on the gates. A full theoretical analysis was carried out in the design office and this indicated basically that the rising gate would not suffer any structural vibrations within itself but that there was a possibility of a mass oscillation of the whole gate about the centre line of its pivots.

66. Tests had already been put in hand by the BHRA and an eighteen month period of testing was carried out to study how these gates could behave under dynamic loading. Mr Young referred to the hydrostatic loading but the gates have to rise up through flowing water and they are subjected, when in an inclined position, to a high degree of overspill. Both the Greater London Council and ourselves have considered the pros and cons of operating the barrier either by an overspill arrangement allowing a certain amount of water to spill over the gates, i.e. maintaining the gates in an inclined position and allowing a surge flood to build up and spill over the crest, or to further incline the gate, taking it through the ninety degree position and allowing under-shot flow. Both these conditions were investigated by BHRA and also by Dr Hardwick, and it was found that the gates behaved more or less as the theoretical calculations predicted.

MR TAPPIN

67. I refer to the torsional question raised by Mr Young. The hydrostatic loading does not cause any twisting of the gate/gate arm assembly about its supports. That is not the same as there being no twist on the gate structure itself. The resultant hydrostatic force which passes through the centre of the trunnion does not pass through the shear centre of the gate span cross section. There is therefore a significant torsional force caused by the hydrostatic load which must be resisted by the gate structure.

MR HOLLOWAY, MR MILLER RICHARDS and MR DRAPER

68. In answer to Mr Young (paragraph 17) theoretical calculations were made to determine the optimum size of all openings in the gate structure to ensure that the rate of flow of water entering or leaving the gate matched the speed of rotation of the gate. The weight of a certain amount of carried water was allowed for in the loads to be catered for by the operating machinery but it was not considered necessary to carry out a model test to confirm these calculations.

69. With reference to Mr Maloy's question on silt deposition the design philosophy adopted for the rising sector gate recognised that silt and debris would collect both on top of the gate and in the gap between the gate and sill unit. In computing the required machinery thrust to move the gate, the force required to shear through consolidated silt was taken into account and is referred to in paragraphs 14 and 15 of Paper 8.

70. Consideration was given to both air and water jetting but it was felt that the provision of permanent water or air vents, in either the sill unit or the gate structure, would be difficult to maintain in a working condition.

71. The accurate determination of the hydrodynamic force acting on the portion of the gate within the sill recess could not be assessed with any degree of certainty either by calculation or by model tests. It was therefore considered prudent to adopt a safe upper bound condition when assessing the hydrodynamic force in this area.

72. The methods employed for fabrication of the gate structure are not described in the Paper, but it is anticipated that subsequent Papers will be given on the construction of the Barrier. However, in answer to Mr Maloy's query, stress

relieving of the gate structure is not being carried out by the fabricator, but his method of fabrication has been determined on the basis that the residual stress locked into the structure shall be minimised and that the distortion of the completed structure shall be within certain dimensional tolerances.

73. In reply to Mr Shipman the effect of allowing water to be drawn from inside the gate into the low pressure area on the flat face just below the leading edge of the gate was investigated during the design period but the benefits were not sufficient to justify its implementation.

74. Dr Hardwick describes a model test (paragraph 7 and Fig. 5 of Paper 12) where ducts were introduced from the curved face to the flat face. This resulted in slightly over a 10% reduction in the maximum hydrodynamic torque on the structure. Ducting from inside the gate would produce a similar but smaller effect since the pressure within the gate is smaller. It can be seen from Fig. 1 of Paper 8 that such a reduction in load would not reduce the design capacity requirements of the operating machinery.

75. Regarding the damping effect referred to by Mr Shipman, because only random excitation of insignificant amplitude occurred in the model tests for normal operating conditions, there would be little justification for the amendment to the existing design.

MR FAIRWEATHER and MR KIRTON

76. In answer to Mr Maloy the effect of a load on the gate, due to a tidal surge or its self weight, is to cause the gate arms to move out of square to the axis of rotation and the driving machinery (see Fig. 3 of Paper 6). To allow for this movement of the gate arms the gate trunnion bearings and the bearings fitted to each end of the link connecting the machinery to the gate arms are of the spherical type (see paragraph 36 of Paper 8).

77. The supply of hydraulic cylinders was not restricted to British manufacturers but suppliers were not nominated in the tender documents, leaving contractors free to select their own suppliers. Prior to letting the contract the supply position was investigated and several companies in the United Kingdom and Europe were considered capable of supplying suitable cylinders.

MR PRATT

78. In reply to Mr Young (paragraph 16) we have calculated that apart from exceptional conditions of ambient humidity it is unlikely that there will be any condensation in the subways. However, we have made space provision for refrigeration plant in the fan chambers if experience indicates that it is necessary to dry the air. Alternatively, on extreme high humidity days, the ventilation systems would be switched off.

79. Communicaion with STWS at Bracknell, queried by Mr McMillan, will be through post office networks via public exchanges, by direct lines and also by radio link. Telex facilities will be provided in the communications centre of the control building.

80. Warnings to shipping are the responsibility of the Port of London Authority and illuminated signals giving information of conditions of barrier closure are to be established at strategic positions on river banks upstream and downstream of the barrier. Paper 9 describes the navigation lights to be displayed on the piers.

81. All vessels will be in radio communication with the PLA control room adjacent to the barrier site on the South Bank. Vessels without radio equipment will be issued with portable units on loan before proceeding into PLA controlled zones. In addition all shipping will be under the supervision of PLA patrol launches and under radar surveillance.

82. A period of about two hours is anticipated from the time of indication of conditions possibly leading the barrier closure and pressing the button to operate the machinery. Short time emergency operations will rarely occur.

MR CROW, MR KING and MR PROSSER

83. The sudden change in hydrodynamic torque, referred to by Mr Shipman, which occurs when the flow pattern over the gate changes from "surface jet" to "diving jet", was observed during the first model tests carried out at BHRA (Ref.3 of Paper 11). Various devices intended to influence the flow patterns along the lines suggested by Mr Shipman have been tested by Dr Hardwick and referred to briefly in Paper 12. In the absence of any practical means of preventing this phenomenon, a series of tests was made using the 1/20 scale hydroelastic model, to obtain values of the hydrodynamic torque for the full range of possible operating conditions (Ref.4 of Paper 11).

DR HARDWICK

84. In answer to Mr Shipman the possibility of drawing water from within the gate into the separation zone near the crest was explored. Openings on the flat face of the gate were created near the crest (where the pressure is lower) and near the toe (where the pressure is higher); a flow from toe to crest within the gate took place but its influence on the external flow pattern and the hydrodynamic torque was negligible. The ducts through the gate from downriver to upriver surfaces (as described in Paper 12) were slightly more successful but the structural and operational complications of openings in a gate's skin seemed to outweigh the hydraulic advantages.

MR CARR

85. Mr Nash has discussed certain aspects of temporary pressure relief systems under the pier bases. Of the various alternatives available to the Contractor, he chose a system of pressure relief wells to ensure the stability of the underwater concrete plugs; these in themselves are not heavy enough to counteract the uplift pressures arising when the cofferdams are dewatered. Any variations from the general agreement found at the southern piers between the measured piezometric pressures and those indicated by the finite element analysis which

the Contractor undertook (based on the Laplace equation) can be ascribed to deviations in practice from the ideal conditions assumed.

86. The Contractor used an empirical approach for the Pier 2 arrangement of wells in view of the successful use of the system for the southern piers. The initial poor performance of these wells may demonstrate two problems. The first is the difficulty of ensuring that the drilling operations in the chalk leave an unsmeared wall to the wells. The second is the selective nature that may exist for the fissures in the chalk, and the possibility that relief wells and piezometer locations may happen to be isolated from the more dominant seepage paths. The acidizing can be presumed to establish fissure connections previously sealed.

87. Mr Milton asked whether with hindsight different or additional information would have been given to the tenderers, and whether design changes would have been made. It is difficult to identify any information aspects that could have been significantly changed or augmented. The available site data was of course provided, and this was quite comprehensive. The amount of design detail in the tender documents, whilst being sufficient to permit realistic pricing, could have been further developed in a number of respects, particularly with regard to the detailed layout of the pier structures. However, this would have considerably delayed tender enquiries until more information had become available from the development of the machinery and operating requirements. Apart from a general increase in the complexity of the piers in this respect, there have been remarkably few design changes of importance since the tender stage.

88. The introduction of the revised tender design of the large sills, embodying embedded subway tubes and longitudinal prestressing, has been referred to earlier in the discussion. A number of items, for example pier roofs, had to be dealt with at tender stage by means of outline indications only, and provisional sums.

89. It is a matter for debate whether the alternative foundation design, involving contiguous bored in situ concrete piles, would have proved as suitable and economic as the sheetpiled cofferdam construction, at least for the abutments and inshore piers. This was apparently not the preponderant view of the tenderers.

90. Mr Bernard Clark comments that the barrier does nothing for the country commercially, and he refers to earlier schemes which would have maintained constant high water in the port of London and its approaches. Leaving aside the disastrous effects on the country's commerce which the barrier and its associated works will prevent, it is questionable that the impounding schemes advocated by Mr Clark would have radically altered the trend experienced over the last decades towards the closure of the upriver docks in favour of ro-ro docks downstream and at the coastal ports. As Mr Clark points out, Rotterdam illustrates the advantages of bringing the docks downriver to deep water. It has been noted in Paper 2 that a barrage scheme would present considerable siltation problems, which half-tide operation of the Barrier would reduce but not remove.

91. Mr Clark's reference to the similarity of the barrier gates to those in Holland, assuming the reference is to the Haringvliet sluice, does not appear to take into consideration the fundamental difference that exists in the two concepts, namely the underwater location of the rising sector gate in its open housed position, which provides an unobstructed passage to shipping.

MR MITCHELL

92. Mr Maloy asked about the surface finish for the sill scallops. The selection of a high quality concrete finish, rather than granolithic concrete or epoxy mortar, resulted from the laboratory investigation referred to in Paper 14. This investigation included tests for abrasion, rate of wear and gouge resistance, the tests being repeated after various periods of watersoak of up to nine months. The specification required a standard concrete with a minimum characteristic strength of 30 N/mm^2, and called for special finishing techniques to achieve a dense smooth surface. The effectiveness of these techniques, involving early removal of top shuttering to permit power floating or other tooling processes, had to be demonstrated on trial slabs before any scallop concrete could be cast.

93. In reply to Mr Maloy's question regarding corrosion of the mild steel cover plates at the sill edges, an estimate was made of the rate of thickness loss likely to occur, including allowance for above-average flow rates during barrier operation. It was found that a 20 mm thickness would be quite adequate to survive the 60 year nominal life of the barrier. The need for a material more expensive than mild steel was discounted.

94. Mr Palmer asked about the inclusion in the specification of PFA as a partial replacement for cement in various concrete mixes. Estimated cost savings were not quantified when deciding this specification but were expected to be significant. A difference in basic material price of at least 50% existed at that time, which would have been offset only partially by the additional storage and handling costs.

95. No mix tests were carried out until after the contract was placed. From previous projects, there is wide experience of replacement of up to 30%. Replacement of up to 50% was allowed by the specification for the underwater concrete, and was subsequently adopted after satisfactory trials by the Contractor, which included an underwater placement trial.

96. In addition to the 20 N/mm^2 strength requirement, the facing concrete was specified to have a minimum cementitious content and a maximum water/cement ratio in accordance with the recommendations of CP110 for severe exposure conditions.

MR DAY

97. The very general use over many years of epoxy coal tar coatings, as indicated by Mr Maloy, is evidence enough of their excellence for a variety of environments. Indeed, a type of coal tar epoxy is now being used for the Gate

Arms, a typical tidal situation where such coatings are known to succeed - as long as the specifications for surface preparation and application are properly carried out!

98.   However, as detailed in Paper 16, the gate span situation may be deemed a special case in that it is important to prevent high velocity water plus debris from exposing bare steel of the skinplate to the activity of microbes in the mud in the scallops of the sills.

99.   From the rotating disc tests initially carried out, it was quite evident that the optimum coating would contain no coal tar. As reported by Mr Maloy from his tests, the most abrasion/erosion resistant coatings were found to be based on epoxy resin suitably pigmented with hard fillers.